Advances in Biostimulant Use on Horticultural Crops

Advances in Biostimulant Use on Horticultural Crops

Guest Editors

Mason MacDonald
Lord Abbey
Raphael Ofoe

MDPI

Basel • Beijing • Wuhan • Barcelona • Belgrade • Novi Sad • Cluj • Manchester

Guest Editors
Mason MacDonald
Department of Plant Food and Environmental Sciences
Dalhousie University
Halifax
Canada

Lord Abbey
Department of Plant Food and Environmental Sciences
Dalhousie University
Halifax
Canada

Raphael Ofoe
Department of Plant Food and Environmental Sciences
Dalhousie University
Halifax
Canada

Editorial Office
MDPI AG
Grosspeteranlage 5
4052 Basel, Switzerland

This is a reprint of the Special Issue, published open access by the journal *Plants* (ISSN 2223-7747), freely accessible at: https://www.mdpi.com/journal/plants/special_issues/24PC8HKM00.

For citation purposes, cite each article independently as indicated on the article page online and as indicated below:

Lastname, A.A.; Lastname, B.B. Article Title. *Journal Name* **Year**, *Volume Number*, Page Range.

ISBN 978-3-7258-7797-3 (Hbk)
ISBN 978-3-7258-7798-0 (PDF)
https://doi.org/10.3390/books978-3-7258-7798-0

Cover image courtesy of Mason MacDonald

Contents

About the Editors

Mason MacDonald

Mason MacDonald holds a PhD in Plant Biology (Laval University, QC, Canada), an MSc in Agriculture (Dalhousie University, NS, Canada), and a BSc in Agriculture (Nova Scotia Agricultural College, NS, Canada). He is an Assistant Professor of Plant Ecophysiology at the Faculty of Agriculture, Dalhousie University, actively teaching courses related to biology, plant ecophysiology, crop physiology, and environmental sampling. Mason's research program relates to understanding and mitigating physiological changes associated with environmental stress. A large part of his research program is related to testing and developing biostimulants for use on agricultural and horticultural crops, specifically to improve yields during stress. Mason has written over 50 peer-reviewed publications and supervised 23 highly qualified personnel (BSc honors, MSc, and PhD students).

Lord Abbey

Lord Abbey holds a PhD in Plant Science (Cranfield, UK), an MSc in Horticulture and Crop Science (Wye, UK), and a BSc (Hons) in Agricultural Science (Univ. of Ghana) with a research focus on health-based sustainable horticultural production systems. He is currently a Full Professor and the Graduate Coordinator in Dalhousie University Faculty of Agriculture, Department of Plant, Food, and Environmental Sciences. He has received several significant industry, federal, and provincial government grants of over $3.6M for his work on climate-smart agriculture, leading to over 120 publications. He has supervised over 76 highly qualified personnel (undergraduates, graduates, and postdoctoral fellows). His work is multidisciplinary and internationally recognized. He has worked extensively with industry and other scientists on sustainable horticultural production, novel biostimulants, and organic and nano(bio)fertilizers derived from municipal solid waste, plant biomass, and marine wastes, and controlled environment production systems and nutrient management for various horticultural crops. He is the current President of the *Canadian Society for Horticultural Science (CSHS)*, a Council member of the *International Society for Horticultural Science (ISHS)*, and a former Board member of the *Nova Scotia Institute of Agrologists (NSIA) Council* and an Associate Editor of the *Canadian Journal of Plant Science* and Guest Editor of several other reputable peer-reviewed international journals.

Raphael Ofoe

Raphael Ofoe is a Postdoctoral Research Fellow in the Department of Plant, Food, and Environmental Sciences at Dalhousie University. He obtained his BSc in Crop Science from the University of Ghana (Accra, Ghana), an MSc in Molecular Plant Sciences from the University of Bath (Bath, UK), and a PhD in Agricultural Science from Dalhousie University (Halifax, Canada). Dr. Ofoe studies plant responses to environmental stress. His teaching experience in plant science is widely interdisciplinary, spanning areas from plant nutrition to genetics and molecular biology, and encompassing fieldwork as well as laboratory and computational analysis of plant data. His current research focuses on the use of nanotechnology to develop biostimulants from environmental waste streams to enhance plant growth and productivity, as well as to understand their effects on the morpho-physiological, biochemical, metabolic, and molecular responses of horticultural plants to diverse environmental stresses (both biotic and abiotic). His research also involves elucidating how such innovative strategies can be used to improve carbon sequestration and mitigate greenhouse gas emissions in both field and controlled environment systems. Dr. Ofoe has worked with several agricultural startup companies in Nova Scotia as their Chief Research Officer and has been involved in planning and securing grants for business-related projects. He has published over 30 peer-reviewed research articles and has presented his research at national and international conferences. He is a proactive member of professional organizations such as the *Canadian Society of Horticultural Science (CSHS)* and the *American Society of Horticultural Science (ASHS)*. He has served on conference committees and contributed as a peer reviewer for well-regarded academic journals.

plants

MDPI

Editorial

Editorial for Advances in Biostimulant Use on Horticultural Crops, First Edition

Raphael Ofoe *, Mason T. MacDonald * and Lord Abbey *

Department of Plant Food and Environmental Sciences, Faculty of Agriculture, Dalhousie University, Bible Hill, NS B2N 5E3, Canada
* Correspondence: raphael.ofoe@dal.ca (R.O.); mason.macdonald@dal.ca (M.T.M.); labbey@dal.ca (L.A.)

1. Introduction

The role of horticultural crops (fruits, vegetables, and ornamentals) is vital globally, not only for nutrition, but also for economic and cultural value. Horticultural industries are challenged by increasing populations, climate volatility, and a growing demand for sustainable, nutrient-dense horticultural crop production. The use of biostimulants has emerged not only as an agricultural input, but as a crucial strategic tool for researchers and farmers to enhance crop productivity and resilience to diverse environmental stresses.

Biostimulants are defined as any substance or microorganisms applied to plants to enhance nutrient use efficiency, abiotic stress tolerance, and/or crop quality traits, independent of their nutrient composition [1]. Unlike fertilizers that supply plants with nutrients, biostimulants function by modulating physiological, biochemical, and molecular processes in plants [2]. Horticultural crops are particularly receptive to biostimulant strategies due to their sensitivity to environmental and management stresses, their premium quality requirements, and the opportunities for targeted growth improvement.

This Special Issue brings together a timely and diverse collection of original research and reviews focused on the application of biostimulants in horticultural crops, exploring their mechanistic bases, practical efficacy, crop-specific responses, and opportunities to reduce the reliance on synthetic inputs. The response to our call for papers has been robust, yielding a collection of thirteen insightful studies that collectively advance our understanding of biostimulants. The research presented here spans the horticultural value chain, from seed priming and early establishment to postharvest quality management, and investigates a wide range of biostimulant sources and their effects on a diversity of crops. This Special Issue focuses on three thematic areas: novel and sustainable biostimulants, enhanced functional and nutritional quality, and the mitigation of abiotic stress.

1.1. Novel and Sustainable Biostimulants

Novel biostimulants continue to be used in horticulture in innovative ways, which is a major highlight in this Special Issue. For instance, Senthilkumar et al. [3] investigated the potential of pyroligneous acid, a by-product of pyrolysis, to improve postharvest needle retention in balsam fir. Although the direct effect on needle retention was inconsistent, the study revealed that pyroligneous acid could induce protective physiological responses, such as proline accumulation and reduced membrane injury. This represents an unforeseen benefit in balsam fir beyond the researchers' set objective. Similarly, Varga et al. [4] demonstrated that water extracts from industrial hemp inflorescences can serve as effective biostimulants, significantly enhancing early growth, biomass, and antioxidant activity in oilseed pumpkin seedlings. This work opens avenues for valorizing agricultural by-products as novel biostimulant inputs. Other featured research included the exploration of plant extract-, fungi-, and protein hydrolysate-based biostimulants [5–7].

1.2. Enhanced Functional and Nutritional Quality

The ability of biostimulants to enhance the nutritional and functional quality of horticultural crops is another critical area of advancement. Lopes et al. [5] clearly show that foliar applications of seaweed extract (*Ecklonia maxima*) and glycine betaine can considerably regulate the accumulation of bioactive compounds and antioxidant capacity in blueberries. Importantly, these findings emphasize that the efficacy of these biostimulants is cultivar- and dose-dependent, which emphasizes the need for precise, tailored application strategies to maximize benefits. This theme of quality improvement is further supported by Lam et al. [6], who found that moderate concentrations of seaweed extract optimized growth, photosynthetic performance, and the accumulation of health-promoting phenolic compounds, such as rosmarinic acid, in Thai basil grown in controlled plant factories. In fact, most studies in this Special Issue describe results that include phytochemical changes rather than only describing growth-promoting properties. This points to an increased mechanistic interest in biostimulant use.

1.3. Mitigation of Abiotic Stress

Biostimulants have long been known to enhance growth characteristics, even during abiotic stress, thereby mitigating some of the deleterious stress effects. This Special Issue features six papers that specifically examine the stress-mitigating properties of biostimulants [3,8–12]. Featured abiotic stresses include water deficit, salinity, alkalinity, and nitrogen deficiency. Other studies, though not directly assessing stress mitigation, studied increases in antioxidant phytochemicals induced by biostimulants [4,6]. Such antioxidant increases align with a potential ability of associated biostimulants to mitigate stress.

2. Conclusions and Future Research

Collectively, the papers in this Special Issue illuminate several key trajectories for biostimulant use in horticulture. Firstly, they confirm the remarkable diversity of materials with biostimulant action and their potential to improve both productivity and quality. Additionally, they highlight the intricate physiological and molecular mechanisms through which these biostimulants function, moving beyond simple growth promotion to include stress modulation and secondary metabolite production. Most importantly, these papers demonstrate the critical importance of context, where the specific biostimulant, concentration, method and timing of application, and target crop are critical considerations. As we look to the future, the research compiled in this issue provides a strong foundation for continued exploration. We see a clear need for deeper mechanistic studies to fully unravel the signaling pathways and genetic networks modulated by these biostimulants. Furthermore, translating this knowledge into practical, reliable recommendations for growers remains a paramount goal. This requires not only continued discovery of new products, but also rigorous, multi-location field trials that account for the complex environmental variables of commercial production.

Funding: This research received no external funding.

Data Availability Statement: No new data was generated for this article.

Conflicts of Interest: The authors declare no conflicts of interest.

 https://doi.org/10.3390/plants15081139

eferences

1. Du Jardin, P. Plant Biostimulants: Definition, Concept, Main Categories and Regulation. *Sci. Hortic.* **2015**, *196*, 3–14. [CrossRef]
2. Shahrajabian, M.H.; Chaski, C.; Polyzos, N.; Petropoulos, S.A. Biostimulants Application: A Low Input Cropping Management Tool for Sustainable Farming of Vegetables. *Biomolecules* **2021**, *11*, 698. [CrossRef] [PubMed]
3. Senthilkumar, N.; Kasu, R.; Ofoe, R.; Abbey, L.; MacDonald, M.T. Effect of Pyroligneous Acid on Needle Retention and Certain Stress-Related Phytochemicals in Balsam Fir (*Abies balsamea*). *Plants* **2026**, *15*, 261. [CrossRef] [PubMed]
4. Varga, I.; Antunović, M.; Kojić, M.T.; Markulj Kulundžić, A.; Iljkić, D.; Baličević, R.; Ravlić, M. Hemp Inflorescence as a Sustainable Biostimulant Tool to Boost Growth and Antioxidant Capacity in Oilseed Pumpkin. *Plants* **2025**, *14*, 3473. [CrossRef] [PubMed]
5. Lopes, T.; Silva, A.P.; Aires, A.; Carvalho, R.; Ferreira, M.; Vicente, A.A.; Gonçalves, B. Effect of Foliar Biostimulant Application on Bioactive Compounds and Antioxidant Capacity in Blueberry (*Vaccinium corymbosum* L.). *Plants* **2025**, *15*, 92. [CrossRef] [PubMed]
6. Lam, V.P.; Bok, G.; Loi, D.N.; Do, M.C.; Park, J. Seaweed Foliar Biostimulants Improve Growth and Phytochemicals of Thai Basil (*Ocimum basilicum* L.) in a Plant Factory. *Plants* **2025**, *14*, 3271. [CrossRef] [PubMed]
7. Vultaggio, L.; Ciriello, M.; Campana, E.; Bellitto, P.; Consentino, B.B.; Rouphael, Y.; Colla, G.; Mancuso, F.; La Bella, S.; Napoli, S.; et al. Single or Blended Application of Non-microbial Plant-based Biostimulants and *Trichoderma atroviride* as a New Strategy to Enhance Greenhouse Cherry Tomato Performance. *Plants* **2024**, *13*, 3048. [CrossRef] [PubMed]
8. Justamante, M.S.; Larriba, E.; Zavala-González, E.A.; Aranda-Martínez, A.; Pérez-Pérez, J.M. Transcriptional Profiling to Assess the Effects of Biological Stimulant Atlanticell Micomix on Tomato Seedlings Under Salt Stress. *Plants* **2025**, *14*, 1198. [CrossRef]
9. Utgés-Minguell, L.; Sierras-Serra, N.; Marín, C.; Pintó-Marijuan, M. Enhanced Production by Terra-Sorb® Symbiotic Biostimulant in Two Model Species Under Nitrogen Stress. *Plants* **2025**, *14*, 1087. [CrossRef] [PubMed]
10. Nie, W.; He, Q.; Ma, J.; Guo, H.; Shi, Q. Exogenous 2,4-Epibrassinolide Alleviates Alkaline Stress in Cucumber by Modulating Photosynthetic Performance. *Plants* **2024**, *14*, 54. [CrossRef] [PubMed]
11. Francioso, O.; Schiavon, M.; Nardi, S.; Castellani, D.; Ferrari, E.; Estrada, M.T.R.; Della Lucia, M.C.; Zuffi, V.; Ertani, A. Mitigation of Salt Stress in *Lactuca sativa* L. Var. Gentile Rossa Using Microalgae as Priming Agents. *Plants* **2024**, *13*, 3311. [CrossRef] [PubMed]
12. Fedeli, R.; Celletti, S.; Loppi, S. Wood Distillate Promotes the Tolerance of Lettuce in Extreme Salt Stress Conditions. *Plants* **2024**, *13*, 1335. [CrossRef] [PubMed]

https://doi.org/10.3390/plants15081139

plants

MDPI

Article

Wood Distillate Promotes the Tolerance of Lettuce in Extreme Salt Stress Conditions

Riccardo Fedeli [1], Silvia Celletti [1,*] and Stefano Loppi [1,2]

[1] BioAgry Lab, Department of Life Sciences (DSV), University of Siena, 53100 Siena, Italy; riccardo.fedeli@student.unisi.it (R.F.); stefano.loppi@unisi.it (S.L.)

[2] BAT Center—Interuniversity Center for Studies on Bioinspired Agro-Environmental Technology, University of Naples "Federico II", 80138 Napoli, Italy

* Correspondence: silvia.celletti@unito.it

Abstract: Soil salinization is an adverse phenomenon in agriculture that severely affects crop growth and yield. The use of natural products, such as wood distillate (WD, derived from the pyrolysis of woody biomass), could be a sustainable approach to enhance the tolerance of plants cultivated in the saline soils. Hence, this study aimed to evaluate the potential of WD, a foliar sprayed at 0.2% (*v*/*v*), in lettuce plants subjected to grow under both moderate and high soil sodium chloride (NaCl) concentrations (ranging from 0 to 300 mM). The changes in the physiological and biochemical responses of these plants to the varying salt stress conditions allowed the identification of a maximum tolerance threshold (100 mM NaCl), specific to lettuce. Beyond this threshold, levels related to plant defense antioxidant power (antiradical activity) were lowered, while those indicative of oxidative stress (malondialdehyde content and electrolyte leakage) were raised, causing significant losses in leaf fresh biomass. On the other hand, WD significantly improved plant growth, enabling plants to survive high salt conditions >200 mM NaCl. Collectively, these observations highlight that treatments with WD could be of paramount importance in coping with current environmental challenges to have better yields under soil conditions of high salt concentrations.

Keywords: antioxidant power; malondialdehyde; oxidative stress; pyroligneous acid; sodium chloride

1. Introduction

A soil becomes saline when it contains a high amount of dissolved salts in the soil solution, such as sodium (Na^+), chlorine (Cl^-), and potassium (K^+) ions, showing a value of electrical conductivity (EC) ≥ 2 dS^{-1} m^{-1} [1]. This excessive concentration of salts can have a negative impact on plant growth and development, both physiologically and morphologically [2]. Basically, a high salt concentration in soil is harmful to plants because it interferes with the normal activity of leaf transpiration and root water uptake, causing osmotic and ionic stress [3]. As a result, plants can show symptoms such as thick leaves with the accumulation of Na^+ and Cl^- within the cells, wilting, discoloration, premature aging, and leaf drop, as well as slowed or even inhibited growth [4]. High salt concentrations in the external medium can induce the closure of leaf stomata; this is a defense mechanism activated by plants, which, in turn, can cause oxidative stress due to the transfer of excess energy to oxygen (O_2), resulting in the formation of superoxide ($O2^{\bullet-}$), hydrogen peroxide (H_2O_2), hydroxyl radical ($OH^\bullet$), and singlet oxygen ($^1O_2$) [5,6]. These molecules are highly reactive O_2-containing molecules (known as reactive oxygen species—ROS—) and, when in excess, they are harmful to the life of cells [7]. As a result, plants activate a series of antioxidant enzymes and biosynthesize different non-enzymatic antioxidant compounds to eliminate excess ROS or convert ROS into less toxic compounds [8,9]. Furthermore, high salt levels cause not only plant oxidative stress, but also lead to a great accumulation of electrolytes (e.g., Na^+, Cl^-, K^+) within plant cells, consequently, the osmotic pressure increases, leading to the rupture of cell plasma membranes and, finally, to cell death [10].

This can lead to severe economic losses in crop quality and yield [11]. However, the extent of damage depends not only on the concentration of salts in the soil, but also on the ability of plants to tolerate and adapt, which varies not only from species to species, but also according to the plant growth stage [12,13].

Globally, more than 800 million ha of agricultural soils are affected by salinization problems [14]. This is of concern especially in regions characterized by arid and semi-arid climates, where water evaporation is accelerated, causing surface salts to concentrate and form a saline crust on the soil surface [15,16]. Consequently, soils lose their physical, chemical, and biological fertility, becoming infertile for agriculture and are often abandoned [17]. In Europe, soil salinization is a remarkable phenomenon affecting a large area, between 1 and 3 million ha, mainly in countries around the Mediterranean basin [18]. This problem is exacerbated by climate change, which leads to higher temperatures and prolonged and intensified drought periods, along with the adoption of inappropriate irrigation farming practices [19]. This is predicted to lead to more than 50% of agricultural soils becoming uncultivable by the end of 2050 due to excess salts [20,21]. In addition, the United Nations estimated that the world population will reach 9.7 billion people by 2050, significantly increasing the demand for food, despite the decrease in usable agricultural areas [22]. This situation makes it crucial to improve crop yield and quality to limit the problem of food security and malnutrition [23,24].

To achieve the goal of improving crop tolerance to salinized conditions, mainly two approaches have been adopted: agronomic techniques and genetic selection. Agronomic techniques involve the use of specific products, such as biofertilizers, phytohormones, potassium salts, and silicon, to modify the metabolic behavior of plants so that they can better adapt to the salinized environment [25]. Alternatively, attempts have been made to genetically select crops that show a greater salt tolerance [26]. However, so far, genetic manipulation has been limited by the complexity of gene regulation, which varies significantly during the different stages of plant development. As a result, there have been just a few cases in which salt-tolerant plant varieties have been developed using these methods [27].

Current environmental policies globally encourage sustainable agronomic measures to address climate change, including salt stress management for plants [28]. An eco-friendly alternative to synthetic chemicals, such as pesticides and fertilizers, is wood distillate (WD), also known as pyroligneous acid or wood vinegar [29]. This dark amber-brown natural product is obtained by distillation during the pyrolysis of woody biomass [30]. Experimentation has shown that WD is both safe for the environment [31,32], human health [33], and beneficial for plant biodiversity [34]. In Italy, it has even been approved for use in organic farming [35].

In agriculture, promising results have been obtained in the use of WD as a plant corroborant and a biostimulant [36,37]. Dissolved WD contains more than 200 organic compounds including acids, alcohols, aldehydes, ketones, phenols, and minerals (mainly calcium and iron) [38]. The composition of liquid products that are obtained from thermochemical processes, such as WD in our case, obtained from pyrolysis, can be influenced by the operating parameters (mainly, temperature, residence time, and pressure) and the type of feedstock used in the process [39,40]. Specifically, while pressure affects the kinetics and pathways of pyrolysis reactions, temperature, as well as the duration of the process, have a significant impact on the characteristics of the final product; as an example, rapid heating promotes the rapid release of volatile compounds and, as a result, the composition of the product will be different than that of a product obtained at a slower heating rate. It can be applied both to the soil by fertigation and to the leaves by spraying. Its applications include improving soil quality [41], controlling weeds [42], regulating plant growth [36], and reducing the negative effects of the presence of bioplastics in the soil [38]. However, the efficacy of WD depends on the concentration used; if WD is applied between 0.2% and 0.5% (v/v), it stimulates plant growth and yield [36,38,43,44], while if it is applied at dosages higher than 0.5% (v/v), it can have herbicidal effects [42,45,46]. To the best of our knowledge, lettuce irrigated with saline water has not been the subject of WD studies.

In this context, this study aims to investigate whether WD can improve the ability of lettuce plants to tolerate moderate and high soil sodium chloride concentrations. This study provides an understanding of how lettuce plants, treated with WD as a foliar spray, tolerate saline irrigation.

2. Results

2.1. Response of the WD-Untreated Plants to Varying Salt Concentrations

The addition of salt to the soil, irrespective of the NaCl concentration, resulted in a significant reduction in leaf fresh weight of plants, which were not sprayed with 0.2% (v/v) WD (indicated as "−WD"). Specifically, the reductions increased linearly with the increasing NaCl concentration as follows: −22.1% for the 50 mM, −36.9% for the 100 mM, −75.1% for the 200 mM, and −88.6% for the 300 mM NaCl, compared to those plants which were neither treated with NaCl nor with WD (control) (Figure 1A). On the other hand, leaf chlorophyll content in the control plants did not change except at the highest NaCl concentration (300 mM), where it increased significantly ($p < 0.05$), reaching values approximately +23% higher than the control (Figure 1B).

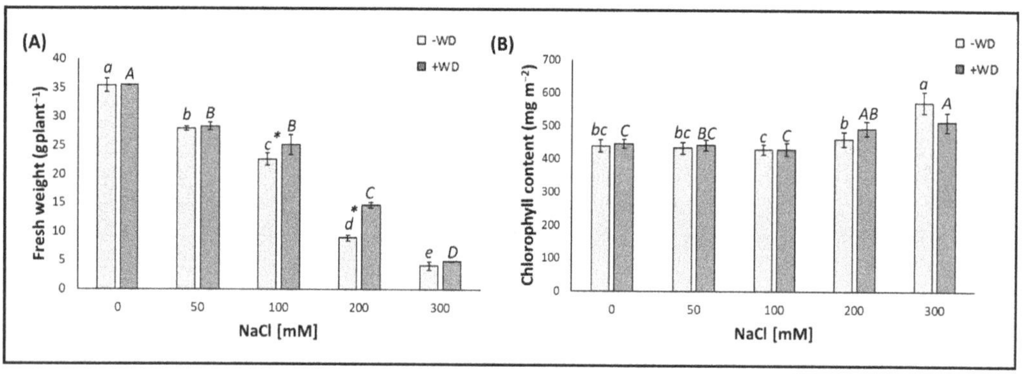

Figure 1. Fresh weight (**A**) and chlorophyll content (**B**) presented as the median ± error. Potting soil treatments with different NaCl concentrations [from 0 (control) to 300 mM] are displayed on the horizontal axis. "−WD" means leaves sprayed with water, while "+WD" means leaves sprayed with 0.2% (v/v) wood distillate (WD). Lowercase letters indicate significant differences ($p < 0.05$) among the different NaCl concentrations within "−WD", whereas uppercase letters indicate significant differences ($p < 0.05$) among the different NaCl concentrations within "+WD". *: indicates significant differences within the same salt treatments evaluated by the Kolmogorov–Smirnov test ($p < 0.05$).

The 50 mM NaCl treatment significantly ($p < 0.05$) lowered the EL by −30%, which is an indicator of osmotic stress. Concentrations higher than 50 mM but less than 300 mM did not change this parameter, but above 300 mM NaCl, a great increase (+308.7%) in EL was observed compared to the control plants (Figure 2A). Amongst the electrolytes, however, Na accumulated abundantly in the leaves of −WD plants, increasing linearly with increasing NaCl concentrations in the following order: +453.4% (at 50 mM) < +613.5% (at 100 mM) < +1128.5% (at 200 mM) < +1269.0% (at 300 mM), with respect to the control. As an example, at 300 mM, the Na content was about 100 g kg_{DW}^{-1} (Figure 2B). Similarly to EL, the content of MDA, a cytotoxic target indicative of the oxidative damage of cell membrane lipids, decreased significantly ($p < 0.05$) (−24.2%) at 50 mM. NaCl remained stable at 100 mM and then increased strongly above this concentration. Particularly, the increases were +55.3% and +65.2% for 200 and 300 mM NaCl, respectively, in comparison to the control (Figure 2C). For carbohydrates, no variation was observed between the contents of the three sugars (such as fructose, glucose, and sucrose) and the respective controls in the various treatments; instead, the starch content varied with an almost opposite trend to those of EL and MDA. In fact, the accumulation of starch increased at the lowest NaCl concentrations (at 50 mM by +71.6%; at 100 mM by +69.3%; and at 200 mM by +77.4%), whereas at the highest NaCl concentration, no change was evident when compared to the control (Figure 2D).

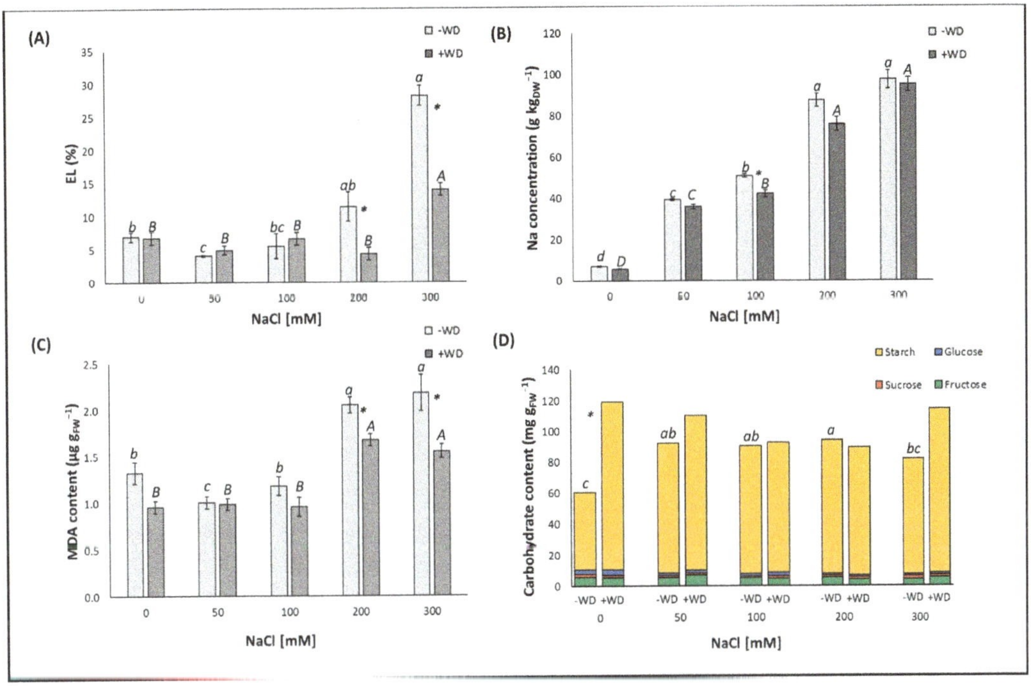

Figure 2. Electrolyte leakage (EL) (**A**), Na concentration (**B**), malondialdehyde (MDA) content (**C**), and carbohydrate content (i.e., glucose, fructose, sucrose, and starch) (**D**) presented as the median ± error. Potting soil treatments with different NaCl concentrations [from 0 (control) to 300 mM] are displayed on the horizontal axis. "−WD" means leaves sprayed with water, while "+WD" means leaves sprayed with 0.2% (v/v) wood distillate (WD). Lowercase letters indicate significant differences ($p < 0.05$) among the different NaCl concentrations within "−WD", whereas uppercase letters indicate significant differences ($p < 0.05$) among the different NaCl concentrations within "+WD". *: indicates significant differences within the same salt treatments evaluated by the Kolmogorov–Smirnov test ($p < 0.05$).

The ARA, expressing the total antioxidant power of −WD-leaves for NaCl concentrations of 50 and 100 mM was higher than the control by +37.4% and +28.6%, respectively, while for NaCl concentrations of 200 and 300 mM, it was less than the control by −22.3% and −40.3%, respectively (Figure 3A). Differently, the antioxidant compounds, in terms of TPC, decreased progressively as the NaCl concentration increased, ranging from values of 2 mg g_{FW}^{-1} (control) to approximately 0.1 mg g_{FW}^{-1} for the highest NaCl concentration (Figure 3B).

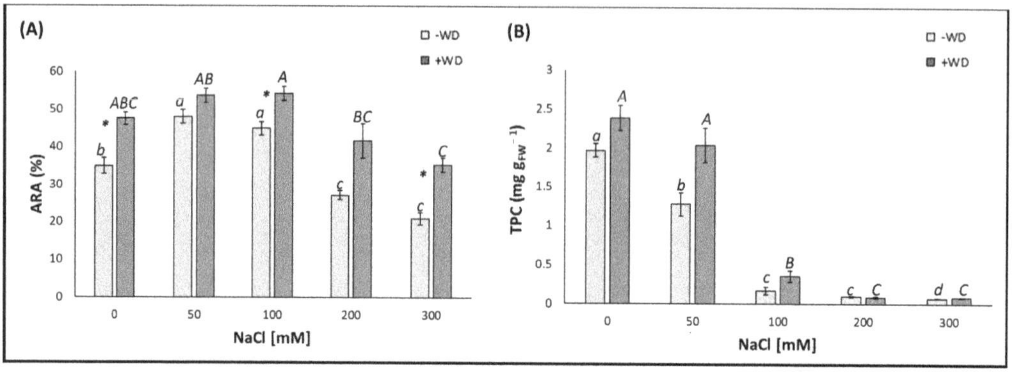

Figure 3. Antiradical activity (ARA) (**A**) and total phenol content (TPC) (**B**) presented as the median ± error. Potting soil treatments with different NaCl concentrations [from 0 (control) to 300 mM] are displayed on the horizontal axis. "−WD" means leaves sprayed with water, while "+WD" means leaves sprayed with 0.2% (*v*/*v*) wood distillate (WD). Lowercase letters indicate significant differences ($p < 0.05$) among the different NaCl concentrations within "−WD", whereas uppercase letters indicate significant differences ($p < 0.05$) among the different NaCl concentrations within "+WD". *: indicates significant differences within the same salt treatments evaluated by the Kolmogorov–Smirnov test ($p < 0.05$).

2.2. Response of the WD-Treated Plants to Varying Salt Concentrations

The effect of NaCl on the leaves of lettuce plants sprayed with 0.2% (*v*/*v*) WD (indicated as "+WD") was evident in terms of reducing the fresh weight of the aerial part of these plants. These reductions were around lighter losses of approximately −25% on average for the 50 mM and 100 mM NaCl concentrations and heavier losses of −58.9% and −86.2% for the respective 200 mM and 300 mM NaCl concentrations, compared to those plants which were not treated with NaCl but with WD (control) (Figure 1A). On the other hand, in the WD-treated plants, the chlorophyll content only varied in the leaves of those plants treated with NaCl at the two highest concentrations (200 and 300 mM), with increases of +11.0% and +14.6%, respectively, when compared to the control (Figure 1B).

The osmotic stress index, i.e., EL, remained stable in the WD-plants treated with NaCl concentrations between 50 and 200 mM, while above 200 mM NaCl a significant increase ($p < 0.05$) was observed, which was about twofold higher than that of the control (Figure 2A). Sodium was abundantly accumulated in the WD-treated leaves, reaching approximately 95 g at the highest NaCl concentration. As expected, the Na content increased linearly with increasing NaCl concentrations as follows: +527.0% (at 50 mM) < +633.5% (at 100 mM) < +1223.1% (at 200 mM) < +1555.8% (at 300 mM), compared to the control (Figure 2B). In the WD plants, the index of cellular oxidative damage, i.e., MDA content, exhibited a similar trend as EL. Specifically, the MDA content showed a significant increase ($p < 0.05$) at 200 mM and 300 mM (+75.8% and +63.2%, respectively), while EL showed a significant increase at 300 mM (+109.0%) (Figure 2C). The content of the analyzed carbohydrates (both simple and complex) was not affected by the combined treatment of WD and NaCl, regardless of the NaCl concentration in comparison to the respective controls (Figure 2D).

Also, in the case of the ARA of the lettuce leaves, no statistical difference was detected between the WD, treated at any NaCl concentration, and the control (Figure 3A). Looking at the TPC, all of the NaCl concentrations, except for the 50 mM, which showed no significant difference to the control, resulted in a severe decline in these antioxidant compounds: −85.3% for the 100 mM and −96.5 for both 200 and 300 mM NaCl (Figure 3B).

2.3. *Differences in the Response between WD-Untreated and WD-Treated Plants at Various Salt Concentrations*

The leaf fresh weight varied between plants without WD and with WD (indicated as "−WD" and "+WD", respectively) for concentrations 100 and 200 mM NaCl. In the presence of WD, an increase, albeit significant ($p < 0.05$), of only +11.1% was found for the 100 mM NaCl concentration, whereas it was more pronounced (+64.0%) for the 200 mM NaCl concentration (Figure 1A). In contrast, chlorophyll content was unaffected by the presence or absence of WD spraying, under any of the specific salt stress conditions tested (Figure 1B).

A significant slowdown of EL was markedly evident with WD solely at the two highest NaCl concentrations. Precisely, at 200 mM NaCl, the reduction was −66.9% and at 300 mM NaCl, it was −50.4% (+WD vs. −WD) (Figure 2A). For the specific electrolyte, namely Na, its accumulation at the leaf level was reduced by about −17.1% with the addition of WD, exclusively at the 100 mM NaCl concentration (Figure 2B). Similar to EL, a reduction in the MDA content was observed, though to a lesser extent, upon the addition of WD to the two highest NaCl concentrations. Specifically, the reduction was −18.5% at 200 mM NaCl, and −28.9% at 300 mM NaCl, when comparing +WD to −WD (Figure 2C). While foliar spraying with WD induced no change in the content of simple carbohydrates (i.e., fructose, glucose, and sucrose) in this plant tissue, WD did allow these plants to accumulate a higher starch content in their leaves, but only when either salt stress was not imposed (+122.8) (Figure 2D).

The ARA was also enhanced by WD treatment if no NaCl was present in the medium. However, this parameter also increased if the NaCl concentration was 100 and 300 mM, showing more modest increases at 0 and 100 mM (around 30% on average for both), and more accentuated increases (of almost 70%) at 300 mM (Figure 3A). In contrast, TPC showed no significant variation between the leaves of −WD and +WD lettuce plants at any NaCl concentration considered (Figure 3B).

3. Discussion

Several studies have shown the important role of natural products in improving nutritional quality, growth and the yield of crop plants under several abiotic stresses, including salt stress [47–54]. Among these products, WD has proved to be very effective [36,55], although its efficacy in mitigating the damage caused by salt stress is not yet well explored and understood.

To the best of our knowledge, only the recent study by Ma et al. [48] evaluated the oxidative damage in rapeseed (*Brassica napus* L.) plants when grown hydroponically in the presence of 70 and 150 mM NaCl, and the effect of WD on the protection of photosystem II in this crop plant. The results obtained by Ma et al. [48] suggest that WD would have the potential to efficiently reduce Na accumulation in the leaves of plants grown in the presence of 150 mM NaCl. Nevertheless, this is the first study investigating the effectiveness of WD in plants cultivated in soils with extreme salt stress concentrations above 150 mM NaCl. It is generally known that, beyond a concentration of 100 mM NaCl, the phytotoxic effects of salt are increasingly detrimental to the functioning of the biochemical and physiological processes of the lettuce plant [56], initially showing stunted growth, and leaf symptoms such as thickening, wilting, yellowing, discoloration, and dropping, ultimately leading, in the most severe cases, to plant death.

In this context, we aimed to investigate whether WD could be an effective ally in enhancing the resistance of lettuce plants to high salt stress conditions, (i.e., 200 mM and 300 mM NaCl), since up to date, this research topic has not yet been explored.

Our results showed that as the NaCl concentration increased, the leaf fresh biomass decreased significantly. This is known to be a sign of stress by the plants since the leaf biomass accumulation is crucial for the survival of the plants [57]. In contrast, the +WD plants were stimulated to produce leaf fresh biomass, except at the highest NaCl concentration (i.e., 300 mM). Wood distillate is to be recognized for its biostimulating action on plant growth, thanks to the presence of various compounds (such as acids, alcohols, aldehydes, ketones, phenols, and nutrients, specifically calcium and iron), that help plants to grow healthily and respond to various environmental stresses [58]. Our results are consistent with those of Ma et al. [48], showing a significant increase in the leaf fresh biomass of rapeseed plants when these plants were foliarly sprayed with 0.2% (*v*/*v*) WD and grown with 150 mM NaCl.

As mentioned above, and also in our case, lettuce plants grown at NaCl concentrations above 200 mM exhibited visible symptoms of leaf thickening, yellowing, and finally leaf loss. The fact that chlorophyll content increased upon exceeding certain salt conditions is not evidence of improved photosynthetic performance by the plants, but a result that corroborates the positive correlation between the increase in measured chlorophyll content and the increase in leaf thickness resulting from exposure to salt stress.

The uncontrolled uptake of Na ion in salt-enriched soils during plant growth has been identified as a primary factor contributing to the occurrence of Na phytotoxicity phenomenon, causing nutrient imbalances as well as disruptions in plant biochemical and physiological processes [59,60]. Consistent with the results obtained by Ma et al. [48], which a decrease in Na in the leaves of rapeseed plants subjected to growth under NaCl of 70 and 150 mM, we observed a significant reduction in Na by WD only in lettuce leaves exposed to 100 mM NaCl. Therefore, these results might indicate that when the external conditions are no longer favorable for plant growth due to high salt levels, the use of WD could be beneficial because it would seem to help salt-stressed plants to reduce Na accumulation, and thus increase plant tolerance to salt by surviving despite such stressful conditions.

Plants are able to increase the level of carbohydrates to balance the salt-induced osmotic disorders [61]. One of the most common adaptive responses involves a reduction in soluble sugars and, simultaneously, an increase in starch synthesis [62–64] to modulate the osmotic potential and cellular turgor [65]. Indeed, our results also revealed that all −WD plants exposed to NaCl, except those grown at the highest NaCl concentration, significantly accumulated starch in the leaves. However, the effect of WD on leaf starch accumulation remains relatively unexplored, as only a few studies have reported a substantial (>100%) increase in starch content in lettuce leaves when sprayed with WD [66].

In this study, we assessed the leaf oxidative stress level through the evaluation of: (i) EL, whose value is linked to the estimation of the integrity of cell plasma membrane; (ii) MDA content, which is correlated to the cell plasma membrane lipid peroxidation level; (iii) ARA, which provides an estimation of the total antioxidant power, and thus of the overall plant capability to scavenge ROS; (iv) TPC, a specific group of powerful plant antioxidants. In particular, we have assisted an increase in EL and MDA for -WD plants as the exposure to NaCl increased, and a decrease in both these parameters for +WD plants only at the two highest NaCl concentrations (i.e., 200 and 300 mM). When these parameters increase, it means that the cell plasma membrane is damaged due to osmotic and ionic stresses (high EL content) or oxidative stress (high MDA content) [67]. For MDA content, our findings are in line with those published by Ma et al. [48], who found a substantial drop in the MDA content of rapeseed plants treated with 0.2% (*v*/*v*) WD at 70 and 150 mM NaCl compared to untreated plants. On the plant antioxidant response capacity, it can be emphasized that salt, in general, induced the activation of the plant response by increasing ARA. However, salt had a great impact in lowering the production and the accumulation of phenolic compounds in lettuce leaves, already noticeable at very low salt concentrations. The actions of WD can be useful in enhancing the non-enzymatic antioxidant compounds to resist not only high salt levels, but also in increasing the levels of phenols in the edible parts of plants grown under non-stressful conditions [68,69]. The use of natural products that induce stress in plants, known as eustress, by activating defense mechanisms and stimulating the increased production of antioxidants [70] is a commonly adopted practice in horticulture to improve the quality of horticultural plants [71]. Wood distillate would also appear to act as an eustressor, strengthening plant defenses and increasing their antioxidant power and the content of antioxidant compounds.

4. Materials and Methods

4.1. Plant Growth and Treatments

Lettuce (*Lactuca sativa* L., cv. Salanova) seedlings were purchased from a local nursery (Cerretani, Siena, Italy). In our laboratory, seedlings were transplanted into plastic pots (10 × 10 × 10 cm), containing 120 g of a commercial growing medium (VigorPlant Italia srl; the characteristics are as follows: moisture content = 43%; porosity = 92%; pH = 5.30 ± 0.03; EC = 1.12 ± 0.01 mS cm^{-1}; and cation exchange capacity = 56.9 ± 2.7 meq100 g DW^{-1}). Seedlings were grown for 5 weeks in a climate chamber with a day/night cycle of 16/8 h and 23/20 °C, 60% relative humidity, and 220 µmol s^{-1} m^{-2} PAR (photosynthetically active radiation). Immediately after transplantation, plants were irrigated using different sodium chloride (NaCl) solutions (JT Baker—Fisher Scientific, Milan, Italy) dissolved in water [0 mM (pH: 7.19; EC: 0.61 mS cm^{-1}), 50 mM (pH: 7.25; EC: 5.78 mS cm^{-1}), 100 mM (pH: 7.34; EC: 11.16 mS cm^{-1}), 200 mM (pH: 7.38; EC: 20.60 mS cm^{-1}), and 300 mM (pH: 7.42; EC: 29.61 mS cm^{-1})]. These salt concentrations were chosen based on our previous published work [72]. For each of the five treatments, five replicates were prepared. During the experimental growth period, the plants were irrigated with these solutions, when necessary, by maintaining the soil at 70% water holding capacity. Half of the plants from each condition were foliarly sprayed with 0.2% (v/v) wood distillate (WD), while the remaining half was foliarly sprayed with water once a week for 5 weeks. The WD used (BioDea®, BioEsperia, Arezzo, Italy) was obtained from the pyrolysis process using sweet chestnut (*Castanea sativa* Mill.) pruning as the feedstock. The wood distillate had the physics-chemical characteristics reported in Table 1, already showed in Celletti et al. [38].

Table 1. Main physico-chemical characteristics of the wood distillate (Celletti et al. [38]).

Parameter	Value	Method
TOC (% DW)	58.03	CHNS Elemental Analysis
TN (% DW)	1.06	CHNS Elemental Analysis
H (% DW)	7.27	CHNS Elemental Analysis
S (% DW)	0.07	CHNS Elemental Analysis
pH	4	UNI EN ISO 10523:2012
Density (g mL^{-1})	1.05	
Flash point (°C)	>60	ASTM D6450-16a
Total organic compounds (g L^{-1})	33.8	
Acidity (mg L^{-1})	1289	APAT CNR IRSA 2010 B Man 29 2003
Organic acids (mg L^{-1})	32.3	
Acetic acid (mg L^{-1})	21.5	
Polyphenols (g L^{-1})	24.5	
Phenols (g L^{-1})	3	
PCBs (mg L^{-1})	<0.2	CNR IRSA 24b Q 64 Vol 3 1988
Hydrocarbons C < 12 (mg L^{-1})	<0.1	EPA 5021A 2014 + EPA 8015D 2003
Hydrocarbons C10–C40 (mg L^{-1})	<0.1	UNI EN ISO 9377-2:2002
16 US-EPA PAHs (mg L^{-1})		EPA 3550C 2007 + EPA 8310 1986
Acenaphthene	<0.05	
Acenaphthylene	<0.05	
Anthracene	<0.05	
Benzo[a]anthracene	<0.05	
Benzo[a]pyrene	<0.05	
Benzo[b]fluoranthene	<0.05	
Benzo[g,h,i]perylene	<0.05	
Benzo[k]fluoranthene	<0.05	
Chrysene	<0.05	
Dibenz[a,h]anthracene	<0.05	
Fluoranthene	<0.05	
Fluorene	<0.05	
Indeno[1,2,3-cd]pyrene	<0.05	
Naphthalene	<0.05	
Phenanthrene	<0.05	
Pyrene	<0.05	

Table 1. *Cont.*

Parameter	Value	Method
Macronutrients (mg L^{-1})		Alkaline melting + ICP-MS analysis
Ca	325.50	
K	23.49	
Mg	6.79	
P	7.28	
Micronutrients (mg L^{-1})		Alkaline melting + ICP-MS analysis
Cu	0.18	
Fe	21.16	
Mn	0.58	
Mo	0.0007	
Zn	3.22	
Other nutrients		Alkaline melting + ICP-MS analysis
Al	1.96	
Ba	0.06	
Cr	0.03	
Na	103.59	

TOC: total organic carbon. **TN**: total nitrogen. **PCBs**: polychlorinated biphenyls. **16 US-EPA PAHs**: list of 16 priority polycyclic aromatic hydrocarbons as classified by the United State Environmental Protection Agency. **Al**: aluminum; **Ba**: barium; **C**: carbon; **Ca**: calcium; **Cr**: chromium; **Cu**: copper; **Fe**: iron; **K**: potassium; **Mg**: magnesium; **Mn**: manganese; **Mo**: molybdenum; **N**: nitrogen; **Na**: sodium; **Zn**: zinc.

At the end of the growth period (corresponding to 5 weeks of growth under the 5 different conditions), the above-ground part of lettuce plants was harvested, weighed for leaf fresh biomass, and stored at −20 °C for subsequent physiological and biochemical analyses.

4.2. *Leaf Analysis*

4.2.1. Chlorophyll

The chlorophyll content was measured using the portable non-destructive device (CCM-300, Opti-Science Inc., Hudson, NH, USA). The measurements were taken, avoiding the nerves of the leaves, on the three youngest and fully expanded leaves [66]. The results were reported as the quantity of chlorophyll on a surface basis (mg m^{-2}) [73].

4.2.2. Electrolyte Leakage

Uniform circular sections (Ø 5 mm) of the youngest and fully expanded leaf were obtained by means of a circular stainless-steel tip, and were carefully rinsed three times with deionized water (dH_2O). Subsequently, the leaf samples were transferred to 50-mL plastic tubes containing 20 mL of dH_2O and, after 2 h at room temperature, the EC1 of the solutions was measured using a conductivity meter (Crison Basic 30, Crison Instruments, S.A., Barcelona, Spain). Then, the sample tubes were placed in an oven at 90 °C for 25 min. Next, in the cooled solutions, the EC2 was measured for the second time. The following formula was used to calculate the electrolyte leakage (EL) [72]:

$$EL\ (\%) = (EC1/EC2) \times 100$$

4.2.3. Sodium

Lettuce leaves were oven-dried at 60 °C and pulverized. Approximately 150 mg DW of this material were microwave-digested (Milestone Ethos 900, Gladeville, Metrohm, Australia) with 3 mL of 70% (*v*/*v*) HNO_3, 0.2 mL of 50% (*v*/*v*) HF, and 0.5 mL of 30% (*v*/*v*) H_2O_2. Inductively coupled plasma-mass spectrometry (ICP-MS, Perkin Elmer NexION 350, Hopkiton, MA, USA) was used to determine the leaf concentration of Na. The analytical quality was assessed using the NCS DC 73350 certified standard reference material, "Poplar leaves", with recoveries ranging from 96 to 111%. The precision of the analysis was calculated using the coefficient of variation of 5 repetitions and was always >97%. The results are presented on a dry weight basis (g kg_{DW}^{-1}).

4.2.4. Malondialdehyde

The method proposed by Celletti et al. [74] was used to assess the level of lipid peroxidation, expressed as malondialdehyde (MDA) content, as MDA is considered to be a biomarker of oxidative stress. An amount of 0.5 g FW of lettuce leaves were homogenized in 5 mL of a pre-chilled reagent prepared by dissolving 0.25 g of 2-thiobarbituric acid (TBA) (Merck KGaA, Darmstadt, Germany) in 100 mL of 10% (*w*/*v*) trichloroacetic acid (Panreac, Castellar del Vallès, Barcelona, Spain). The samples were first incubated at 95 °C for 30 min and then ice-cooled to stop the reaction. After centrifugation at 5000 rpm for 20 min, the absorbance of the supernatants was measured at 532 nm and 600 nm using a UV-Vis spectrophotometer (8453, Agilent, Santa Clara, CA, USA). The non-specific turbidity adjustment was derived by subtracting the absorbance value observed at 600 nm. The lipid peroxidation level was estimated using the molar extinction coefficient (155 mM^{-1} cm^{-1}) of the MDA–TBA complex.

4.2.5. Carbohydrates

The content of three soluble sugars (fructose, glucose, and sucrose) was determined according to Fedeli et al. [66]. In total, 1 g of the leaf sample was homogenized in 2 mL dH_2O before a centrifugation step at 15000 rpm for 5 min. The supernatant was filtered at 0.45 μm (LLG-Syringe filters SPHEROS, Petaluma, CA, USA) with a syringe before being analyzed with HPLC (Waters 600 system, Milford, MA, USA), equipped with a Waters 2410 refractive index detector. Sugars were partitioned with dH_2O as the mobile phase, eluted at 0.5 mL min^{-1}, with a Waters Sugar-Pak I ion-exchange column (6.5 × 300 mm) kept at 90 °C using an external temperature controller (Waters Column Heater Module, Milford, MA, USA). Sugar contents were determined using calibration curves prepared by dissolving analytical sugars (Merck KGaA, Darmstadt, Germany) in dH_2O at a concentration ranging from 0.1 to 20 mg mL^{-1}. The results were expressed on a dry weight basis (mg $g_{DW}{}^{-1}$).

The content of the complex carbohydrate, i.e., starch, in the leaf samples was determined using the method described by Loppi et al. [75]. About 50 mg of dried sample was homogenized in 2 mL of dimethyl sulfoxide (DMSO) (≥99.9%, Carlo Erba, Cornaredo, MI, Italy), and then 0.5 mL of 8 M HCl was added. The sample tubes were placed at 60 °C for 30 min and then on ice to stop the reaction. In total, 0.5 mL of 8 M NaOH and 7 mL of dH_2O was added to the samples. Subsequently, they were centrifuged at 4000 rpm for 5 min. A total of 0.5 mL of supernatant was added to 2.5 mL of Lugol's solution (previously prepared by mixing 0.05 M HCl, 0.03% (*w*/*v*) I_2, and 0.06% (*w*/*v*) KI). After 15 min, samples were read at 605 nm using a UV-Vis spectrophotometer (8453, Agilent, Santa Clara, CA, USA). A calibration curve of pure starch (Merck KGaA, Darmstadt, Germany) (in the range of 10–400 g mL^{-1}) was used for the quantification of starch in the samples. The results are expressed on a dry weight basis (mg $g_{DW}{}^{-1}$).

4.2.6. Antioxidants

The total antioxidant power was assessed using the method described by Fedeli et al. [76]. After homogenizing 0.5 g FW of lettuce leaves in 2 mL of 80% (*v*/*v*) ethanol, the mixture was centrifuged at 15,000 rpm for 5 min. An aliquot of 200 μL of supernatant was taken and added to 1 mL of 2,2-diphenyl-1-picrylhydrazyl (DPPH) solution (Merck KGaA, Darmstadt, Germany), which was prepared by dissolving 3.9 mg of DPPH in 100 mL of 80% (*v*/*v*) methanol. To quantify the antioxidant capacity of the samples, a blank and a control were prepared. After an incubation of 1 h in the dark, the absorbance of the samples was measured at 517 nm using a UV-Vis spectrophotometer (Agilent 8453, Santa Clara, CA, USA). The total antioxidant power was expressed as the antiradical activity (ARA, %), using the following formula:

$$ARA\ (\%) = 100 \times [1 - (sample\ absorbance)/(control\ absorbance)]$$

The total phenol content (TPC) was determined using the method described by Lamaro et al. [77]. Briefly, 0.5 mg FW of lettuce leaves were homogenized in 4 mL of 70% (*v*/*v*) acetone (Merck KGaA, Darmstadt, Germany). The homogenate was centrifuged at 4000 rpm for 5 min. Afterwards, 0.5 mL of the collected supernatant was added to 3.95 mL dH_2O, 0.125 mL Folin–Ciocalteu reagent (Merck KGaA, Darmstadt, Germany), 0.75 mL saturated Na_2CO_3. After incubation at 37 °C for 30 min, the samples were again centrifuged at 4000 rpm for 5 min and the absorbance of the supernatant was measured at 765 nm using a UV-Vis spectrophotometer (Agilent 8453; Santa Clara, CA, USA). A calibration curve of gallic acid (GA, >98%, Merck KGaA, Darmstadt, Germany) (in the range of 5–20 g mL^{-1}) was used for the quantification of TPC in the samples. The results are expressed as mg of the GA equivalent on a dry weight basis (mg GAE $g_{DW}{}^{-1}$).

4.3. Statistical Analysis

Due to the non-normality of the data (Shapiro–Wilk test, $p < 0.05$), a non-parametric approach was followed [78]. The parameters analyzed were expressed as their median value and the associated error was expressed as the interquartile range divided by the square root of the number of observations from five biological replicates (n = 5). The significance of differences ($p < 0.05$) between treatments was assessed using the Kruskal–Wallis test and pairwise differences within the same salt treatments were assessed using the Kolmogorov–Smirnov test. All calculations were performed with the R software (v 4.4.0) [79].

5. Conclusions

This study revealed the varying thresholds of salt tolerance in lettuce plants, delineating their resilience up to 100 mM NaCl and their subsequent decline in defense mechanisms beyond this threshold. The observed reductions in leaf fresh weight and antioxidant responses, coupled with increased oxidative stress, underscore the critical importance of understanding plant responses to environmental stressors.

In addition, the application of 0.2% (*w*/*w*) WD via foliar spraying emerges as a promising agronomic practice, effectively supporting plant growth and enabling tolerance to extreme NaCl conditions, even from high NaCl concentrations such as 200 mM.

Thus, in the face of worsening environmental challenges, the use of WD is a novel approach that offers a feasible solution for cultivating plants in soils contaminated by high salt levels, traditionally deemed unsuitable, and contributes to sustainable agricultural practices.

Author Contributions: Conceptualization, S.C. and S.L.; Investigation, R.F.; Data curation, R.F. and S.C.; Methodology, R.F. and S.C.; Writing—original draft preparation, R.F. and S.C.; Writing—review and editing, R.F., S.C. and S.L.; Supervision, S.C. and S.L.; Resources, S.C. and S.L. All authors have read and agreed to the published version of the manuscript.

Funding: This research received no external funding.

Data Availability Statement: Data will be made available on reasonable request by the corresponding author.

Acknowledgments: We thank Francesco Barbagli (BioDea and BioEsperia Srl) for kindly providing the wood distillate.

Conflicts of Interest: The authors declare no conflicts of interest.

References

1. Hassani, A.; Azapagic, A.; Shokri, N. Global predictions of primary soil salinization under changing climate in the 21st century. *Nat. Commun.* **2021**, *12*, 6663. [CrossRef] [PubMed]
2. Chourasia, K.N.; Lal, M.K.; Tiwari, R.K.; Dev, D.; Kardile, H.B.; Patil, V.U.; Pramanik, D. Salinity stress in potato: Understanding physiological, biochemical and molecular responses. *Life* **2021**, *11*, 545. [CrossRef] [PubMed]
3. Aroca, R.; Porcel, R.; Ruiz-Lozano, J.M. Regulation of root water uptake under abiotic stress conditions. *J. Exp. Bot.* **2012**, *63*, 43–57. [CrossRef] [PubMed]

4. Sévin, D.C.; Stählin, J.N.; Pollak, G.R.; Kuehne, A.; Sauer, U. Global metabolic responses to salt stress in fifteen species. *PLoS ONE* **2016**, *11*, e0148888. [CrossRef]
5. Petretto, G.L.; Urgeghe, P.P.; Massa, D.; Melito, S. Effect of salinity (NaCl) on plant growth, nutrient content, and glucosinolate hydrolysis products trends in rocket genotypes. *Plant Physiol. Biochem.* **2019**, *141*, 30–39. [CrossRef]
6. Raza, A.; Ashraf, F.; Zou, X.; Zhang, X.; Tosif, H. Plant adaptation and tolerance to environmental stresses: Mechanisms and perspectives. In *Plant Ecophysiology and Adaptation under Climate Change: Mechanisms and Perspectives I: General Consequences and Plant Responses*; Springer: New York, NY, USA, 2020; pp. 117–145.
7. Waszczak, C.; Carmody, M.; Kangasjärvi, J. Reactive oxygen species in plant signaling. *Annu. Rev. Plant Biol.* **2018**, *69*, 209–236. [CrossRef] [PubMed]
8. Ahmad, R.; Hussain, S.; Anjum, M.A.; Khalid, M.F.; Saqib, M.; Zakir, I.; Ahmad, S. Oxidative stress and antioxidant defense mechanisms in plants under salt stress. In *Plant Abiotic Stress Tolerance: Agronomic, Molecular and Biotechnological Approaches*; Springer: New York, NY, USA, 2019; pp. 191–205.
9. Giordano, M.; Petropoulos, S.A.; Rouphael, Y. Response and defence mechanisms of vegetable crops against drought, heat and salinity stress. *Agriculture* **2021**, *11*, 463. [CrossRef]
10. McNeil, P.L.; Steinhardt, R.A. Loss, restoration, and maintenance of plasma membrane integrity. *J. Cell Biol.* **1997**, *137*, 1–4. [CrossRef] [PubMed]
11. Imadi, S.R.; Shah, S.W.; Kazi, A.G.; Azooz, M.M.; Ahmad, P. Phytoremediation of saline soils for sustainable agricultural productivity. In *Plant Metal Interaction*; Elsevier: Amsterdam, The Netherlands, 2016; pp. 455–468.
12. Hasanuzzaman, M.; Nahar, K.; Fujita, M. Plant response to salt stress and role of exogenous protectants to mitigate salt-induced damages. In *Ecophysiology and Responses of Plants under Salt Stress*; Springer: New York, NY, USA, 2013; pp. 25–87.
13. Hossain, M.S.; Dietz, K.J. Tuning of redox regulatory mechanisms, reactive oxygen species and redox homeostasis under salinity stress. *Front. Plant Sci.* **2016**, *7*, 548. [CrossRef]
14. Ullah, A.; Bano, A.; Khan, N. Climate change and salinity effects on crops and chemical communication between plants and plant growth-promoting microorganisms under stress. *Front. Sustain. Food Syst.* **2021**, *5*, 618092. [CrossRef]
15. Kovda, V.A. Origin and regime of saline soils. *Soil Sci.* **1949**, *67*, 71. [CrossRef]
16. Rengasamy, P. World salinization with emphasis on Australia. *J. Exp. Bot.* **2006**, *57*, 1017–1023. [CrossRef] [PubMed]
17. Amini, S.; Ghadiri, H.; Chen, C.; Marschner, P. Salt-affected soils, reclamation, carbon dynamics, and biochar: A review. *J. Soils Sediments* **2016**, *16*, 939–953. [CrossRef]
18. European Commission 2008. Available online: https://esdac.jrc.ec.europa.eu/themes/soil-salinization (accessed on 12 October 2023).
19. Ondrasek, G. Irrigation in Agroecosystems. BoD–Books on Demand. 2019. Available online: https://www.intechopen.com/books/6952 (accessed on 6 January 2024).
20. Hasanuzzaman, M.; Nahar, K.; Alam, M.M.; Bhowmik, P.C.; Hossain, M.A.; Rahman, M.M. Potential use of halophytes to remediate saline soils. *BioMed. Res. Int.* **2014**, *2014*, 589341. [CrossRef] [PubMed]
21. Pitman, M.G.; Läuchli, A. Global impact of salinity and agricultural ecosystems. In *Salinity: Environment-Plants-Molecules*; Springer: Dordrecht, The Netherlands, 2022; pp. 3–20.
22. GAP Report. Global Agricultural Productivity Report§ (GAP Report§) Global Harvest Initiative, Washington. 2018. Available online: https://globalagriculturalproductivity.org/wp-content/uploads/2019/01/GHI_2018-GAP-Report_FINAL-10.03.pdf (accessed on 21 August 2022).
23. Dhankher, O.P.; Foyer, C.H. Climate resilient crops for improving global food security and safety. *Plant Cell Environ.* **2018**, *41*, 877–884. [CrossRef]
24. Wen-bin, Z.; Feng-ying, D. Closing crop yield and efficiency gaps for food security and sustainable agriculture. *J. Integr. Agric.* **2021**, *20*, 343.
25. Arora, S.; Singh, A.K.; Sahni, D. Bioremediation of salt-affected soils: Challenges and opportunities. In *Bioremediation of Salt Affected Soils: An Indian Perspective*; Springer: New York, NY, USA, 2017; pp. 275–301.
26. Shahbaz, M.; Ashraf, M.; Al-Qurainy, F.; Harris, P.J. Salt tolerance in selected vegetable crops. *Crit. Rev. Plant Sci.* **2012**, *31*, 303–320. [CrossRef]
27. Foolad, M.R. Recent advances in genetics of salt tolerance in tomato. *Plant Cell Tissue Organ. Cult.* **2004**, *76*, 101–119. [CrossRef]
28. Del Buono, D. Can biostimulants be used to mitigate the effect of anthropogenic climate change on agriculture? It is time to respond. *Sci. Total Environ.* **2021**, *751*, 141763. [CrossRef]
29. Grewal, A.; Abbey, L.; Gunupuru, L.R. Production, prospects and potential application of pyroligneous acid in agriculture. *J. Anal. Appl. Pyrolysis* **2018**, *135*, 152–159. [CrossRef]
30. Fačkovcová, Z.; Vannini, A.; Monaci, F.; Grattacaso, M.; Paoli, L.; Loppi, S. Effects of wood distillate (pyroligneous acid) on sensitive bioindicators (lichen and moss). *Ecotoxicol. Environ. Saf.* **2020**, *204*, 111117. [CrossRef]
31. Fačkovcová, Z.; Vannini, A.; Monaci, F.; Grattacaso, M.; Paoli, L.; Loppi, S. Uptake of trace elements in the water fern Azolla filiculoides after short-term application of chestnut wood distillate (Pyroligneous Acid). *Plants* **2020**, *9*, 1179. [CrossRef] [PubMed]
32. Filippelli, A.; Ciccone, V.; Loppi, S.; Morbidelli, L. Characterization of the safety profile of sweet chestnut wood distillate employed in agriculture. *Safety* **2021**, *7*, 79. [CrossRef]

33. Fanfarillo, E.; Fedeli, R.; Fiaschi, T.; de Simone, L.; Vannini, A.; Angiolini, C.; Loppi, S.; Maccherini, S. Effects of wood distillate on seedling emergence and first-stage growth in five threatened arable plants. *Diversity* **2022**, *14*, 669. [CrossRef]
34. Italian Ministerial Decree 6793 18/07/2018. 2018. Available online: https://www.gazzettaufficiale.it/eli/id/2018/09/05/18A05693/sg (accessed on 18 July 2023).
35. Fedeli, R.; Vannini, A.; Celletti, S.; Maresca, V.; Munzi, S.; Cruz, C.; Loppi, S. Foliar application of wood distillate boosts plant yield and nutritional parameters of chickpea. *Ann. Appl. Biol.* **2023**, *182*, 57–64. [CrossRef]
36. Ofoe, R.; Qin, D.; Gunupuru, L.R.; Thomas, R.H.; Abbey, L. Effect of pyroligneous acid on the productivity and nutritional quality of greenhouse tomato. *Plants* **2022**, *11*, 1650. [CrossRef] [PubMed]
37. Ofoe, R.; Mousavi, S.M.N.; Thomas, R.H.; Abbey, L. Foliar application of pyroligneous acid acts synergistically with fertilizer to improve the productivity and phytochemical properties of greenhouse-grown tomato. *Sci. Rep.* **2024**, *14*, 1934. [CrossRef]
38. Celletti, S.; Fedeli, R.; Ghorbani, M.; Aseka, J.M.; Loppi, S. Exploring sustainable alternatives: Wood distillate alleviates the impact of bioplastic in basil plants. *Sci. Total Environ.* **2023**, *900*, 166484. [CrossRef] [PubMed]
39. Wagner, A.; Kaupenjohann, M. Suitability of biochars (pyro- and hydrochars) for metal immobilization on former sewage-field soils. *Eur. J. Soil Sci.* **2014**, *65*, 139–148. [CrossRef]
40. Celletti, S.; Lanz, M.; Bergamo, A.; Benedetti, V.; Basso, D.; Baratieri, M.; Cesco, S.; Mimmo, T. Evaluating the Aqueous Phase From Hydrothermal Carbonization of Cow Manure Digestate as Possible Fertilizer Solution for Plant Growth. *Front. Plant Sci.* **2021**, *12*, 687434. [CrossRef]
41. Najafi-Ghiri, M.; Mirsoleimani, A.; Boostani, H.R.; Amin, H. Influence of wood vinegar and potassium application on soil properties and Ca/K ratio in citrus rootstocks. *J. Soil Sci. Plant Nutr.* **2022**, *22*, 334–344. [CrossRef]
42. Chu, L.; Liu, H.; Zhang, Z.; Zhan, Y.; Wang, K.; Yang, D.; Yu, J. Evaluation of Wood Vinegar as an Herbicide for Weed Control. *Agronomy* **2022**, *12*, 3120. [CrossRef]
43. Fedeli, R.; Vannini, A.; Grattacaso, M.; Loppi, S. Wood distillate (pyroligneous acid) boosts nutritional traits of potato tubers. *Ann. Appl. Biol.* **2023**, *183*, 135–140. [CrossRef]
44. Fedeli, R.; Cruz, C.; Loppi, S.; Munzi, S. Hormetic Effect of Wood Distillate on Hydroponically Grown Lettuce. *Plants* **2024**, *13*, 447. [CrossRef] [PubMed]
45. Fedeli, R.; Fiaschi, T.; Angiolini, C.; Maccherini, S.; Loppi, S.; Fanfarillo, E. Dose-Dependent and Species-Specific Effects of Wood Distillate Addition on the Germination Performance of Threatened Arable Plants. *Plants* **2023**, *12*, 3028. [CrossRef] [PubMed]
46. Aguirre, J.L.; Baena, J.; Martín, M.T.; González, S.; Manjón, J.L.; Peinado, M. Herbicidal effects of wood vinegar on nitrophilous plant communities. *Food Energy Secur.* **2020**, *9*, e253. [CrossRef] [PubMed]
47. Theerakulpisut, P.; Kanawapee, N.; Panwong, B. Seed priming alleviated salt stress effects on rice seedlings by improving Na+/K+ and maintaining membrane integrity. *IJPB* **2017**, *7*, 6402. [CrossRef]
48. Ma, J.; Islam, F.; Ayyaz, A.; Fang, R.; Hannan, F.; Farooq, M.A.; Zhou, W. Wood vinegar induces salinity tolerance by alleviating oxidative damages and protecting photosystem II in rapeseed cultivars. *Ind. Crops Prod.* **2022**, *189*, 115763. [CrossRef]
49. Wu, S.; Yuan, Y.; Liu, Q. Effects of wood vinegar on the growth of oil sunflower (Helianthus annuus L.) seedlings in a salt-affected soil of Yellow River Delta, China. In *Advances in Renewable Energy and Sustainable Development*; CRC Press: Boca Raton, FL, USA, 2022; pp. 79–83.
50. Shannon, M.C.; Grieve, C.M. Tolerance of vegetable crops to salinity. *Sci. Hortic.* **1998**, *78*, 5–38. [CrossRef]
51. Xu, C.; Mou, B. Evaluation of lettuce genotypes for salinity tolerance. *HortScience* **2015**, *50*, 1441–1446. [CrossRef]
52. Borella, M.; Baghdadi, A.; Bertoldo, G.; Della Lucia, M.C.; Chiodi, C.; Celletti, S.; Nardi, S. Transcriptomic and physiological approaches to decipher cold stress mitigation exerted by brown-seaweed extract (BSE) application in tomato. *Front. Plant Sci.* **2023**, *14*, 1232421. [CrossRef]
53. Zulfiqar, F.; Moosa, A.; Ferrante, A.; Nafees, M.; Darras, A.; Nazir, M.M.; Soliman, T.M. Exogenous foliar application of melatonin mitigates salt induced oxidative stress and promotes growth in Gerbera jamosonii. *S. Afr. J. Bot.* **2023**, *161*, 678–684. [CrossRef]
54. Zuzunaga-Rosas, J.; González-Orenga, S.; Tofei, A.M.; Boscaiu, M.; Moreno-Ramón, H.; Ibáñez-Asensio, S.; Vicente, O. Effect of a biostimulant based on polyphenols and glycine betaine on tomato plants' responses to salt stress. *Agronomy* **2022**, *12*, 2142. [CrossRef]
55. Fedeli, R.; Marotta, L.; Frattaruolo, L.; Panti, A.; Carullo, G.; Fusi, F.; Saponara, S.; Gemma, S.; Butini, S.; Cappello, A.R.; et al. Nutritionally-enriched tomatoes (*Solanum lycopersicum* L.) grown with wood distillate: Chemical and biological characterization for quality assessment. *J. Food Sci.* **2023**, *88*, 5324–5338. [CrossRef] [PubMed]
56. Sardar, H.; Khalid, Z.; Ahsan, M.; Naz, S.; Nawaz, A.; Ahmad, R.; Abou Fayssal, S. Enhancement of salinity stress tolerance in lettuce (*Lactuca sativa* L.) via foliar application of nitric oxide. *Plants* **2023**, *12*, 1115. [CrossRef] [PubMed]
57. Zhu, J.K. Plant salt tolerance. *Trends Plant Sci.* **2001**, *6*, 66–71. [CrossRef]
58. Wei, Q.; Ma, X.; Dong, J. Preparation, chemical constituents and antimicrobial activity of pyroligneous acids from walnut tree branches. *J. Anal. Appl. Pyrolysis* **2010**, *87*, 24–28. [CrossRef]
59. Khan, I.; Raza, M.A.; Awan, S.A.; Shah, G.A.; Rizwan, M.; Ali, B.; Huang, L. Amelioration of salt induced toxicity in pearl millet by seed priming with silver nanoparticles (AgNPs): The oxidative damage, antioxidant enzymes and ions uptake are major determinants of salt tolerant capacity. *Plant Physiol. Biochem.* **2020**, *156*, 221–232. [CrossRef]
60. Mostofa, M.G.; Saegusa, D.; Fujita, M.; Tran, L.S.P. Hydrogen sulfide regulates salt tolerance in rice by maintaining Na+/K+ balance, mineral homeostasis and oxidative metabolism under excessive salt stress. *Front. Plant Sci.* **2015**, *6*, 1055. [CrossRef]

61. Hayashi, H.; Alia Mustardy, L.; Deshnium, P.; Ida, M.; Murata, N. Transformation of Arabidopsis thaliana with the cod A gene for choline oxidase; accumulation of glycinebetaine and enhanced tolerance to salt and cold stress. *Plant J.* **1997**, *12*, 133–142. [CrossRef]
62. Adams, C.A.; Rinne, R.W.; Fjerstad, M.C. Starch deposition and carbohydrate activities in developing and germinating soybean seeds. *Ann. Bot.* **1980**, *45*, 577–582. [CrossRef]
63. Glavier, V.; Perez-Alfocea, F.; Bourgeais-Chaillou, P.; Guerrier, G. Influence of cotyledons upon sucrose metabolism in axis of germinating pea. *J. Plant Physiol.* **1991**, *138*, 421–428. [CrossRef]
64. Perez-Alfocea, F.; Bolarin, M.C.; Guerrier, G. Sucrose Metabolism in NaCl-Treated Calli from Lycopersicon esculentum L. pennellii and their Interspecific Hybrid. *J. Plant Physiol.* **1995**, *145*, 161–167. [CrossRef]
65. Kerepesi, I.; Galiba, G. Osmotic and salt stress-induced alteration in soluble carbohydrate content in wheat seedlings. *Crop Sci.* **2000**, *40*, 482–487. [CrossRef]
66. Fedeli, R.; Vannini, A.; Guarnieri, M.; Monaci, F.; Loppi, S. Bio-based solutions for agriculture: Foliar application of wood distillate alone and in combination with other plant-derived corroborants results in different effects on lettuce (*Lactuca sativa* L.). *Biology* **2022**, *11*, 404. [CrossRef] [PubMed]
67. Bajji, M.; Kinet, J.M.; Lutts, S. Osmotic and ionic effects of NaCl on germination, early seedling growth, and ion content of Atriplex halimus (Chenopodiaceae). *Can. J. Bot.* **2002**, *80*, 297–304. [CrossRef]
68. Benzon, H.R.L.; Lee, S.C. Potential of wood vinegar in enhancing fruit yield and antioxidant capacity in tomato. *Korean J. Plant Resour.* **2016**, *29*, 704–711. [CrossRef]
69. Mohd Amnan, M.A.; Teo, W.F.A.; Aizat, W.M.; Khaidizar, F.D.; Tan, B.C. Foliar application of oil palm wood vinegar enhances Pandanus amaryllifolius tolerance under drought stress. *Plants* **2023**, *12*, 785. [CrossRef]
70. Baenas, N.; Garcia-Viguera, C.; Moreno, D.A. Elicitation: A tool for enriching the bioactive composition of foods. *Molecules* **2014**, *19*, 13541–13563. [CrossRef]
71. Marchica, A.; Ascrizzi, R.; Flamini, G.; Cotrozzi, L.; Tonelli, M.; Lorenzini, G.; Pellegrini, E. Ozone as eustress for enhancing secondary metabolites and bioactive properties in Salvia officinalis. *Ind. Crops Prod.* **2021**, *170*, 113730. [CrossRef]
72. Fedeli, R.; Vannini, A.; Djatouf, N.; Celletti, S.; Loppi, S. Can lettuce plants grow in saline soils supplemented with biochar? *Heliyon* **2024**, *10*, e26526. [CrossRef] [PubMed]
73. Gitelson, A.A.; Buschmann, C.; Lichtenthaler, H.K. The chlorophyll fluorescence ratio F735/F700 as an accurate measure of the chlorophyll content in plants. *Remote Sens. Environ.* **1999**, *69*, 296–302. [CrossRef]
74. Celletti, S.; Fedeli, R.; Ghorbani, M.; Loppi, S. Impact of starch-based bioplastic on growth and biochemical parameters of basil plants. *Sci. Total Environ.* **2023**, *856*, 159163. [CrossRef] [PubMed]
75. Loppi, S.; Fedeli, R.; Canali, G.; Guarnieri, M.; Biagiotti, S.; Vannini, A. Comparison of the Mineral and Nutraceutical Profiles of Elephant Garlic (*Allium ampeloprasum* L.) Grown in Organic and Conventional Fields of Valdichiana, a Traditional Cultivation Area of Tuscany, Italy. *Biology* **2021**, *10*, 1058. [CrossRef] [PubMed]
76. Fedeli, R.; Celletti, S.; Loppi, S.; Vannini, A. Comparison of the Effect of Solid and Liquid Digestate on the Growth of Lettuce (*Lactuca sativa* L.) Plants. *Agronomy* **2023**, *13*, 782. [CrossRef]
77. Helsel, D.R. Advantages of nonparametric procedures for analysis of water quality data. *Hydrol. Sci. J.* **1987**, *32*, 179–190. [CrossRef]
78. Lamaro, G.P.; Tsehaye, Y.; Girma, A.; Vannini, A.; Fedeli, R.; Loppi, S. Evaluation of Yield and Nutraceutical Traits of Orange-Fleshed Sweet Potato Storage Roots in Two Agro-Climatic Zones of Northern Ethiopia. *Plants* **2023**, *12*, 1319. [CrossRef]
79. R Core Team. *R: A Language and Environment for Statistical Computing*; R Foundation for Statistical Computing: Vienna, Austria, 2023. Available online: https://www.R-project.org/ (accessed on 15 December 2023).

plants

MDPI

Article

Single or Blended Application of Non-Microbial Plant-Based Biostimulants and *Trichoderma atroviride* as a New Strategy to Enhance Greenhouse Cherry Tomato Performance

Lorena Vultaggio [1], Michele Ciriello [2], Emanuela Campana [2], Pietro Bellitto [1], Beppe Benedetto Consentino [1,*], Youssef Rouphael [2], Giuseppe Colla [3], Fabiana Mancuso [1], Salvatore La Bella [1,4], Simona Napoli [1,*] and Leo Sabatino [1]

[1] Dipartimento Scienze Agrarie, Alimentari e Forestali, Università di Palermo, Viale delle Scienze, 90128 Palermo, Italy; lorena.vultaggio@unipa.it (L.V.); pietro.bellitto@unipa.it (P.B.); fabiana.mancuso@you.unipa.it (F.M.); salvatore.labella@unipa.it (S.L.B.); leo.sabatino@unipa.it (L.S.)

[2] Department of Agricultural Sciences, University of Naples Federico II, 80055 Portici, Italy; michele.ciriello@unina.it (M.C.); emanuela.campana@unina.it (E.C.); youssef.rouphael@unina.it (Y.R.)

[3] Department of Agriculture and Forest Sciences, University of Tuscia, 01100 Viterbo, Italy; giucolla@uinitus.it

[4] Research Consortium for the Development of Innovative Agro-Environmental Systems (Corissia), Via della Libertà 203, 90143 Palermo, Italy

* Correspondence: beppebenedetto.consentino@unipa.it (B.B.C.); simona.napoli@unipa.it (S.N.)

Abstract: The need to increase yield and enhance the sustainability of crop production systems has led to the development and employment of natural products, such as plant biostimulants. In recent years, a number of reports have researched the effects of biostimulants on plant performance; however, few studies have focused on the mutual application of microbial and/or non-microbial biostimulants. This research, conducted in the framework of the SO.MI.PR.O.N regional project, aimed to investigate the single or mutual application of three biostimulants, a tropical plant extract (PE), a vegetal protein hydrolysate (PH), and *Trichoderma atroviride*, on 'Creativo' F1 cherry tomato plants cultivated during two growing cycles (2022–2023 and 2023–2024). Our results showed that plants treated with the combination Tricho + PE + PH had statistically significant higher fresh shoot biomass (+64.2%, 1647.0 g plant^{-1}), total fruit production (+37.9%, 1902.5 g plant^{-1}), marketable fruit production (+52.9%, 1778.5 g plant^{-1}), and average weight of marketable fruits (+53.1%, 17.0 g) compared to control plants (untreated plants). Furthermore, biostimulant treatments, especially *T. atroviride*, variably enhanced cherry tomato fruits' qualitative traits, such as firmness, total soluble solids, ascorbic acid, lycopene, and total polyphenols compared to control plants. Overall, the best combinations to increase tomato fruit qualitative features were PE + PH, Tricho + PE, and Tricho + PH. From an economic point of view, the best treatment for achieving the highest net return was PE. This study underlines that biostimulant features (yield, qualitative aspects, and economic profitability) can be supported through the application of specific biostimulant combinations.

Keywords: biostimulant synergy; greenhouse cultivation; microbial biostimulants; non-microbial biostimulants; *Solanum lycopersicum* L.

1. Introduction

Tomato (*Solanum lycopersicum* L.) is one of the most consumed vegetables at the global level and its fruits are consumed fresh or processed (soups, juices, sauces, etc.) [1]. Tomato is a notable source of bioactive components efficient of granting appropriate health benefits in humans, such as carotenoids (primarily lycopene and β-carotene), ascorbic acid, flavonoids, and phenols, displaying antioxidant and antiradical activity [2–5]. In the Italian diet, tomato fruits can be considered as a primary source of lycopene, β-carotene, and polyphenols [6–8], and, accompanied by a restricted caloric supply, tomato fruits can be considered as a natural functional food. As reported by the FAO Statistical Database [9] (http://www.fao.org, accessed on 17 July 2024), the estimated total world production of tomato is approximately 190 million tonnes, with China contributing 31.6% to world production. In Europe, Italy is the main tomato producer (6,000,000 t in 2021), while the Apulia region is the leading producer in Italy (more than 2,000,000 t), followed by Sicily (632,038 t) [10].

Tomato is commonly grown in intensive monoculture systems, especially when it is cultivated under a protected or controlled environment [11–13]. As greenhouse cultivation has a prominent impact on agro-ecosystems, efforts are currently focused on finding sustainable and pioneering agronomic tools that could provide good yield and crop quality and, simultaneously, decrease the horticultural sector's impact on the environment [14–21]. Concomitantly, consumers' growing concern about ecological and well-being topics drives them to eat healthy and accessible vegetables cultivated via eco-friendly and sustainable cultivation practices [22–25]. Thus, contemporary horticulture is exposed to the strict reduction in synthetic pesticide and fertilizer use [26,27]. There are reports on an eclectic assortment of vegetables that, once exposed to the foliar or root supply of non-microbial biostimulants, such as vegetal-based protein hydrolysates (V-PHs) or plant extracts (PEs), manifest in enhanced plant metabolism, enhanced mineral uptake through the stimulation and accumulation of the phytochemical's biosynthesis, and increased abiotic distress tolerance [28,29]. This in turn increases the yield of vegetables [30,31].

The upsurge in crop yield encouraged by non-microbial biostimulants (V-PHs and PEs) under favorable, sub- or supra-optimal, or unfavorable growing environments could be related to various direct and indirect collaborative physiological phenomena, comprising the elicitation of enzymatic actions, triggering of hormone-like activity, and modulation of the plant root system, which leads to advanced absorption capacity of water and nutrients [32,33]. Simultaneously, within the biostimulants panorama, microbial biostimulants, such as plant growth-promoting rhizobacteria (PGPR), arbuscular mycorrhizal fungi (AM), or *Trichoderma* spp., substantially promote plant growth and productivity, in addition to nutritional and functional quality [34–38]. *Trichoderma* spp. are versatile fungi with diverse effective tasks, making them a suggested tool to boost horticultural sustainability. These acknowledgments may include the capacity of *Trichoderma* spp. as biocontrol agents, plant growth elicitors, and biofertilizers [39–44]. Likewise, *Trichoderma* spp. might reduce plant abiotic stress by triggering endogenous mechanisms controlled by plant hormones and alterations in the host plant's metabolism [45–47].

Notwithstanding relevant studies on the influences of microbial and non-microbial biostimulants on vegetables, an improved comprehension of the merging application of microbial and/or non-microbial biostimulants is essential to enhance horticultural sustainability and resilience [48–51]. In this regard—to the best of our knowledge—there is no scientific indication of the effect of the joint application of *T. atroviride* and two non-microbial biostimulants, namely a V-PH and a PE, on the plant growth, yield, and fruit quality features of cherry tomato. The present research wants to fill the research gap regarding the effects of microbial and non-microbial biostimulants' mutual application (additive, synergistic, or antagonistic) on tomato plants. Thus, a study was conducted with the aim of explaining the influences of three biostimulants (*T. atroviride*, a V-PH, and a PE)—supplied alone or in a blend—on the performance and economic profitability of an F_1 cherry tomato genotype. The results of this study will help to understand the possible interactions between different classes of biostimulants for the modulation of cherry tomato plant performance.

2. Results

2.1. Analysis of Variance (ANOVA)

Table 1 shows the Analysis of Variance (ANOVA) output for all recorded parameters.

For all parameters, ANOVA did not point out a statistically significant influence of year; indeed, the p-values ranged from 0.398 to 0.959. Contrariwise, biostimulant treatments had a highly significant effect (p-value ≤ 0.001) on tomato productive and qualitative traits. Something that is noteworthy, as shown in Table 1, is the significant influence of the interaction between year and biostimulant recorded for fresh shoot biomass (p-value ≤ 0.001), titratable acidity (p-value ≤ 0.001), ascorbic acid (p-value = 0.013), and lycopene (p-value = 0.004).

Table 1. Analysis of variance output for all recorded parameters in response to year, biostimulant, and their interaction.

Parameters	Year (Y)	Biostimulant (B)	Y × B
		Significance (p-value)	
Total fruit production	n.s. (0.737)	*** (0.001)	n.s. (0.823)
Marketable fruit production	n.s. (0.768)	*** (0.001)	n.s. (0.507)
Number of marketable fruits	n.s. (0.778)	*** (0.001)	n.s. (0.787)
Average weight of marketable fruits	n.s. (0.901)	*** (0.001)	n.s. (0.632)
Fresh shoot biomass	n.s. (0.398)	*** (0.001)	*** (0.001)
Fruit dry matter	n.s. (0.592)	*** (0.001)	n.s. (0.955)
Firmness	n.s. (0.937)	*** (0.001)	n.s. (0.743)
Total soluble solids	n.s. (0.959)	*** (0.001)	n.s. (0.609)
Titratable acidity	n.s. (0.831)	*** (0.001)	*** (0.001)
Ascorbic acid	n.s. (0.597)	*** (0.001)	* (0.013)
Total polyphenols	n.s. (0.937)	*** (0.001)	n.s. (0.740)
Lycopene	n.s. (0.774)	*** (0.001)	** (0.004)

Biostimulant (B) effect and Y × B interaction were determined using Tukey's HSD test ($p \leq 0.05$) while the mean Y effect was compared using the t-test. n.s.: not significant; *: significant at $p \leq 0.05$; **: significant at $p \leq 0.01$; ***: significant at $p \leq 0.001$.

2.2. Greenhouse Temperatures, Fresh Shoot Biomass, and Fruit Production Traits

Overall, during the first cultivation year (2022–2023), lower maximum and minimum temperatures were recorded as compared to the second year; in particular, lower maximum temperatures were recorded at the beginning of the growing cycle (first month) and in the final phase of the cultivation period (last two months), whereas lower minimun temperatures were observed from mid-January to mid-February.

The biostimulant × year interaction statistically influenced fresh shoot biomass (Figure 1).

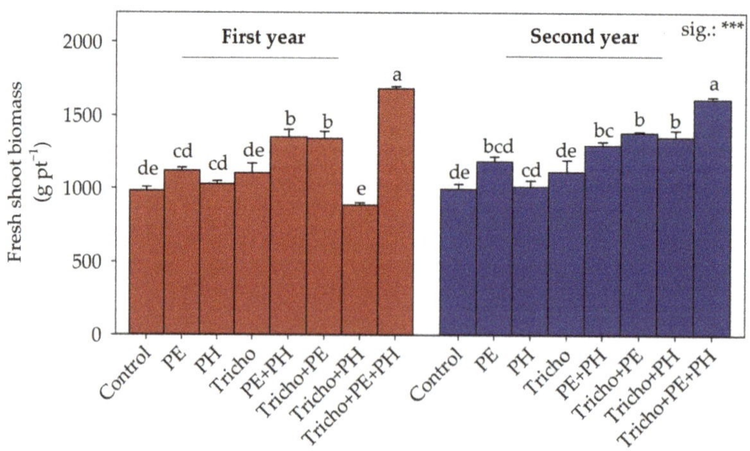

Figure 1. Fresh shoot biomass as influenced by cultivation year and biostimulant interaction. Separation of Y × B interaction was performed using Tukey's HSD test ($p \leq 0.05$). ***: significant at $p < 0.001$. First year: 2022–2023; second year: 2023–2024; Control: non −treated plants; PE: tropical plant extract; PH: legume-derived protein hydrolysate; Tricho: *Trichoderma atroviride*. Different letters indicate significantly different means. Bars indicate the standard error.

Plants supplied with Tricho + PE + PH cultivated both during the first or second year had the highest values, followed by those recorded in the plants treated with Tricho + PE grown in the first or in the second year, those supplied with PE + PH grown during the first year, and those treated with Tricho + PH cultivated during the second year. The lowest fresh shoot biomass was found in tomato plants supplied with Tricho + PH grown in the first year. ANOVA showed that year (Y) did not statistically affect productive features (Table 2).

Table 2. Total fruit production, marketable fruit production, number of marketable fruits, and average weight of marketable fruit as influenced by cultivation year, biostimulant, and their interaction.

P.T.	Total Fruit Production	Marketable Fruit Production	Number of Marketable Fruits	Average Weight of Marketable Fruit
	[g pt^{-1}]		[n° pt^{-1}]	[g]
Year (Y)				
I	1648.7	1513.8	109.6	13.9
II	1667.2	1531.0	110.9	14.0
Biostimulant (B)				
Control	1379.1 f	1163.0 d	105.2 b	11.1 c
PE	1904.2 a	1727.0 a	141.5 a	12.2 bc
PH	1490.5 e	1369.4 c	108.5 b	12.8 bc
Tricho	1581.5 cd	1475.7 b	106.2 b	14.0 b
PE + PH	1800.9 b	1686.3 a	98.1 b	17.3 a
Tricho + PE	1548.6 de	1454.3 bc	108.1 b	13.5 bc
Tricho + PH	1656.6 c	1524.9 b	108.9 b	14.1 b
Tricho + PE + PH	1902.5 a	1778.5 a	105.7 b	17.0 a
Significance				
Y	n.s	n.s	n.s	n.s
B	***	***	***	***
Y × B	n.s	n.s	n.s	n.s

Data are presented as the mean ± SD. Separation of the mean biostimulant (B) effect was performed using Tukey's HSD test ($p \leq 0.05$) while the mean year (Y) effect was compared using the *t*-test. ***: significant at $p \leq 0.001$; n.s: not statistically significant. Different letters indicate significantly different means. First year: 2022–2023; second year: 2023–2024; Control: non-treated plants; PE: tropical plant extract; PH: legume-derived protein hydrolysate; Tricho: *Trichoderma atroviride*.

The application of PE, PH, Tricho, and all of their combinations positively affected total fruit production, marketable fruit production, and average weight of marketable fruit (Table 2). Cherry tomato plants treated with PE or with Tricho + PE + PH had the highest total fruit production, followed by those sprayed with PE + PH. The lowest total fruit production values were recorded in control plots (Table 2). The biostimulant treatments, accomplishing the highest increments in terms of marketable fruit production were PE, PE + PH, or the combinations of all biostimulants (Tricho + PE + PH). Control plots showed the lowest values. Plants from plots treated with PE had the highest number of marketable fruits. The PE + PH and Tricho + PE + PH combinations led to a significant higher average weight of marketable fruits compared to the control plants (Table 2). In comparison, control plants had the lowest average weight of marketable fruits.

2.3. *Fruit Quality Features*

For firmness, TSS, and fruit dry matter, ANOVA did not highlight a significant effect of year (Y) (Table 3).

Contrariwise, the results underline that biostimulant supply statistically affects the firmness, TSS, and dry matter of tomato fruits. Overall, *T. atroviride*-treated plots exhibited the highest firmness values (+33% compared to the control plots). The lowest firmness was obtained in tomatoes from control plants and in those from plants treated with all biostimulants (Table 3). For fruit TSS, our data showed that fruits from PH or Tricho plots showed an increase of 38.6% compared to tomato fruits from control plants, which in turn exhibited the lowest value. The highest fruit dry matter was recorded in tomato fruits from control and Tricho + PH-treated plots, whereas plants treated with the combinations of the three biostimulants (Tricho + PE + PH) showed the lowest values (Table 3).

Table 3. Firmness, TSS, and fruit dry matter as influenced by cultivation year, biostimulant, and their interaction.

P.T.	Firmness	TSS	Fruits Dry Matter
	N	°Brix	%
Year (Y)			
I	26.9	8.7	9.0
II	26.8	8.7	9.1
Biostimulant (B)			
Control	22.5 d	7.0 d	9.7 a
PE	25.3 c	8.7 b	8.7 de
PH	27.3 bc	9.7 a	9.3 abc
Tricho	29.9 a	9.7 a	9.4 ab
PE + PH	28.9 ab	8.1 c	8.9 cde
Tricho + PE	29.6 ab	8.7 b	8.8 de
Tricho + PH	28.7 ab	8.9 b	9.6 a
Tricho + PE + PH	22.7 d	8.8 b	8.5 e
Significance			
Y	n.s	n.s	n.s
B	***	***	***
Y × B	n.s	n.s	n.s

Data are presented as the mean ± SD. Separation of the mean biostimulant (B) effect was performed using Tukey's HSD test ($p \leq 0.05$) while the mean year (Y) effect was compared using the *t*-test.; n.s: not statistically significant; ***: significant at $p \leq 0.001$. Different letters indicate significantly different means. First year: 2022–2023; second year: 2023–2024; Control: non-treated plants; PE: tropical plant extract; PH: legume-derived protein hydrolysate; Tricho: *Trichoderma atroviride*.

ANOVA underlined that the biostimulant × year interaction significantly influenced titratable acidity and ascorbic acid (Figure 2). Concerning titratable acidity, fruits from Tricho-inoculated plants cultivated in the first year had the highest values, followed by those treated with the same biostimulant and cultivated during the second year (Figure 2a). Plants supplied with the PH revealed the lowest values, independently of the cultivation year.

Fruits from Tricho-treated plants grown during the second year had the highest ascorbic acid concentration, followed by those harvested from plants treated with the PE + PH combination grown during the second year. Fruits from control plots cultivated in the first year had the lowest values (Figure 2b).

For lycopene, ANOVA evidenced a significant influence of Y × B interaction. The highest values were found in fruits from plants grown in the second year and supplied with Tricho, whereas the lowest values were observed in control or PE-treated plots cultivated during the first or second year (Figure 3).

Statistical analysis revealed that year did not affect total polyphenol concentration (Figure 4).

Inversely, biostimulant supply statistically affected total polyphenols in tomato fruits (Table 3). Fruits from Tricho-treated plants showed the highest polyphenol content (+78%), followed by those treated with Tricho + PE + PH (+49%). The lowest polyphenol concentration was found in fruits from control plots.

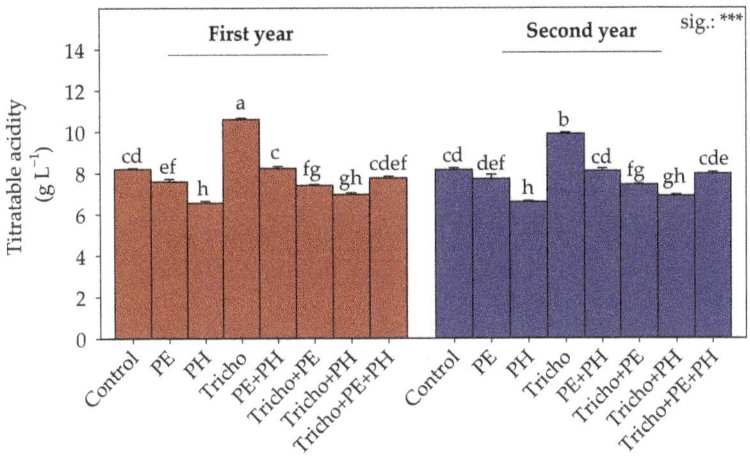

(**a**)

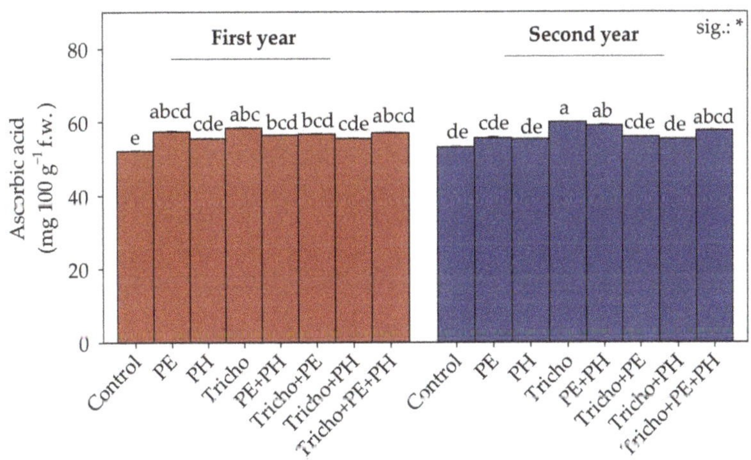

(**b**)

Figure 2. Titratable acidity (**a**) and ascorbic acid (**b**) as influenced by cultivation year and biostimulant interaction. Separation of Y × B interaction was performed using Tukey's HSD test ($p \leq 0.05$). First year: 2022–2023; second year: 2023–2024; Control: non-treated plants; PE: tropical plant extract; PH: legume-derived protein hydrolysate; Tricho: *Trichoderma atroviride*. Bars indicate standard error. Different letters indicate significantly different means. *: significant at $p \leq 0.05$; ***: significant at $p \leq 0.001$.

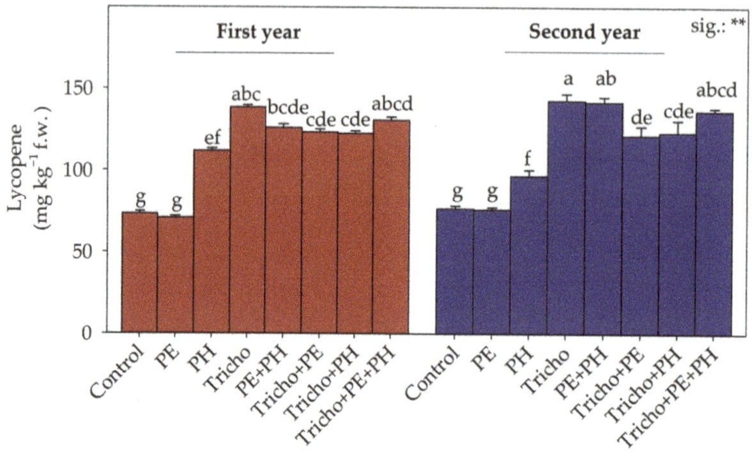

Figure 3. Lycopene as influenced by cultivation year and biostimulant interaction. Separation of Y × B interaction was performed using Tukey's HSD test ($p \leq 0.05$). First year: 2022–2023; second year: 2023–2024; Control: non-treated plants; PE: tropical plant extract; PH: legume-derived protein hydrolysate; Tricho: *Trichoderma atroviride*. Bars indicate the standard error. Different letters indicate significantly different means. **: significant at $p \leq 0.01$.

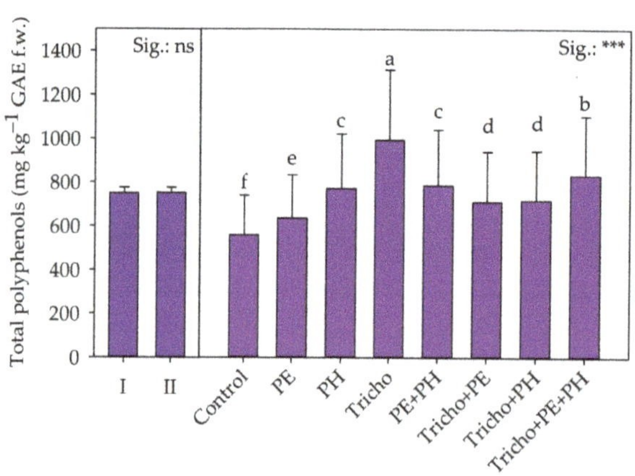

Figure 4. Total polyphenols as influenced by cultivation year and biostimulants. Separation of the mean biostimulant (B) effect was performed using Tukey's HSD test ($p \leq 0.05$) while the mean Y effect was compared using the *t*-test. ***: significant at $p \leq 0.001$. First year: 2022–2023; second year: 2023–2024; Control: non-treated plants; PE: tropical plant extract; PH: legume-derived protein hydrolysate; Tricho: *Trichoderma atroviride*. Bars indicate the standard error. Different letters indicate significantly different means.

2.4. Heat-Map Analysis

To summarize the impact of all experimental factors on cherry tomato plants, a heat-map analysis, involving the entire data set, was performed (Figure 5).

The heat-map analysis showed two dendrograms (Dendrogram 1 and Dendrogram 2). The first, located on the top, contains the combinations between year (Y) and biostimulant (B). Dendrogram 2, located on the left side of the graphical analysis, includes the dependent variables, which affected the distribution.

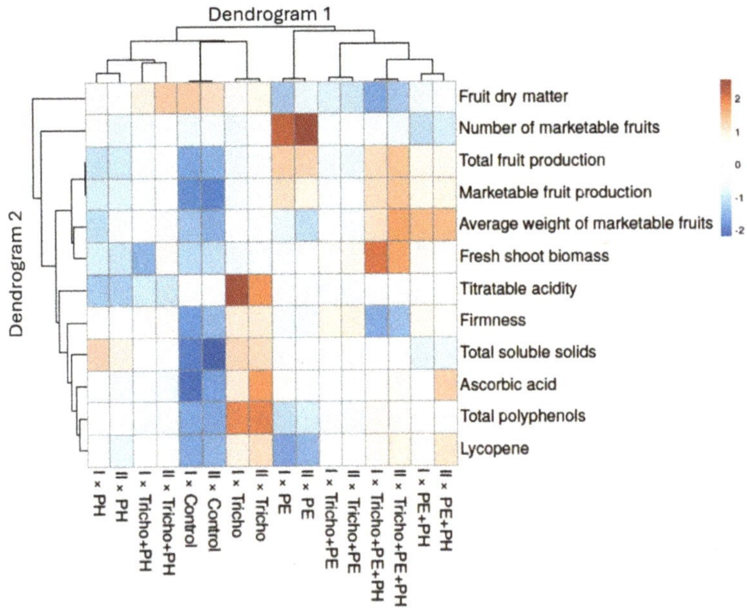

Figure 5. Heat-map analysis including all cherry tomato plant traits in response to cultivation year and biostimulant treatments. The heat-map figure was created using the https://biit.cs.ut.ee/clustvis/ online program (accessed on 1 August 2024) package. I: first cultivation year; II: second cultivation year; Control: non-treated plants; PE: tropical plant extract; PH: legume-derived protein hydrolysate; Tricho: *Trichoderma atroviride*.

For fruit dry matter, values similar to the control were detected in Tricho + PH and Tricho treatments. Regarding the number of marketable fruits, all treatments showed similar values to the control, with the exception of PE treatment. Contrariwise, for total fruit production, marketable fruit production, and the average weight of marketable fruits, all treatments revealed dissimilar values compared with the control, excluding PH treatment. For fresh shoot biomass, PH, Tricho, and PE treatments had similar values compared to the control. Moreover, for titratable acidity and firmness, the heat-map analysis revealed that Tricho + PE + PH showed analogous values to the control. However, for firmness, PE + PH treatment also had values comparable to the control. The treatment PE + PH—for total soluble solids—gave similar values to the control, whereas, for ascorbic acid, the heat-map detected that all treatments were different from the control. Lastly, for total polyphenols and lycopene, data similar to the control were detected in PE treatment.

2.5. Partial Budget Analysis

The biostimulant effects on the partial budget analysis of "Creativo F" cherry tomato are presented in Table 4.

All biostimulants, applied alone or in combination, increased the profitability of the cherry tomato crop. The economic assessment showed that the highest added gross return was observed in plants treated with PE (28,154.4 EUR ha^{-1}), followed by those treated with PE + PH (23,963.6 EUR ha^{-1}), which in turn was higher than those exposed to PE + PH + Tricho (20,009.5 EUR ha^{-1}). Overall, our results showed that the greatest net economic benefit was related to yield increase. In particular, the use of PE was the most profitable treatment.

Table 4. Added returns of cherry tomato grown under greenhouse conditions achieved by tropical plant extract (PE), plant protein hydrolysate (PH), *Trichoderma atroviride* (Tricho), or their combinations compared to the untreated control.

Biostimulant	Yield Increase (kg ha^{-1})	Price (EUR kg^{-1})	Added Gross Return (EUR ha^{-1})	Added Variable Cost (EUR ha^{-1})			Total (EUR ha^{-1})	Added Net Return (EUR ha^{-1})
				Biostimulant Treatment	Application	Harvest		
PE	18,612.0	2.0	37,224.0	4875.0	100	4094.6	9069.6	28,154.4
PH	6811.2	2.0	13,622.4	1800.0	100	1498.5	3398.5	10,223.9
Tricho	10,319.1	2.0	20,638.2	9270.0	100	2270.2	11,640.2	8998.0
PE + PH	17,268.9	2.0	34,537.8	6675.0	100	3799.2	10,574.2	23,963.6
Tricho + PE	9612.9	2.0	19,225.8	14,145.0	200	2114.8	16,459.8	2766.0
Tricho + PH	11,942.7	2.0	23,885.4	11,070.0	200	2627.4	13,897.4	9988.0
Tricho + PE + PH	20,311.5	2.0	40,623.0	15,945.0	200	4468.5	20,613.5	20,009.5

The costs of biostimulants were defined by the suppliers (PE = 65 EUR L^{-1}; PH = 12 EUR L^{-1}; Tricho = 150 EUR kg^{-1}); the costs of biostimulant applications were considered reputable based on the information provided by local agricultural contractors.

3. Discussion

Overall, data on most tomato productive and growth traits revealed a positive effect of the three biostimulants. This is in accordance with the results of Colla et al. [32], who studied the effect of different biostimulants (plant extract and protein hydrolysate) on tomato plants and reported an increase in productive features. Moreover, our data corroborate those of Sani et al. [51], who stated that tomato plants treated with *Trichoderma* spp. and grown under organic conditions significantly enhanced growth and yield. Colla et al. [52] reported that the mechanism behind the stimulatory effect of PE might be related to its phytohormone and signaling compound content, which increases the plant photosynthesis rate and, consequently, productive features. Furthermore, the same authors reported that the beneficial effect of PE on plant growth and yield could be linked to its ability to increase plant nutrient uptake via the modulation of plant root expansion. The positive effect on tomato fresh shoot biomass and productive parameters recorded in plants treated with the PH can also be associated with well-known hormone-like activity, which in turn stimulates the expression of some genes involved in cell division and expansion [53]. In comparison, De Palma et al. [54] reported that *Trichoderma* genera are implicated in the upregulation of many proteins involved in photosynthesis, with repercussions for photosynthetic rate and, accordingly, plant yield. *Trichoderma* spp. are also implicated in numerous biological and chemical activities in the rhizosphere, such as the increase in nutrient availability and the increase in root growth [55]. In addition, various authors [56,57] state that *Trichoderma* spp. can produce secondary metabolites with hormone-like activity, small peptides, and volatile organic compounds, which are known for their biostimulatory effects. Remarkably, our results showed a positive effect of biostimulants on productive traits, especially when they were applied in combination. For total fruit production, marketable fruit production, and the average weight of marketable fruits, all biostimulants (Tricho + PE + PH) had an additive effect. However, the number of marketable fruits did not show a similar trend. Moreover, the positive additive effect between PE and PH was also documented for marketable fruit production and the average weight of marketable fruits. Even for fresh shoot biomass, the combination of all biostimulants showed synergistic action. However, in this case, the year of cultivation also had a significant effect on plant response to biostimulant application.

Fruit firmness was enhanced by biostimulant application. Likewise, Vultaggio et al. [34], investigating the effect of biostimulants on woodland strawberry, revealed an increase in firmness in fruits from plants inoculated with *Trichoderma*. Contrariwise, Soteriou et al. [56] reported that a vegetal PH did not have a significant effect on watermelon fruit firmness. Since contrasting results are reported in the literature, we can speculate that the variation in this parameter to PH application depends on the plant species. However, the positive effect recorded on fruit firmness in plants treated with *T. atroviride* can be associated with its capacity to make calcium more available for plants [57]. We also recorded a synergistic effect of biostimulants on fruit firmness, with no statistical differences among PE + PH, Tricho + PE, and Tricho + PH. However, when all biostimulants were combined (Tricho + PE + PH), no significant variation from the control was recorded. This behavior can be linked to an excessive effect of biostimulants in increasing plant water uptake, which, in turn, can cause a reduction in fruit firmness. Thus, in this case, the mutual application of all biostimulants had a counterproductive effect.

Our total soluble solids (TSS) data concur with those of Sani et al. [51], who found an increase in fruit TSS when plants were inoculated with *Trichoderma* spp. The effect exerted by PE and PH on fruit TSS can be linked to their amino acid and peptide content, which have a positive effect on plant primary metabolism [58]. Furthermore, it seems that PE and PH biostimulants promote photosynthesis rates, with a positive impact on soluble solids biosynthesis [52]. The *T. atroviride* effect on TSS could be the consequence of increased plant mineral uptake, which in turn elicited the production of soluble solids [58]. However, no synergistic effect among the biostimulants was recorded for TSS.

Fruit dry matter was reduced by biostimulant application, especially when all of them were combined. Since it has been reported that PE, PH, and *T. atroviride* application can increase plant water uptake [52,54], we can assume that the decrease in fruit dry matter could be the consequence of higher water content in fruit tissues compared to the control. These data may also be partially related to firmness since we noticed that fruit dry matter decreases when the biostimulant applications were combined.

Our data on ascorbic acid corroborate those of Sani et al. [51], who reported an upsurge of ascorbic acid in fruits from plants inoculated with *T. atroviride*. The higher values of ascorbic acid recorded in fruits from plants cultivated in the second-year experiment can be related to the higher temperature recorded. Indeed, since ascorbic acid production increases for ROS detoxification when plants are under stress [59] and considering that the temperatures of the second year were higher than those of the first cultivation cycle, we can hypothesize that this response is a direct temperature-related effect. Regarding the biostimulants' effect on ascorbic acid, there are reports on the influence of PE, PH, and *Trichoderma* spp. applications on plant secondary metabolism elicitation [55].

Titratable acidity was reduced by PE and PH application and enhanced by *T. atroviride* inoculation. Our data are not in line with those of Abdelkader et al. [60], who reported no significant effect of biostimulants on tomato acidity. Since the genotype used in this study is different from that employed by Abdelkader et al. [60] and considering that the growing conditions were different, we can hypothesize that the effects of these biostimulants on tomato acidity are both environmental and plant genetic dependent. However, our data on *T. atroviride*'s impact on titratable acidity corroborate those of Ruiz-Cisneros et al. [61], who noted an improvement in fruits from inoculated plants. The outcome may be related to the fruit size increase recorded in inoculated plants since, as reported by Tigist et al. [62], acidity increases with the mass of the fruits.

Lycopene in tomato fruits was increased by PH and *T. atroviride* application. Our data agree with those of Choi et al. [63], who, studying the impact of a PH on greenhouse lettuce and tomato performance, reported an increase in fruit lycopene concentration when plants were supplied with the biostimulant. This could be the result of the activation of physiological and molecular mechanisms [64]. Furthermore, our data agree with those of Sani et al. [51], who reported an increase in fruit lycopene concentration when plants were inoculated with *T. atroviride* compared to control plants. As suggested by López-Bucio et al. [58], *T. atroviride* inoculation can increase plant stress tolerance via an increase in antioxidant production, such as lycopene. Interestingly, a high additive effect among the biostimulants was recorded. Stand-alone PE application did not statistically influence lycopene concentration, whereas, when the PE treatments were combined with other biostimulants, an increase in lycopene concentration compared to the control was recorded.

Biostimulants promoted fruit total polyphenols. As reported by Colla et al. [32] and López-Bucio et al. [58], the application of microbial and non-microbial biostimulants has a positive effect on plant secondary metabolism activity and, consequently, the biosynthesis of polyphenols. The small peptides and amino acids comprising the PE and PH had a hormone-like effect on plants, with positive outcomes on the stimulation of secondary metabolites [5]. Contrariwise, *T. atroviride* can establish a relationshup with plant roots and can produce secondary metabolites and phytohormone signaling molecules, which elicit plant secondary metabolism [36]. Finally, for total polyphenols, no synergistic effect was attributed to the biostimulants.

4. Materials and Methods

4.1. Plant Material and Experimental Site

This study was performed in Comiso (Ragusa province, RG) on an experimental farm (36°99′64.8″ N, 14°59′88.3″ E) belonging to the lead partner of the SO.MI.PR.O.N. project, at 189 m a.s.l. The trial was carried out, for two successive cultivation cycles (2022–2023 and 2023–2024), in an unheated greenhouse (width 52 m and length 72 m) covered with a transparent polyethylene film, during the autumn–spring period. The experiment was carried out in a sandy clay soil (45.5% sand, 16.0% silt, and 38.5% clay) at pH 7.8, containing 1.8% organic matter, 0.09% total nitrogen, 6.2 mg kg^{-1} dw of P_2O_5, and 410.0 mg kg^{-1} dw of potassium. On 19 September (first cultivation cycle) and 25 September (second cultivation cycle), seedlings of the 'Creativo' F_1 hybrid cherry tomato (*Solanum lycopersicum* L.) (HM Clause, Davis, CA, USA) were transplanted, obtaining a plant density of 3 plants m^{-2}. Prior to fertilization, soil samples were collected from the experimental field and analyzed for their physico-chemical properties. All tomato plants were fertigated with 230 kg ha^{-1} of N, 150 kg ha^{-1} of P_2O_5, and 250 kg ha^{-1} of K_2O, according to the average withdrawls of tomato crops grown in Mediterranean environment conditions and the soil source. A two-branch pruning system was adopted. Throughout the growing cycles, all tomato plants were managed according to the method of Tesi [65]. Maximum and minimum temperatures inside the greenhouse were recorded via a data logger (BL30, Trotec, Heinsberg, Germany) (Figure 6).

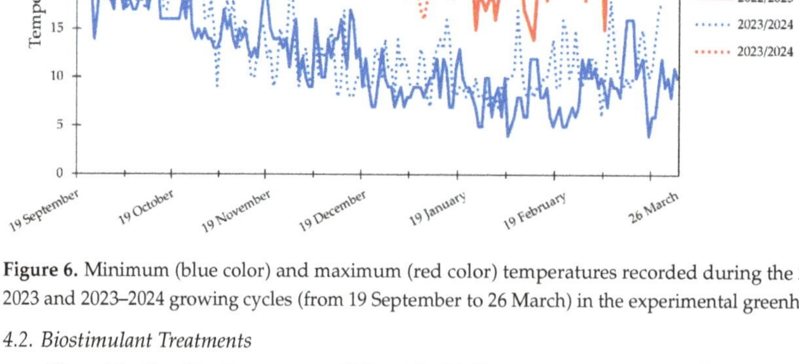

Figure 6. Minimum (blue color) and maximum (red color) temperatures recorded during the 2022–2023 and 2023–2024 growing cycles (from 19 September to 26 March) in the experimental greenhouse.

4.2. Biostimulant Treatments

Three biostimulant treatments, all from the Hello Nature company (Hello Nature Italia srl, Rivoli Veronese, Italy), a *Trichoderma atroviride* (Tricho) Condor®, a tropical plant extract (PE) Auxym®, and a legume-derived protein hydrolysate (PH) Trainer® were used to evaluate the single and combined effects. The day before transplantation, the tomato plants were inoculated, with the tomato root system soaked for 15 min using a dose of 6.25 g L^{-1} of Condor®. Root inoculation was repeated 15 and 45 days after transplantation(DAT), with

100 mL of solution supplied per plant. Every 15 days, from the beginning of the tomato flowering stage to the end of the growing cycle, the tomato plants were sprayed with PE and PH using a dosage of 1.5 mL L^{-1} and 3 mL L^{-1}, respectively. For each treatment, a solution of 0.5 L m^2 was supplied. The control plants were sprayed with only water.

4.3. Fruit Production and Fresh Shoot Biomass

Fruit production traits and fresh shoot biomass were recorded for all plants. Total fruit production was instantly recorded after each harvest, and it was parted into marketable and non-marketable fruit production and presented as g $plant^{-1}$. The number of marketable fruits, presented as number $plant^{-1}$, was recorded, and the average weight of marketable fruits, expressed as g $fruit^{-1}$, was calculated. In comparison, fresh shoot biomass (g $plant^{-1}$) was recorded at the end of the growing cycle.

4.4. Proximal Fruit Quality

At the third harvest, qualitative analyses were performed on five randomly collected fruits for each treatment. Fruit dry matter content was assessed using a ventilated oven at a temperature of 80 °C, with 500 g of the fruit sample dried until constant weight. Data were expressed as percentages.

Fruit firmness was measured using a Trsnc digital penetrometer (Forlì, Italy). On two opposite sides of the fruits' equatorial zone, with a 3 mm stainless steel cylindrical probe, the fruits' resistance to penetration was evaluated. The values were expressed as newtons (N). After obtaining the juice by centrifuging the tomatoes and filtering, a digital refractometer (MTD-045nD, Three-In-One Enterprises Co., Ltd., New Taipei, Taiwan) was used for the evaluation of the soluble solids content (TSS) values. The data were reported as °Brix.

Titratable acidity was calculated with Han et al.'s [66] method, using a 10 g aliquot of fruits mixed in 50 mL of distilled H_2O and titrated to an endpoint of pH 8.1, with 0.1 N NaOH. Titratable acidity was expressed as mg L^{-1}. For determination of ascorbic acid content, tomato fruits were squeezed and then measurements were performed with a reflectometer (Merck RQflex* 10, Darmstadt, Germany), using Reflectoquant Ascorbic Acid Test Strips. The value was expressed as mg of ascorbic acid per 100 g of fresh weight (mg 100 g fw^{-1}).

4.5. Fruits' Functional Components

Total polyphenols were estimated according to the method of Meda et al. [67], using a spectrophotometric assay, and the results were reported as mg kg^{-1} gallic acid fresh weight. Briefly, 2 g of each sample was weighed out and extracted with 50 mL of methanol. Extraction was conducted under stirring for 60 min at 60 °C. The mixture was filtered through filter paper (Whatman No. 3), filled into a 50 mL volumetric flask, and allowed to set in the dark until analysis. The standard or tomato sample extract (100 µL; in triplicate) was mixed with 0.4 mL Folin-C reagent (Sigma–Aldrich Chemie, Steinheim, Germany). After 5 min, 0.8 mL of 10% sodium carbonate (Na_2CO_3) was added to promote rapid oxidation of the phenolic compound, resulting in colorimetric variation. The absorbance (at 765 nm) of the solution was recorded after 2 h of incubation at room temperature and compared with a calibration curve using gallic acid (Sigma–Aldrich Chemie, Steinheim, Germany) (0–200 mg/L) as standard to produce the calibration curve. The lycopene content was established according to the protocol used by Sadler et al. [68] and expressed in mg kg^{-1} of fresh weight. Briefly, 5 g of homogenized sample was extracted by adding 50 mL of a mixture of hexane/acetone/ethanol (2:1:1, $v/v/v$) and stirred. The value was evaluated by measuring the absorbance (at 472 nm) of the lycopene hexane fraction and comparing it with a calibration curve using pure lycopene (Sigma, St. Louis, MO, USA) as standard.

4.6. Experimental Setup and Statistics

Eight biostimulant treatments (the control, PE, PH, Tricho, PE + PH, PE + Tricho, PH + Tricho, and PE + PH + Tricho) were employed in a randomized complete block design (RCBD), with three replications per treatment (9 plants per replicate), rendering a total of 216 plants. This experimental setup was repeated for two consecutive years (Y). All statistical analyses were performed with SPSS V. 28 software, using ANOVA and Student's *t*-test. Separation of the mean biostimulant (B) and Y × B interaction was performed using Tukey's HSD test ($p \leq 0.05$) while the mean Y effect was compared using the *t*-test. Percentage data were subjected to arcsin conversion prior to ANOVA analysis ($Ø = \arcsin (p/100)^{1/2}$). The entire data set was also subjected to heat-map analysis through the online program package clustvis (https://biit.cs.ut.ee/clustvis/), accessed on 3 June 2024.

4.7. Partial Budget Analysis

A partial budget analysis was performed to evaluate the net economic benefits that may encourage cherry tomato cultivation using microbial and/or non-microbial biostimulants. The economic approach of Giordano et al. [69] and Vultaggio et al. [31] was adopted. The biostimulants' added costs and gross returns were considered. The added net return was calculated as follows:

$$\text{Added net return} = \text{Added gross return} - \text{Added variable cost}$$

5. Conclusions

This study explored the application, alone or combined, of three biostimulants (tropical plant extracts, vegetal protein hydrolysate, and *Trichoderma atroviride*) on the 'Creativo' F1 cherry tomato hybrid. Overall, we found that the use of biostimulants increased plant performance in terms of growth, yield, and quality and boosted the nutraceutical traits of tomatoes, such as lycopene and total polyphenols. The combined application of all biostimulants (Tricho + PE + PH) was able to increase growth and yield traits, total soluble solids, ascorbic acid, lycopene, and total polyphenols, whereas it reduced fruit firmness, fruit dry matter, and titratable acidity compared to the control. In comparison, the best combinations for increasing tomato qualitative features were PE + PH, Tricho + PE, and Tricho + PH. Thus, the current study highlights that precise biostimulant combinations can promote specific traits (e.g., yield, quality, functional traits, economic profitability, etc.). The results of this study could be the basis for the definition of new biostimulant products and agronomic protocols, using microbial and non-microbial products. However, to better understand the physiological and biochemical mechanisms behind these responses, further studies including omics analyses are needed.

Author Contributions: Conceptualization, L.V., B.B.C., Y.R., G.C., S.L.B., and L.S.; methodology, Y.R., G.C.; B.B.C., S.N., and L.S.; software, M.C., P.B., and F.M.; validation, Y.R., G.C., and L.S.; formal analysis, L.V., M.C., E.C., P.B., B.B.C., F.M., S.N., and L.S.; investigation, L.V., M.C., E.C., P.B., B.B.C., F.M., S.N., and L.S.; resources, Y.R., G.C., and S.L.B.; data curation, L.V., B.B.C., and L.S.; writing—original draft preparation, L.V., B.B.C., S.N., and L.S.; writing—review and editing, L.V., M.C., E.C., B.B.C., Y.R., G.C., S.L.B., S.N., and L.S.; visualization, Y.R., G.C., and L.S.; supervision, Y.R., G.C., S.L.B., and L.S.; project administration, G.C.; funding acquisition, Y.R., G.C., and S.L.B. All authors have read and agreed to the published version of the manuscript.

Funding: This research was funded by the "SO.MI.PR.O.N." project (Natural products and microorganisms in horticulture for a sustainable vegetables production of high nutraceutical value), Sicilian Region, Programma di Sviluppo Rurale Sicilia (PSR-Sicilia) 2014-2020, CUP G46D20000150009.

Data Availability Statement: The data produced in this research are available upon request due to privacy restrictions.

Conflicts of Interest: The authors declare no conflicts of interest.

References

1. Ali, M.Y.; Sina, A.A.I.; Khandker, S.S.; Neesa, L.; Tanvir, E.M.; Kabir, A.; Khalil, M.I.; Gan, S.H. Nutritional composition and bioactive compounds in tomatoes and their impact on human health and disease: A Review. *Foods* **2020**, *10*, 45. [CrossRef] [PubMed]
2. Del Giudice, R.; Raiola, A.; Tenore, G.C.; Frusciante, L.; Barone, A.; Monti, D.M.; Rigano, M.M. Antioxidant bioactive compounds in tomato fruits at different ripening stages and their effects on normal and cancer cells. *J. Funct. Foods* **2015**, *18*, 83–94. [CrossRef]
3. Ibrahim, D.M.S.; Mujtaba, D.A.; Hassan, D.I.; Nadeem, D.M.; Sohail, D.A.; Abbasi, D.K.S.; Samee, A.; Azeem, W.; Waqar, M.; Fatima, I. Antioxidant and nutraceutical potential of tomato for health and life quality: A review: Tomato as functional fruit. *Food Sci. Appl. Microbiol. Rep.* **2023**, *2*, 8–25. [CrossRef]
4. Salas-Pérez, L.; García-Hernández, J.L.; Márquez-Hernández†, C.; Fortis-Hernández, M.; Estrada-Arellano, J.R.; Esparza-Rivera, J.R.; Preciado-Rangel, P. Yield and nutraceutical quality of tomato fruits in organic substrates. *Ecosistemas Y Recur. Agropecu.* **2017**, *4*, 169–175. [CrossRef]
5. Frusciante, L.; Carli, P.; Ercolano, M.R.; Pernice, R.; Di Matteo, A.; Fogliano, V.; Pellegrini, N. Antioxidant nutritional quality of tomato. *Mol. Nutr. Food Res.* **2007**, *51*, 609–617. [CrossRef]
6. Martí, R.; Roselló, S.; Cebolla-Cornejo, J. Tomato as a source of carotenoids and polyphenols targeted to cancer prevention. *Cancers* **2016**, *8*, 58. [CrossRef]
7. Tudor-Radu, M.; Vîjan, L.E.; Tudor-Radu, C.M.; Tița, I.; Sima, R.; Mitrea, R. Assessment of ascorbic acid, polyphenols, flavonoids, anthocyanins and carotenoids content in tomato fruits. *Not. Bot. Horti Agrobot. Cluj-Napoca* **2016**, *44*, 477–483. [CrossRef]
8. Sabatino, L.; Palazzolo, E.; D'Anna, F. Grafting suitability of sicilian eggplant ecotypes onto *Solanum torvum*: Fruit composition, production and phenology. *J. Food Agric. Environ.* **2013**, *11*, 1195–1200.
9. FAO Statistical Database. 2021. Available online: http://www.Fao.Org (accessed on 17 July 2024).
10. ISTAT. Available online: http://Dati.Istat.It/Index.Aspx?QueryId=33703# (accessed on 17 July 2024).
11. Carricondo-Martínez, I.; Berti, F.; Salas-Sanjuán, M.d.C. Different organic fertilisation systems modify tomato quality: An opportunity for circular fertilisation in intensive horticulture. *Agronomy* **2022**, *12*, 174. [CrossRef]
12. Pane, C.; Celano, G.; Piccolo, A.; Villecco, D.; Spaccini, R.; Palese, A.M.; Zaccardelli, M. Effects of on-farm composted tomato residues on soil biological activity and yields in a tomato cropping system. *Chem. Biol. Technol. Agric.* **2015**, *2*, 4. [CrossRef]
13. Giordano, M.; El-Nakhel, C.; Carillo, P.; Colla, G.; Graziani, G.; Di Mola, I.; Mori, M.; Kyriacou, M.C.; Rouphael, Y.; Soteriou, G.A.; et al. Plant-derived biostimulants differentially modulate primary and secondary metabolites and improve the yield potential of red and green lettuce cultivars. *Agronomy* **2022**, *12*, 1361. [CrossRef]
14. D'Anna, F.; Sabatino, L. Morphological and agronomical characterization of eggplant genetic resources from the Sicily area. *J. Food Agric. Environ.* **2013**, *11*, 401–404.
15. Bulgari, R.; Franzoni, G.; Ferrante, A. Biostimulants application in horticultural crops under abiotic stress conditions. *Agronomy* **2019**, *9*, 306. [CrossRef]
16. Shahrajabian, M.H.; Chaski, C.; Polyzos, N.; Petropoulos, S.A. Biostimulants application: A low input cropping management tool for sustainable farming of vegetables. *Biomolecules* **2021**, *11*, 698. [CrossRef] [PubMed]
17. Bashir, M.A.; Rehim, A.; Raza, Q.U.A.; Zhai, L.; Liu, H.; Wang, H. Biostimulants as plant growth stimulators in modernized agriculture and environmental sustainability. In *Technology in Agriculture*; IntechOpen: London, UK, 2021.
18. Hamid, B.; Zaman, M.; Farooq, S.; Fatima, S.; Sayyed, R.Z.; Baba, Z.A.; Sheikh, T.A.; Reddy, M.S.; El Enshasy, H.; Gafur, A.; et al. Bacterial plant biostimulants: A sustainable way towards improving growth, productivity, and health of crops. *Sustainability* **2021**, *13*, 2856. [CrossRef]
19. Mannino, G.; Campobenedetto, C.; Vigliante, I.; Contartese, V.; Gentile, C.; Bertea, C.M. The application of a plant biostimulant based on seaweed and yeast extract improved tomato fruit development and quality. *Biomolecules* **2020**, *10*, 1662. [CrossRef] [PubMed]
20. Schiattone, M.I.; Viggiani, R.; Di Venere, D.; Sergio, L.; Cantore, V.; Todorovic, M.; Perniola, M.; Candido, V. Impact of irrigation regime and nitrogen rate on yield, quality and water use efficiency of wild rocket under greenhouse conditions. *Sci. Hortic.* **2018**, *229*, 182–192. [CrossRef]
21. Ullah, F.; Saqib, S.; Khan, W.; Zhao, L.; Khan, W.; Li, M.Y.; Xiong, Y.C. Sodium Nitroprusside and Melatonin Improve Physiological Vitality and Drought Acclimation via Synergistically Enhancing Antioxidant Response in Dryland Maize. *J. Plant Growth Regul.* **2024**, *43*, 1–18. [CrossRef]
22. Consentino, B.B.; Sabatino, L.; Mauro, R.P.; Nicoletto, C.; De Pasquale, C.; Iapichino, G.; La Bella, S. Seaweed extract improves *Lagenaria siceraria* young shoot production, mineral profile and functional quality. *Horticulturae* **2021**, *7*, 549. [CrossRef]
23. Trudel, R. Sustainable Consumer Behavior. *Consum. Psychol. Rev.* **2019**, *2*, 85–96. [CrossRef]
24. Yue, C.; Lai, Y.; Wang, J.; Mitchell, P. Consumer preferences for sustainable product attributes and farm program features. *Sustainability* **2020**, *12*, 7388. [CrossRef]
25. Lamichhane, J.R.; Dachbrodt-Saaydeh, S.; Kudsk, P.; Messéan, A. Toward a reduced reliance on conventional pesticides in european agriculture. *Plant Dis.* **2016**, *100*, 10–24. [CrossRef] [PubMed]
26. Expósito, A.; Velasco, F. Exploring environmental efficiency of the european agricultural sector in the use of mineral fertilizers. *J. Clean. Prod.* **2020**, *253*, 119971. [CrossRef]
27. Sytar, O.; Olšovská, K. Plant-based proteins as a food source and plant growth biostimulants. *Discov. Food* **2024**, *4*, 78. [CrossRef]

28. Francesca, S.; Cirillo, V.; Raimondi, G.; Maggio, A.; Barone, A.; Rigano, M.M. A novel protein hydrolysate-based biostimulant improves tomato performances under drought stress. *Plants* **2021**, *10*, 783. [CrossRef]
29. Xu, C.; Leskovar, D.I. Effects of *A. Nodosum* seaweed extracts on spinach growth, physiology and nutrition value under drought stress. *Sci. Hortic.* **2015**, *183*, 39–47. [CrossRef]
30. Mejias, L.; Martínez-Avila, O.; Mora, M.; Castaño, O.; Kharytonov, M. The prospects of the waste-based biostimulants production for agriculture. In Proceedings of the International Symposium, ISB-INMA-TEH', Agricultural and Mechanical Engineering, Bucharest, Romania, 6–8 October 2022.
31. Vultaggio, L.; Allevato, E.; Consentino, B.B.; Bellitto, P.; Napoli, S.; Cannata, C.; Ntatsi, G.; Vasto, S.; Baldassano, S.; La Bella, S.; et al. Joint action of *Trichoderma atroviride* and a vegetal derived-protein hydrolysate improves performances of woodland strawberry in Italy. *Horticulturae* **2024**, *10*, 459. [CrossRef]
32. Colla, G.; Nardi, S.; Cardarelli, M.; Ertani, A.; Lucini, L.; Canaguier, R.; Rouphael, Y. Protein hydrolysates as biostimulants in horticulture. *Sci. Hortic.* **2015**, *196*, 28–38. [CrossRef]
33. Battacharyya, D.; Babgohari, M.Z.; Rathor, P.; Prithiviraj, B. Seaweed extracts as biostimulants in horticulture. *Sci. Hortic.* **2015**, *196*, 39–48. [CrossRef]
34. Vultaggio, L.; Allevato, E.; Sabatino, L.; Ntatsi, G.; Rouphael, Y.; Torta, L.; La Bella, S.; Consentino, B.B. Modulation of cherry tomato performances in response to molybdenum biofortification and arbuscular mycorrhizal fungi in a soilless system. *Heliyon* **2024**, *10*, e33498. [CrossRef]
35. Srivastava, S.N.; Singh, V.; Awasthi, S.K. Trichoderma induced improvement in growth, yield and quality of sugarcane. *Sugar Tech* **2006**, *8*, 166–169. [CrossRef]
36. Vukelić, I.D.; Prokić, L.T.; Racić, G.M.; Pešić, M.B.; Bojović, M.M.; Sierka, E.M.; Kalaji, H.M.; Panković, D.M. Effects of *Trichoderma harzianum* on photosynthetic characteristics and fruit quality of tomato plants. *Int. J. Mol. Sci.* **2021**, *22*, 6961. [CrossRef] [PubMed]
37. Jamal Uddin, A.F.M.; Ahmad, H.; Hasan, M.R.; Mahbuba, S.; Roni, M.Z.K. Effects of *Trichoderma* spp. on growth and yield characters of Bari tomato—14. *Int. J. Bus. Soc. Sci. Res.* **2016**, *4*, 112–117.
38. Wahab, A.; Batool, F.; Muhammad, M.; Zaman, W.; Mikhlef, R.M.; Qaddoori, S.M.; Ullah, S.; Abdi, G.; Saqib, S. Unveiling the complex molecular dynamics of arbuscular mycorrhizae: A comprehensive exploration and future perspectives in harnessing phosphate-solubilizing microorganisms for sustainable progress. *EEB* **2023**, *219*, 105633. [CrossRef]
39. Ghazanfar, M.U.; Raza, M.; Raza, W.; Qamar, M.I. *Trichoderma* as potential biocontrol agent, its exploitation in agriculture: A review. *Plant Prot.* **2018**, *2*, 3.
40. Sood, M.; Kapoor, D.; Kumar, V.; Sheteiwy, M.S.; Ramakrishnan, M.; Landi, M.; Araniti, F.; Sharma, A. *Trichoderma*: The "secrets" of a multitalented biocontrol agent. *Plants* **2020**, *9*, 762. [CrossRef] [PubMed]
41. Andrzejak, R.; Janowska, B. *Trichoderma* spp. improves flowering, quality, and nutritional status of ornamental plants. *Int. J. Mol. Sci.* **2022**, *23*, 15662. [CrossRef]
42. Sachdev, S.; Singh, R.P. Isolation, characterisation and screening of native microbial isolates for biocontrol of fungal pathogens of tomato. *Clim. Chang. Environ. Sustain.* **2018**, *6*, 46. [CrossRef]
43. Sachdev, S.; Singh, A.; Singh, R.P. Optimization of culture conditions for mass production and bio-formulation of *Trichoderma* using response surface methodology. *3 Biotech* **2018**, *8*, 360. [CrossRef]
44. Pacheco-Trejo, J.; Aquino-Torres, E.; Reyes-Santamaría, M.I.; Islas-Pelcastre, M.; Pérez-Ríos, S.R.; Madariaga-Navarrete, A.; Saucedo-García, M. Plant defensive responses triggered by *Trichoderma* spp. as tools to face stressful conditions. *Horticulturae* **2022**, *8*, 1181. [CrossRef]
45. Z Zaidi, N.W.; Dar, M.H.; Singh, S.; Singh, U.S. Trichoderma species as abiotic stress relievers in plants. In *Biotechnology and Biology of Trichoderma*; Gupta, V.K., Schmoll, M., Herrera-Estrella, A., Upadhyay, R.S., Druzhinina, I., Tuohy, M.G., Eds.; Elsevier: Amsterdam, The Netherlands, 2014; pp. 515–525. ISBN 978-0-444-59576-8.
46. Mastouri, F. Use of *Trichoderma* spp. to Improve Plant Performance under Abiotic Stresses. Ph.D. Thesis, Cornell University, Ithaca, NY, USA, 2010.
47. Macías-Rodríguez, L.; Contreras-Cornejo, H.A.; Adame-Garnica, S.G.; del-Val, E.; Larsen, J. The interactions of *Trichoderma* at multiple trophic levels: Inter-kingdom communication. *Microbiol. Res.* **2020**, *240*, 126552. [CrossRef] [PubMed]
48. Sachdev, S.; Singh, R.P. *Trichoderma*: A multifaceted fungus for sustainable agriculture. In *Ecological and Practical Applications for Sustainable Agriculture*; Bauddh, K., Kumar, S., Singh, R.P., Korstad, J., Eds.; Springer: Singapore, 2020; pp. 261–304. ISBN 9789811533723.
49. Benito, P.; Celdrán, M.; Bellón, J.; Arbona, V.; González-Guzmán, M.; Porcel, R.; Yenush, L.; Mulet, J.M. The combination of a microbial and a non-microbial biostimulant increases yield in lettuce (*Lactuca sativa*) under salt stress conditions by up-regulating cytokinin biosynthesis. *J. Integr. Plant Biol.* **2024**, *66*, 2140–2157. [CrossRef] [PubMed]
50. Tabacchioni, S.; Passato, S.; Ambrosino, P.; Huang, L.; Caldara, M.; Cantale, C.; Hett, J.; Del Fiore, A.; Fiore, A.; Schlüter, A.; et al. Identification of beneficial microbial consortia and bioactive compounds with potential as plant biostimulants for a sustainable agriculture. *Microorganisms* **2021**, *9*, 426. [CrossRef] [PubMed]
51. Sani, N.H.; Islam, N.; Uddain, J.; Chowdhury, S.N.; Subramaniam, S. Synergistic effect of microbial and nonmicrobial biostimulants on growth, yield, and nutritional quality of organic tomato. *Crop. Sci.* **2020**, *60*, 2102–2114. [CrossRef]
52. Colla, G.; Cardarelli, M.; Bonini, P.; Rouphael, Y. Foliar applications of protein hydrolysate, plant and seaweed extracts increase yield but differentially modulate fruit quality of greenhouse tomato. *HortScience* **2017**, *52*, 1214–1220. [CrossRef]

53. Malécange, M.; Sergheraert, R.; Teulat, B.; Mounier, E.; Lothier, J.; Sakr, S. Biostimulant properties of protein hydrolysates: Recent advances and future challenges. *Int. J. Mol. Sci.* **2023**, *24*, 9714. [CrossRef]
54. De Palma, M.; Docimo, T.; Guida, G.; Salzano, M.; Albrizio, R.; Giorio, P.; Ruocco, M.; Tucci, M. Transcriptome modulation by the beneficial fungus *Trichoderma longibrachiatum* drives water stress response and recovery in tomato. *Environ. Exp. Bot.* **2021**, *190*, 104588. [CrossRef]
55. Sofo, A.; Tataranni, G.; Xiloyannis, C.; Dichio, B.; Scopa, A. Direct effects of *Trichoderma harzianum* strain t-22 on micropropagated shoots of gisela6® (*Prunus cerasus* × *Prunus canescens*) rootstock. *Environ. Exp. Bot.* **2012**, *76*, 33–38. [CrossRef]
56. Pelagio-Flores, R.; Esparza-Reynoso, S.; Garnica-Vergara, A.; López-Bucio, J.; Herrera-Estrella, A. *Trichoderma*-induced acidification is an early trigger for changes in arabidopsis root growth and determines fungal phytostimulation. *Front. Plant Sci.* **2017**, *8*, 822. [CrossRef]
57. Soteriou, G.A.; Rouphael, Y.; Emmanouilidou, M.G.; Antoniou, C.; Kyratzis, A.C.; Kyriacou, M.C. Biostimulatory action of vegetal protein hydrolysate and the configuration of fruit physicochemical characteristics in grafted watermelon. *Horticulturae* **2021**, *7*, 313. [CrossRef]
58. López-Bucio, J.; Pelagio-Flores, R.; Herrera-Estrella, A. *Trichoderma* as biostimulant: Exploiting the multilevel properties of a plant beneficial fungus. *Sci. Hortic.* **2015**, *196*, 109–123. [CrossRef]
59. Celi, G.E.A.; Gratão, P.L.; Lanza, M.G.D.B.; Reis, A.R.D. Physiological and Biochemical Roles of Ascorbic Acid on Mitigation of Abiotic Stresses in Plants. *Plant Physiol. Biochem.* **2023**, *202*, 107970. [CrossRef] [PubMed]
60. Abdelkader, M.M.; Gaplaev, M.S.; Terekbaev, A.A.; Puchkov, M.Y. The influence of biostimulants on tomato plants cultivated under hydroponic systems. *J. Hort. Res.* **2021**, *29*, 107–116. [CrossRef]
61. Ruiz-Cisneros, M.F.; Ornelas-Paz, J.D.J.; Olivas-Orozco, G.I.; Acosta-Muñiz, C.H.; Sepúlveda-Ahumada, D.R.; Pérez-Corral, D.A.; Rios-Velasco, C.; Salas-Marina, M.Á.; Fernández-Pavía, S.P. Efecto de *Trichoderma* spp. y hongos fitopatógenos sobre el crecimiento vegetal y calidad del fruto de jitomate. *Rev. Mex. Fitopatol.* **2018**, *36*, 444–456. [CrossRef]
62. Tigist, M.; Workneh, T.S.; Woldetsadik, K. Effects of variety on the quality of tomato stored under ambient conditions. *J. Food Sci. Technol.* **2013**, *50*, 477–486. [CrossRef] [PubMed]
63. Choi, S.; Colla, G.; Cardarelli, M.; Kim, H.-J. Effects of plant-derived protein hydrolysates on yield, quality, and nitrogen use efficiency of greenhouse grown lettuce and tomato. *Agronomy* **2022**, *12*, 1018. [CrossRef]
64. Ertani, A.; Pizzeghello, D.; Francioso, O.; Sambo, P.; Sanchez-Cortes, S.; Nardi, S. *Capsicum chinensis* L. growth and nutraceutical properties are enhanced by biostimulants in a long-term period: Chemical and metabolomic approaches. *Front. Plant Sci.* **2014**, *5*, 375. [CrossRef] [PubMed]
65. Tesi, R. *Orticoltura Mediterranea Sostenibile*, 1st ed.; Pàtron: Bologna, Italy, 2010; ISBN 978-88-555-3062-0.
66. Han, C.; Zhao, Y.; Leonard, S.W.; Traber, M.G. Edible coatings to improve storability and enhance nutritional value of fresh and frozen strawberries (*Fragaria* × *Ananassa*) and raspberries (*Rubus ideaus*). *Postharvest Biol. Technol.* **2004**, *33*, 67–78. [CrossRef]
67. Meda, A.; Lamien, C.E.; Romito, M.; Millogo, J.; Nacoulma, O.G. Determination of the total phenolic, flavonoid and proline contents in burkina fasan honey, as well as their radical scavenging activity. *Food Chem.* **2005**, *91*, 571–577. [CrossRef]
68. Sadler, G.; Davis, J.; Dezman, D. Rapid extraction of lycopene and β-carotene from reconstituted tomato paste and pink grapefruit homogenates. *J. Food Sci.* **1990**, *55*, 1460–1461. [CrossRef]
69. Giordano, M.; El-Nakhel, C.; Caruso, G.; Cozzolino, E.; De Pascale, S.; Kyriacou, M.C.; Colla, G.; Rouphael, Y. Stand-alone and combinatorial effects of plant-based biostimulants on the production and leaf quality of perennial wall rocket. *Plants* **2020**, *9*, 922. [CrossRef] [PubMed]

plants

Article

Mitigation of Salt Stress in *Lactuca sativa* L. var. Gentile Rossa Using Microalgae as Priming Agents

Ornella Francioso [1], Michela Schiavon [2,*], Serenella Nardi [3], Davide Castellani [1,3], Erika Ferrari [4], Maria Teresa Rodriguez Estrada [1], Maria Cristina della Lucia [3], Veronica Zuffi [1] and Andrea Ertani [2]

[1] Department of Agricultural and Food Sciences, Alma Mater Studiorum, University of Bologna, Viale G. Fanin 40, 40127 Bologna, Italy; ornella.francioso@unibo.it (O.F.); dca@algaenergy.com (D.C.); maria.rodriguez@unibo.it (M.T.R.E.); veronica.zuffi3@unibo.it (V.Z.)

[2] Department of Agricultural, Forest and Food Sciences (DISAFA), University of Turin, Largo Paolo Braccini 2, Grugliasco, 10095 Turin, Italy; andrea.ertani@unito.it

[3] Department of Agronomy, Animals, Food, Natural Resources and Environment (DAFNAE), University of Padova, Viale dell'Università 16, Legnaro, 35020 Padova, Italy; serenella.nardi@unipd.it (S.N.); mcri.dellalucia@gmail.com (M.C.d.L.)

[4] Department of Chemical and Geological Sciences, University of Modena and Reggio Emilia, Via G. Campi 103, 41125 Modena, Italy; erika.ferrari@unimore.it

* Correspondence: michela.schiavon@unito.it

Abstract: Using renewable biomass in agriculture, particularly microalgae as a biostimulant, offers economic and environmental sustainability benefits by reducing costs, improving nutrient cycling, and enhancing water use efficiency. Microalgae contain bioactive compounds that boost crop tolerance to environmental stresses, including salinity. Saline soils, characterized by elevated sodium chloride (NaCl) levels, negatively impact many crops, resulting in low productivity and high remediation costs. Therefore, this study evaluates the biostimulant properties of a microalgae-based commercial preparation (MR) on lettuce (*Lactuca sativa* L.) plants grown hydroponically and exposed to saline stress. The extract was chemically characterized through elemental analysis, lipid composition (gas chromatography with flame ionization detector—GC-FID), the determination of functional groups (Fourier Transformed Infrared—FT-IR), structure (^{1}H,^{13}C Nuclear Magnetic Resonance—NMR), with their hormone-like activity also assessed. Lettuce plants were treated with or without the microalgae blend, in combination with 0, 50 mM, or 100 mM NaCl. The contents of nutrients, soluble proteins, chlorophylls, and phenols, as well as the lipid peroxidation, antioxidants and root traits of lettuce plants, were estimated. The microalgae applied to salt-stressed plants resulted in a significant increase in biomass, protein, and chlorophyll contents. Additionally, significant effects on the secondary metabolism and mitigation of salinity stress were observed in terms of increased phenol content and the activity of antioxidant enzymes, as well as decreased lipid peroxidation. The potassium (K^+) content was increased significantly in plants treated with 100 mM NaCl after addition of microalgae, while the content of sodium (Na^+) was concurrently reduced. In conclusion, our results demonstrate that using microalgae can be a potent approach for improving the cultivation of *Lactuca sativa* L. under saline stress conditions.

Keywords: microalgae; biostimulant; sustainability; renewable biomass; salt stress

1. Introduction

In recent years, the use of microalgae as a valuable industrial feedstock has attained increasing interest within the agricultural sector [1–4]. The appeal of microalgae lies in their ability to prosper in non-potable waters, including industrial effluents [5]. Furthermore, their capacity to efficiently modulate metabolite biosynthesis in response to various environmental conditions, along with their significant production of antioxidant molecules, enhances their attractiveness [6–9]. Indeed, various microalgae species, including *Spirulina* spp., *Dunaliella salina*, and *Chlorella* spp., have been recently reviewed [10] and acknowledged as renewable sources for food, agrochemicals, and nutraceuticals [11].

The metabolites derived from microalgae exhibit promising potential in augmenting soil fertility, bolstering plant resistance to abiotic stress, eliciting defense responses, and optimizing nutrient uptake [1,12]. However, the precise mechanisms of their action and their impact on plant physiology still lack comprehensive understanding. It is known that the growth-promoting properties associated with microalgae arise from their content in diverse biomolecules, such as amino acids, phenolics, phytohormones, terpenoids, and polysaccharides [13]. For instance, applying extracts obtained from *Chlorella vulgaris* and *Scenedesmus quadricauda* during seed hydration improved the germination rate and increased root length, root diameter, and the number of root tips in sugar beet [14]. Similar positive effects were observed when extracts of cyanobacteria species were applied to seeds of maize and wheat [15]. Moreover, the inoculation of live cyanobacteria suspensions in pepper, coriander, fennel, and cumin seeds promoted germination and enhanced the development of roots and leaves [16].

The utilization of microalgae as biostimulant to mitigate abiotic stress has become popular due to their richness in signaling and bioactive molecules, growth enhancers, antioxidants, and natural mineral chelators that support plant growth [17]. The application of extracts from *D. salina* has been shown to enhance salinity tolerance in pepper and tomato plants by increasing the activity of antioxidant enzymes [18,19]. Similarly, applying polysaccharides derived from *Spirulina platensis, D. salina, Porphyridium* spp., and *Phaeodactylum tricornutum* to tomato plants has been demonstrated to enhance phenylalanine lyase and chitinase enzyme activity, total polyphenol content, and reactive oxygen species (ROS) scavenging activity [18].

Among environmental stresses that could be mitigated by applying biostimulants to crops, salinity poses a major concern, as the amount of salinized land is progressively increasing on a global scale [20]. Areas characterized by high concentrations of sodium chloride (NaCl) salt in saline (sodic) soils pose a considerable threat to crop productivity by causing osmotic stress and toxicity in plants [21,22]. An excess of NaCl can restrict water uptake, imbalance the uptake of nutrients—particularly potassium (K^+) and calcium (Ca^{2+})—and induce changes in protein conformation that result in alterations to plant metabolism [23,24].

There are several methods for reclaiming saline soils for crop cultivation, including deep tillage, sub-seeding, horizon mixing [25], profile inversion, washing, and leaching. The most common method for salt removal is leaching, which involves applying large volumes of water. However, due to limited water resources, alternative strategies are needed to reduce remediation costs and prevent crop productivity loss in saline environments. Eustress, a form of moderate and beneficial stress, has been increasingly studied for its potential to enhance the quality of horticultural crops, particularly vegetables. Unlike distress, which severely impacts plant health, eustress can trigger positive physiological responses, leading to improved plant performance and quality traits [26]. Among the various types of eustress, mild salt stress has gained significant attention for its role in enhancing the nutritional and functional quality of vegetables by increasing the biosynthesis of secondary metabolites, such as phenolics, flavonoids, and antioxidants [27]. For instance, moderate salinity levels have been shown to enhance the concentration of bioactive compounds in leafy vegetables like lettuce and spinach, improving their antioxidant properties without compromising yield [26,27]. Moreover, controlled saline eustress can increase the content of essential minerals such as potassium and magnesium, contributing to a better taste and nutritional profile of the produce [28–30].

Lettuce, a crop highly sensitive to salt, plays a crucial role in the agricultural sector [31]. Recent studies have shown that salinity can significantly disrupt lettuce's physiological processes, particularly those associated with electron transfer in photosynthesis [32]. This disruption often leads to ionic imbalance at the cellular level and ultimately results in reduced yields. Furthermore, increasing concentrations of NaCl applied to lettuce have been reported to induce proline accumulation while simultaneously causing declines in total phenol concentration (TPC), total flavonoid concentration (TFC), ascorbic acid levels, chlorophyll content, and antioxidant activities [33]. Its cultivation and distribution are critical for supporting the livelihoods of farmers, farmworkers, and other stakeholders across the agricultural supply chain. The lettuce trade has witnessed notable growth in recent years, which has been driven by increased demand for both fresh and processed forms, resulting in expanded cultivation and export opportunities for many nations [34]. Given this scenario, contemplating the application of biostimulants in the form of microalgae blends to lettuce emerges as a potential sustainable strategy to alleviate salt stress in this horticulture crop.

So far, various microalgae extracts have been used as biostimulants to enhance crop nutrition and productivity, but limited investigation into the specific effects they induce on plant growth and productivity under salinity stress has been conducted. For these reasons, the aim of this study was to assess the capacity of a microalgae-based product to mitigate salt stress in hydroponically cultivated *Lactuca sativa* L. plants. Previous testing involved a thorough dissection and analysis of the chemical characteristics of this product.

2. Results

2.1. Characteristics of the Microalgae-Based Product

2.1.1. Elemental Analysis and Lipid Content

The elemental analyses revealed the occurrence of several macro- and microelements in the microalgae-based product (MR) (Table 1). In particular, relatively high contents of N, P, and K were determined. Among the micronutrients, the most abundant were Ca, Mg, and Fe.

Table 1. Elemental composition of microalgae-based product (mean value, ±SD, n = 3).

Parameters	Concentration
TOC (g kg^{-1})	30 ± 2.90
TKN (g kg^{-1})	70 ± 1.78
P (mg kg^{-1})	749 ± 37.84
K (mg kg^{-1})	2452 ± 38.8
Ca (mg kg^{-1})	190 ± 8.70
Mg (mg kg^{-1})	231 ± 9.81
Cu (mg kg^{-1})	1.76 ± 0.38
Zn (mg kg^{-1})	2.23 ± 0.18
Fe (mg kg^{-1})	63 ± 3.82
Mn (mg kg^{-1})	2.10 ± 0.11
pH	3.8 ± 0.06
Moisture (%)	8.84 ± 0.07
Ash (%)	6.75 ± 0.20

Abbreviations. TOC: total organic carbon; TKN: total nitrogen.

The total lipid content accounted for 11% DW (dry weight), as shown in Table 2. Free fatty acids (FFAs) and diglycerides (DAGs) represented 42.30% DW and 38.50% DW of the total lipids, respectively. Additionally, hydrocarbons and carotenoids (HYCAs) were found to be 14.6% DW in the sample; in contrast, triglycerides (TAGs) and di-acylglycerols (PSTOCs) were less than 1% DW.

Table 2. Total lipid contents (% dry weight, DW) and lipid class distribution (% total lipids) of microalgae-based product (mean value, ±SD, n = 3).

Parameters	Concentration
Total lipid	11.11 ± 0.10
MAG	2.91 ± 0.02

Table 2. *Cont.*

Parameters	Concentration
DAG	38.51 ± 0.18
TAG	0.84 ± 0.08
FFA	42.31 ± 0.17
HYCA	14.67 ± 0.11
PSTOC	0.75 ± 0.01
Total lipid	11.11 ± 0.10

Abbreviations. MAG: monoacylglycerols; DAG: di-acylglycerols; TAG: triacylglycerols; FFA: phytosterols + tocopherols; HYCA: hydrocarbons + carotenes; PSTOC: free fatty acids.

2.1.2. Content in Amino Acids

The MR contained a varied amino acid composition, with a predominance of certain specific ones (Table 3). Glutamic acid and leucine were among the most abundant amino acids, followed by alanine and valine. Other amino acids present in significant quantities included lysine, isoleucine, and threonine. Serine, histidine, asparagine, and glycine were also well represented in the formulation, although in slightly lower amounts. The presence of phenylalanine, tryptophan, and methionine completed the range of essential amino acids. Additionally, the product contained aspartic acid, arginine, tyrosine, proline, cysteine, and a small amount of glutamine.

Table 3. Amino acid content of microalgae-based product (MR). Values are the average of three replicates (average ± SD).

Amino Acid	% (g/g)
Glutamic acid	0.43
Leucine	0.42
Alanine	0.38
Valine	0.31
Lysine	0.29
Isoleucine	0.26
Threonine	0.26
Serine	0.22
Histidine	0.19
Asparagine	0.17
Glycine	0.14
Phenylalanine	0.12
Tryptophan	0.10
Methionine	0.10
Aspartic acid	0.09
Arginine	0.07
Tyrosine	0.07
Proline	0.07
Cysteine	0.04
Glutamine	0.03

2.1.3. Spectroscopic Characterization of Microalgal-Based Product (MR)

In this study, different spectroscopic techniques were employed with the objective of providing a comprehensive characterization of the microalgal based product. In particular, FTIR spectroscopy is a time-efficient technique due to the simplicity of the sample preparation process, the relatively short time required for spectral analysis, and the ability to analyze only minute quantities of the sample. Furthermore, FTIR provides significant advantages such as a higher signal-to-noise ratio, high energy throughput, and high accuracy and stability. The aforementioned properties of FTIR are effective in the characterization of complex samples, such as microalgal-based products.

The FTIR spectrum of the microalgae-based product (MR) (Figure 1) showed distinct bands corresponding to different biomolecules such as lipids, carbohydrates, and aromatic compounds. In particular, the broad band at 3272 cm^{-1} was assigned to O–H stretching, while a slight shoulder at 3065 cm^{-1} was assigned to C=C–H stretching vibration and N-H stretching. The region between 2954 cm^{-1} and 2853 cm^{-1} was typical of aliphatic lipid chains [35], as well as that between 1368–1329 cm^{-1}, which corresponded to the twisting vibration of -CH_2 in C14:0–C18:0 [36]. The strong bands at 1624 cm^{-1} and the weak band at 1245 cm^{-1} were due to C=O and C–OH stretching vibration of acidic groups, respectively. The high content of FFAs might support the presence of these bands. The strong band at 1516 cm^{-1} corresponded to aromatic C=C stretching. The bands around 1037 cm^{-1} could be assigned to C–C–H and O–C–H deformation, C–O stretching, and C–O and C–C stretching vibrations of pyranose rings [37], with the contribution of mineral compounds.

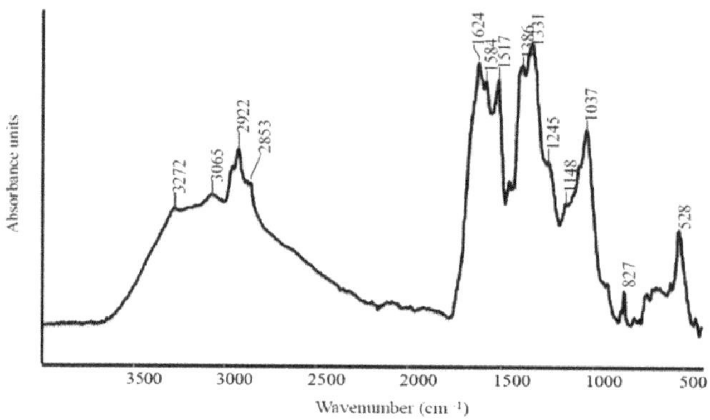

Figure 1. FT-IR spectrum of microalgae-based product (MR).

In addition to FT-IR analysis, NMR spectroscopy was used to better characterize the MR. Indeed, NMR spectroscopy is an efficient tool to investigate the chemical composition of algae and microalgae, as reported by Stirk et al. [38]. Indeed, this technique allows us to evaluate all those species that have even slight solubility in the selected solvent, D_2O for this study, and that behave differently according to their mobility, molar mass, and wettability. As shown in the ^1H NMR spectra (Figure 2), minimal differences were observed using 1D pulse sequences for water suppression; particularly, *cpmg* permitted the focus on "sharper" peaks due to highly mobile protons with long relaxation time, while information was lost for those protons in a fast-relaxing environment. In all cases, resonances were consistent with the presence of proteins, amino acids (AAs), carbohydrates, and lipids. The ^1H NMR spectrum can be chunked into typical regions, as reported in Figure 2. Although overcrowded signals prevented a quantitative NMR analysis, a semi-quantitative analysis pinpointed prevalent amounts of amino acids and peptides over carbohydrates and lipids. Specifically, the integrated area corresponding to terminal aliphatic methyl groups (region F, Figure 2) was close to that for methylene groups, with an integrated area ratio approximately near 1, which suggested the presence of a little amount of long-chain lipids in comparison to amino acids. The peak around 0.9 ppm benefitted from the contribution of leucine, isoleucine, and valine—in addition to terminal lipid CH_3. In the aromatic/double-bond regions (A + B), the presence of tyrosine (7.10–6.8 ppm), histidine (8.4–8.2 ppm), phenylalanine (7.4 ppm), and nucleobases, such as uracil (5.9/7.8 ppm), was detected and confirmed by ^1H,^1H COSY spectrum (Figure S1). The broad weak resonance at 5.3 ppm indicated the presence of diacylglycerol. Similarly, ^{13}C NMR spectrum showed significant overlap; however, different regions could be assigned on the basis of the chemical shifts and 2D ^1H/^{13}C correlations due to scalar couplings. Definitely, the applied pulse sequence for HSQC (*hsqcedetgp*) being phase-sensitive using Echo/Antiecho-TPPI gradient selection could discriminate between CH_2 and CH/CH_3 cross-peaks (Figure S2).

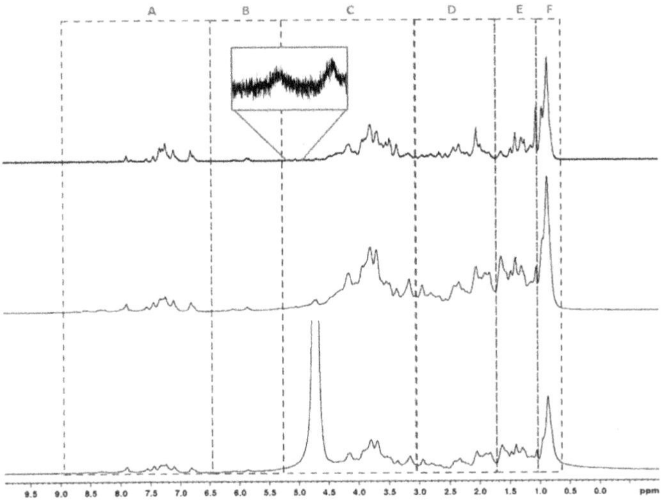

Figure 2. ^{1}H NMR spectra of microalgae-based product MR in D_2O (298K @ 600 MHz). From bottom to top, spectra acquired using the pulse sequences: *zgcppr*, *noesypr1d*, and *cpmgpr1d*, respectively. Six typical regions are highlighted by dashed square boxes: (**A**) aromatic protons from amino acids and nucleobases (9.0–6.5 ppm); (**B**) nucleobases and double-bond (lipids and carotenoids) protons (6.5–5.3 ppm), the expansion in cpmgpr1d spectrum highlights diacylglycerol methylene group; (**C**) amino acids, carbohydrates (5.5–3.0 ppm); (**D**) methylene protons of amino acids/peptides and carboxyl-rich alicyclic molecules (3.0–1.7 ppm); (**E**) methylene protons of lipids and amino acids/peptides (1.7–1.0 ppm); (**F**) methyl protons of amino acids and lipids.

As shown in Figure 3, ^{13}C NMR spectrum can be divided into five main chemical shift regions associated with typically occurring compounds: FAs and protein/amino acid aliphatic side chains (0–40 ppm), protein/amino acid C alfa (40–60 ppm), C\O in carbohydrates (60–100 ppm), aromatic and double-bonded carbons (110–160 ppm), and amide and carbonyl regions (160–180 ppm). As for carbohydrate resonances, specific sub-regions were detected: 90–112 ppm for the anomeric carbon (C1), 70–80 ppm for the ring carbons (C2-5), and 60–70 ppm for the C6 carbons. This latter was also overlapped with lipids, particularly glycerol that forms mono-, di-, and triacylglycerols resonating in the same spectral region. Phase-sensitive $^{1}H,^{13}C$ HSQC analysis (Figure S2) highlighted the presence of two cross-peaks due to methylene groups tentatively attributable to diacetyl-glycerol, respectively, at δ 3.95(^{1}H)/63.9(^{13}C) and 3.8(^{1}H)/66.2(^{13}C) ppm [39]. Triacylglycerols were not observed, probably due to their poor water solubility. Carboxylic groups of free fatty acids were detected around 173 ppm. At higher chemical shifts (180–174 ppm) the contribution of amide groups was observed, while at lower chemical shifts (~170 ppm), esters were present. The signals in the range 150–160 ppm were possibly due to tyrosine and arginine residues.

The overlapping of different species hindered quantitative analysis. Nevertheless, the low intensity of anomeric carbons suggested a minimal contribution of carbohydrates, with proteins and amino acids being the main constituents. The absence of a very intense and broad peak around 20 ppm accounted for a low content of long-chain hydrocarbons and lipids, although their presence could be underestimated because of their low solubility in the aqueous medium.

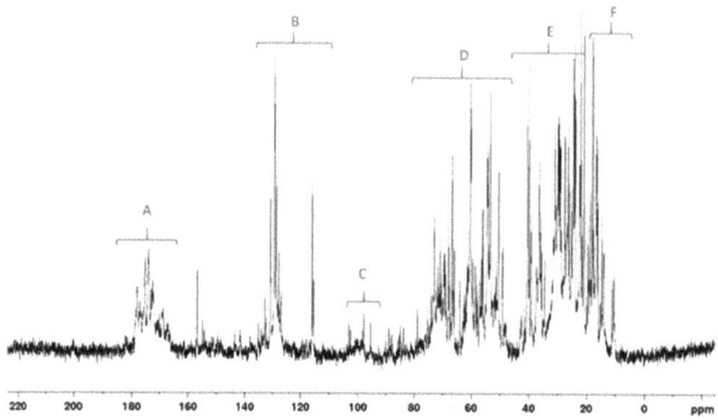

Figure 3. ^{13}C NMR spectrum of microalgae-based product MR in D_2O (298K @ 600 MHz). Six typical regions are highlighted by dashed square boxes: (**A**) carbonyl (amides/carboxylic/esters); (**B**) aromatic carbons from amino acids and nucleobase double-bonds (lipids and carotenoids), (**C**) anomeric carbons; (**D**) amino acids/peptides/carbohydrates/diacylglycerols; (**E**) CH_2/CH carbons of lipids and amino acids/peptides; (**F**) methyl carbons of amino acids and lipids.

2.2. Hormone-like Activities

The hormone-like activity of the product was assayed to determine if the presence of certain compounds endowed in the formulation could exhibit the capacity to stimulate plant growth through this hormone-regulated mechanisms. The IAA concentrations applied during the bioassay inhibited the root elongation of watercress in a dose-dependent manner (Figure 4A). In contrast, this dose-dependent inhibition did not occur when the microalgae-based product—MR—was applied at watercress seedlings. Conversely, GA concentrations promoted the elongation of lettuce shoots in a dose-dependent manner (Figure 4B), and a similar trend was observed when the microalgae-based product was applied, suggesting an effect on shoot elongation consistent with the GA-like activity of the MR.

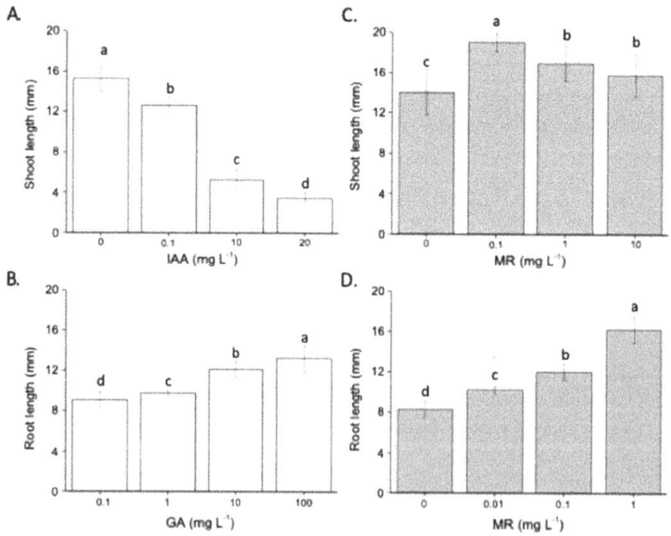

Figure 4. Auxin-like activity of Indoleacetic acid (IAA) (**A**) and MR (**C**), and gibberellin-like activity of gibberellic acid (**B**) and MR (**D**) supplied at increasing doses. Data represent the means of three measurements with ten plants in each. Values above bars following the same letter are not statistically different at $p < 0.05$ according to the Student–Newman–Keuls test.

2.3. Effect of MR on Root Morphology

To evaluate the effectiveness of the MR in stimulating root growth in lettuce, various root morphological parameters were analyzed and compared between MR-treated and untreated plants (Table 4). The results revealed overall stimulation of the root system. The projected area (cm^2), which is classified as the two-dimensional area measurement of a three-dimensional object projecting its shape onto an arbitrary plane, indicated a stimulation increase of about 13% in the MR-treated plants compared to the control. A similar rate of increase was calculated for the surface area, which takes into account the entire three-dimensional surface of the root and showed a 13% increase compared to the control. The surface area of a solid object is mathematically defined as a measure of the total area that the root's surface occupies in a given space. The length per volume, referred to as the ratio of root length (cm) to the volume of soil (m^3), indicated a considerably higher growth rate (+47%) in plants treated with the MR. On average, the data recorded for average diameter (mm) and root volume (cm^3) showed lower values than the control conditions. The number of root tips, forks, and crossings showed increases of 26%, 28%, and 66%, respectively, in MR-treated plants compared to untreated plants.

Table 4. Total root length, surface area, number of root tips, forks, crossings, and length of fine roots of untreated plants and MR treated plants. Values are the average of three replicates (average ± SD). Different letters in the same column indicate significant differences between treatments ($p < 0.05$).

Parameters	C	MR	% Increase
Projected Area (cm^2)	47.34 ± 12.22 ^{a}	53.49 ± 8.28 ^{a}	113
Surface Area (cm^2)	148.73 ± 24.80 ^{b}	168.05 ± 23.1 ^{a}	114
Average Diameter (mm)	0.74 ± 0.046 ^{b}	0.57 ± 0.05 ^{a}	77
Root Volume (cm^3)	2.74 ± 0.14 ^{b}	2.38 ± 0.09 ^{a}	87
Tips	942 ± 23 ^{b}	1186 ± 38 ^{a}	126
Forks	6254 ± 56 ^{b}	7990 ± 37 ^{a}	128
Crossings	942 ± 57 ^{b}	1568 ± 67 ^{a}	166

2.4. Effects of the Microalgae-Based Product and NaCl on Leaf and Root Biomass, Chlorophyll and Protein Content

Figure 5 illustrates the growth of lettuce plants at the end of the trial before harvesting. A significant increase in leaf dry biomass was observed in plants treated with MR, either alone or in combination with 50 mM or 100 mM NaCl, compared to untreated plants (Figure 6A).

Figure 5. Plants of *Lactuca sativa* L. grown for 14 days in a complete culture solution with 0 (control), 50, and 100 mM NaCl and microalgae-based product (MR).

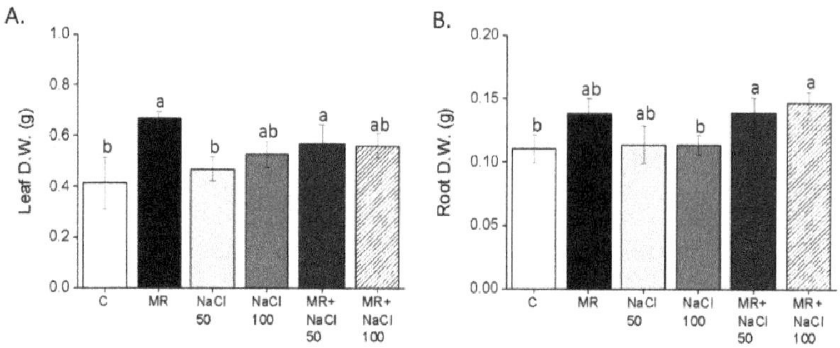

Figure 6. Leaves (**A**) and roots (**B**) dry weight of *Lactuca sativa* L. plants grown for 14 days in a complete culture solution with 0 (control), 50, and 100 mM NaCl and microalgae-based product (MR). All data represent the means of five measurements per treatment with three plants in each (±SD). Different letters in the same column indicate significant differences between treatments ($p < 0.05$) according to the Student–Newman–Keuls test.

Similarly, the addition of MR to plants grown with NaCl concentration increased the leaf and root biomass (Figure 6B). Lettuce plants treated with salt at both concentrations showed no significant changes in leaf and root biomass compared to the control.

The leaf chlorophyll content was substantially decreased by NaCl application compared to the control plants (Figure 7A), while the microalgae-based product (MR) enhanced it. The application of the MR to NaCl-treated plants restored chlorophyll levels, with values similar to those observed in the control group and in MR salt-untreated plants. A significant increase in total protein content was observed in plants treated with MR (Figure 7B), but MR did not improve the protein content in plants exposed to salt stress, which exhibited similar protein levels compared to the control plants.

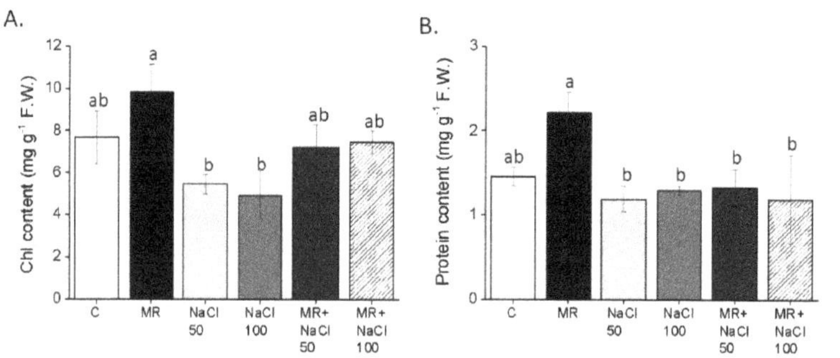

Figure 7. Chlorophyll (**A**) and protein (**B**) content of *Lactuca sativa* L. plants grown for 14 days in a complete culture solution with 0 (control), 50, and 100 mM NaCl and microalgae-based product (MR). All data represent the means of five measurements per treatment with three plants in each (±SD). Different letters in the same column indicate significant differences between treatments ($p < 0.05$) according to the Student–Newman–Keuls test.

2.5. Effects of the Microalgae-Based Product on Total Phenol Content

Overall, all treatments resulted in an increase in phenol content compared to the control, with the higher concentration of salt yielding the greatest effect. Specifically, plants treated with MR, in combination or not with 50 mM NaCl, showed a slight increase in phenol content compared to the control (Figure 8A). The addition of MR to plants treated with 100 mM NaCl decreased the amount of phenols by about 40% compared to plants treated with NaCl alone.

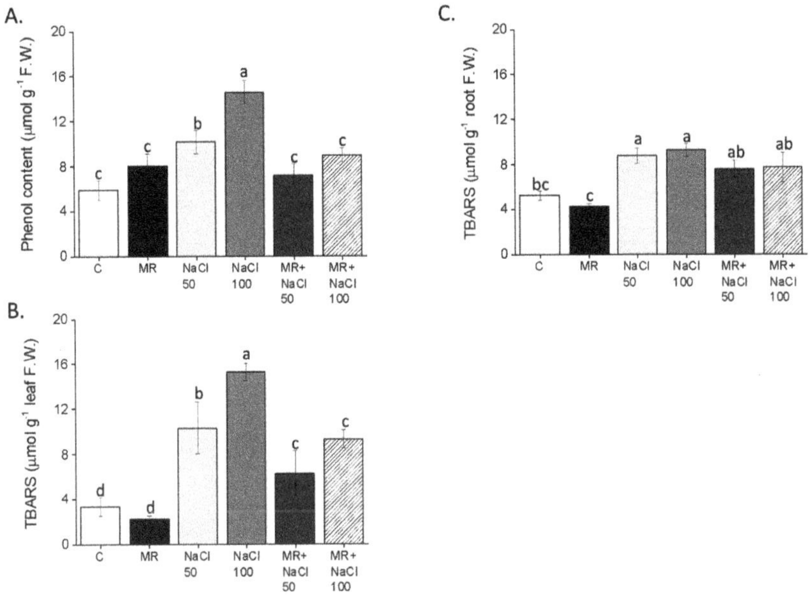

Figure 8. Phenols content in leaves (**A**) and lipid peroxidation in leaves (**B**) and roots (**C**) of *Lactuca sativa* L. plants grown for 14 days in a complete culture solution with 0 (control), 50, and 100 mM NaCl and microalgae-based product (MR). All data represent the means of five measurements per treatment with three plants in each (±SD). Different letters in the same column indicate significant differences between treatments ($p < 0.05$) according to the Student–Newman–Keuls test.

2.6. Effects of the Microalgae-Based Product on the Accumulation of Na^+ and K^+

The K^+ content in the leaves was comparable between control plants and those supplemented with the MR (Table 5). The NaCl treatments led to an increase in leaf K^+ accumulation, which was further amplified by the addition of MR. In the roots, K^+ content was higher in NaCl-treated plants compared to the control and MR groups. Nevertheless, the MR treatment improved K^+ retention in the roots relative to plants treated with NaCl alone. Regarding the Na^+ content, it was significantly elevated in both leaves and roots of plants treated with NaCl (Table 5).

The addition of MR reduced Na^+ accumulation in both leaves and roots compared to NaCl-only treatments. The K^+/Na^+ ratio was highest in control plants and decreased in NaCl-treated plants. The addition of MR improved the K^+/Na^+ ratio in both leaves and roots under salt stress conditions.

Table 5. Concentration and K^+/Na^+ in leaves and roots of *Lactuca sativa* L. plants grown for 14 days in a complete culture solution with 0 (control), 50, and 100 mM NaCl and MR. All data represent the means of five measurements per treatment with three plants in each (±SD). Different letters along each column indicate significant differences between treatments.

	K^+	Na^+	K^+/Na^+
Leaf	mg kg^{-1} D.W.		
C	7.74 ± 0.82 ^{c}	1.96 ± 0.21 ^{c}	3.95 ± 0.20 ^{a}
MR	7.61 ± 0.38 ^{c}	1.87 ± 0.06 ^{c}	4.07 ± 0.31 ^{a}
NaCl 50	8.99 ± 0.11 ^{b}	3.33 ± 0.22 ^{a}	2.70 ± 0.11 ^{d}
NaCl 100	9.42 ± 0.15 ^{b}	3.17 ± 0.20 ^{a}	2.97 ± 0.10 ^{d}
MR + NaCl 50	9.17 ± 0.12 ^{b}	2.54 ± 0.12 ^{b}	3.54 ± 0.23 ^{c}
MR + NaCl 100	10.24 ± 0.55 ^{a}	2.37 ± 0.13 ^{b}	4.32 ± 0.27 ^{a}

Table 5. *Cont.*

	K^+	Na^+	K^+/Na^+
Root			
C	10.03 ± 0.97 [a]	3.02 ± 0.41 [c]	3.33 ± 0.32 [a]
MR	9.60 ± 0.53 [a]	3.03 ± 0.05 [c]	3.17 ± 0.13 [a]
NaCl 50	9.06 ± 0.16 [b]	4.10 ± 0.25 [a]	2.21 ± 0.09 [c]
NaCl 100	8.27 ± 0.12 [c]	4.28 ± 0.19 [a]	1.93 ± 0.23 [c]
MR + NaCl 50	10.29 ± 0.18 [a]	3.56 ± 0.09 [b]	2.89 ± 0.13 [b]
MR + NaCl 100	9.94 ± 0.85 [a]	2.94 ± 0.10 [c]	3.38 ± 0.22 [a]

2.7. Effects of the Microalgae-Based Product and NaCl on Lipid Peroxidation and Antioxidants

The addition of NaCl, particularly at the 100 mM concentration, significantly increased total phenol production, while MR supplementation resulted in only a slight increase compared to the control plants (Figure 8A). However, the MR reduced the phenol content in NaCl-treated plants to levels similar to those in control plants.

Lipid peroxidation, as indicated by the MDA assay, was significantly higher in the leaves of lettuce plants treated with 50 and 100 mM NaCl compared to both control plants and MR-primed plants not exposed to salt (Figure 8B). The MR substantially decreased (by about 60%) lipid peroxidation in the leaves of lettuce exposed to NaCl. In the roots, a similar trend was observed as in the leaves, although the reduction in lipid peroxidation due to MR priming was less pronounced (Figure 8C).

Regarding APX enzyme activity, all treatments showed a significant increase compared to the control (Figure 9A). In particular, the treatments with the MR and 100 mM NaCl resulted in the most significant increases. The combination of the MR and NaCl did not further increase the activity of APX in plants. The trend observed for GPX was the same as for APX (Figure 9B).

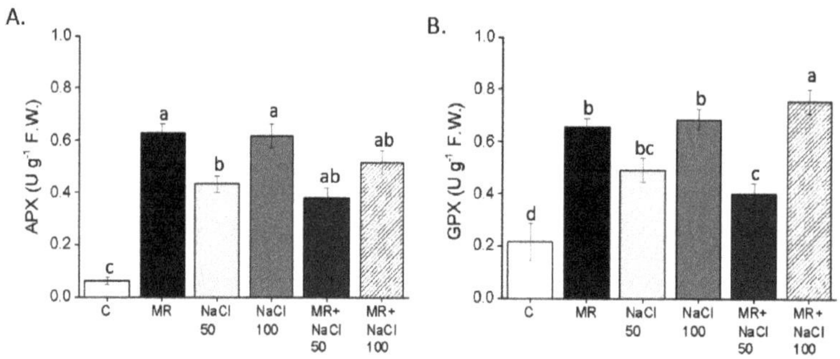

Figure 9. APX (**A**) and GPX (**B**) in roots of *Lactuca sativa* L. plants grown for 14 days in a complete culture solution with 0 (control), 50 and 100 mM NaCl and microalgae-based product (MR). All data represent the means of five measurements per treatment with three plants in each (±SD). Different letters in the same column indicate significant differences between treatments ($p < 0.05$) according to the Student–Newman–Keuls test.

3. Discussion

Plants constantly face various adverse environmental stress conditions, with soil salinity posing a significant global challenge for agricultural production [40,41]. Ongoing efforts aim to develop solutions that help crops mitigate salt-induced stress, including the production of valuable products to enhance plant resilience. In this study, we specifically evaluated the effectiveness of a microalgal blend as a priming agent with biostimulant properties in lettuce plants subjected to two concentrations of NaCl. These concentrations were chosen based on previous studies [32,42]. Overall, the chemical characterization of the microalgal blend confirmed the antioxidant properties of the MR product, being in line with previous reports [43,44]. The MR demonstrated positive effects on various parameters of lettuce. These benefits were likely due to the bioactive compounds contained in the formulation [4,45–47], some of which are gibberellins [48] that may mimic (GA)-like activity, as suggested by the Audus test, thereby improving plant growth and root traits [8,49,50]. In addition, the lipid composition of the blend was characterized by a high presence of DAGs, FFAs, and HYCA, which are all known for their bioactive properties in plants and for their involvement in cellular defense systems against stress. These compounds may synergistically enhance plant resilience and development, promoting improved physiological responses and growth outcomes regardless of environmental stressors. DAGs, for instance, serve as secondary messengers in plant cells, being involved in signaling pathways that regulate growth and stress responses. Algal-derived DAGs may enhance plant signaling mechanisms that control responses to environmental stresses such as salinity, and they may influence the synthesis of hormones, especially abscisic acid, which is required in osmotic stress regulation [47]. FFA, including phytosterols and tocopherols, were the most abundant lipid components in the blend. These compounds are known for their capacity to improve plant cell membrane stability and regulate hormone signaling pathways under stress conditions. Tocopherols, in particular, are potent lipid-soluble antioxidants that protect plant cells from oxidative stress. The HYCA fraction, which was the third most abundant lipid component, contains carotene. This compound has well-known antioxidant properties, protecting plant cells from oxidative damage [51,52]. Regarding the content of amino acids, glutamate, leucine, and alanine were the most abundant in the MR. In recent years, some reports have noted that the exogenous application of amino acids can protect plants from abiotic stresses, including glutamate [53,54] and leucine [55], while the effects of alanine has not yet been clarified in this context [55]. On the other hand, the content of proline, widely recognized as the most significant osmolyte for mitigating osmotic stress under salinity [56], was relatively low in the product. This suggests that its contribution to the product's beneficial properties was likely minimal.

When NaCl was applied to the plants, we did not observe significant negative effects of NaCl on plant growth and protein content compared to the control group (i.e., the non-salt-treated plants). These findings differ from those of other studies, where similar salt concentrations often resulted in reduced plant growth [31,33,57]. One possible explanation is that the lettuce variety used in the study may have been relatively salt-tolerant, as suggested by its capacity to enhance K^+ accumulation in the leaves and roots in response to salt exposure. Additionally, this variety demonstrated a substantial increase in the activity of two key antioxidant enzymes, APX and GPX, with the effect being even more pronounced at higher NaCl concentrations, where lipid peroxidation was more evident. The activity of these enzymes typically serves to counteract NaCl-induced oxidative stress [58]. Thus, while the application of the MR to non-salt-treated lettuce plants significantly improved growth, chlorophyll and protein content, and antioxidant enzyme activity, its addition to salt-treated plants did not further significantly enhance these parameters, except for leaf growth. On the other hand, it is important to note that lipid peroxidation decreased in salt-treated plants supplemented with MR, likely due to the microalgal blend ability to considerably restrict Na^+ accumulation in lettuce and concurrently increase K^+ intake and the K^+/Na^+ ratio. The reduced accumulation of Na^+ in these plants following MR application likely resulted in lower stress levels, which may explain the decreased production of phenolic compounds, typically acting as non-enzymatic cellular antioxidants, as well as the lack of a cumulative effect on antioxidant enzyme activity [59,60]. The reduction in stress was further supported by the lower levels of lipid peroxidation, which, under high-stress conditions, would otherwise compromise cell membrane integrity [47]. We cannot however exclude that some constituents of MRs with antioxidant properties, such as tocopherols and carotenes, may directly contribute to safeguarding lettuce cells against peroxidation and maintaining cellular redox status [48].

4. Materials and Methods

MR*, the commercial biostimulant preparation, was produced and provided by AlgaEnergy ((Madrid, Spain)) as an enzymatic digestion of microalgae extracts stabilized at specific pH values for conservation. The entire production process, from cultivation conditions to the final product, is safeguarded by patent protection. Further insights into its composition and elaboration process, via UPT®, can be obtained upon request to the provider.

4.1. Chemical Characterization of the Microalgae-Based Product

4.1.1. Elemental Analysis, Quantification of Moisture and Ashes

The elemental characterization was performed on the lyophilized microalgae-based product, termed MR. Total carbon (C) and nitrogen (N) contents were determined using a dry combustion procedure inside an element analyzer (CHNS-O EA 1110 Thermo Fisher Scientific, Waltham, MA, USA). The contents of macro- and microelements were quantified via microwave-assisted acid digestion (Milestone, Shelton, CT, USA) using HNO_3 suprapur (Carlo Erba Reagents Srl, Milan, Italy) and H_2O_2 (30% v/v). The mineralized samples were then diluted with ultrapure water, and each element was analyzed by Inductively Coupled Plasma–Optical Emission Spectroscopy (ICP-OES) (Spectro Amatek Arcos II ICP-OES, Kleve, Germany).

Moisture and ash content were determined using a TG-DTA92 instrument (SETARAM, Caluire-et-Cuire, France). A quantity of 4–5 mg of lyophilized sample was accurately measured and placed in an alumina crucible. The sample was hence subjected to heating, ranging from 30 to 700 °C, within a dynamic air atmosphere (air flow: 5 L h^{-1}). The heating rate was 10 °C min^{-1}. All analyses were conducted in triplicate.

4.1.2. Analysis of Functional Groups and Structure via Attenuated Total Reflectance (ATR-IR) and Nuclear Magnetic Resonance (NMR) Spectroscopies

The Fourier Transformed Infrared (FT-IR) spectrum of lyophilized MR was recorded with an Alpha FT-IR instrument (Bruker Optics, Ettlingen, Germany). The spectral range was set between 4000 and 400 cm^{-1}, with a resolution of 4 cm^{-1} and 64 scans. Prior to sample measurement, a background spectrum was recorded against air. Less than 1 mg of freeze-dried sample was used [49].

The NMR spectra were recorded using a Bruker Biospin FT-NMR AVANCE III HD (Bruker, Billericca, MA, USA) (600 MHz) spectrometer equipped with a CryoProbe BBO H&F 5 mm in inverse detection. Nominal frequencies were 150.90 MHz for ^{13}C and 600.13 for ^{1}H, respectively. Fifteen mg of the investigated lyophilized microalgae-based product were suspended in 0.6 mL of D_2O (99.8%) and dropped into a 5 mm NMR tube. All experiments were performed at 298 K. A 90° ($\pi/2$) pulse was calibrated, and standard NMR parameters were used. As a result, 1D ^{1}H NMR data were acquired using standard Bruker pulse sequences: *zg* (1D sequence), *zgcppr* (1D sequence with presaturation and using composite pulse for selection), *noesypr1d*, and *cpmgpr1d* (Carr-Purcell-Meiboom-Gill). These 1D ^{13}C NMR spectra were acquired using Bruker pulse sequences *zgpg_pisp_f2.fas*. Typical 2D homo- and heteronuclear techniques were used for assignment, applying Bruker library pulse sequences, namely, *cosygpprqf* (^{1}H,^{1}H-COSY), *hsqcedetgp* (^{1}H,^{13}C-HSQC), and *hmbcgplpndqf* (^{1}H,^{13}C-HMBC) [61,62].

4.1.3. Lipid Content

A modified version [49] of the Folch method [50] was used for lipids extraction from 8 g of the lyophilized microalgae-based product. The lipid content was quantified gravimetrically and expressed as percentage. Three independent replicates of the sample were analyzed. The qualitative profile of the main lipid classes (hydrocarbons + carotenes, HYCA; free fatty acids, FFA; monoacylglycerols, MAG; phytosterols + tocopherols, PSTOC; di-acylglycerols, DAG; triacylglycerols, TAG) was determined by gas chromatography–flame ionization detection (GC-FID) according to Toschi et al. [51].

An aliquot of 30 mg from the lipid extract was dissolved in 1 mL of n-hexane and injected into a GC-FID Shimadzu GC 2010 plus (Kyoto, Japan), which was coupled with a fused silica capillary column (Mega SE52 (Legnano, Milan, Italy); 10 m × 0.32 mm i.d. × 0.1 μm film thickness). The temperature program of the oven ranged from 100 to 355 °C at 5 °C min^{-1}, which was held at 355 °C for 15 min. Both injector and FID temperatures were set at 355 °C. The injection was performed in the split mode (1:25), with a constant helium flow (carrier gas) of 2.02 mL min^{-1}. The lipid classes were identified using different commercial standards, and the data are expressed as area percentage values. Two independent replicates of the sample were analyzed.

4.1.4. Amino Acid Composition

The free amino acid content in plant samples was analyzed following the method outlined by [52]. Two grams of sample was transferred into a centrifuge tube. An internal standard (1.7-diaminoheptane solution, Merck, Darmstadt, Germany) was added to the sample to facilitate accurate quantification. Subsequently, 40 mL of trichloroacetic acid (5% *w:v*, Avantor Performance Materials, Gliwice, Poland) was added to the sample, which was homogenized by agitation. After centrifugation at 10,000× *g* for 10 min, 100 μL was added with 2.5 mL of di-sodium tetraborate (3% water solution, J.T. Baker, Pol-Aura, Gliwice, Poland) and 2.5 mL of dansyl-chloride (20 mM in acetonitrile, abcr GmbH, Karlsruhe, Germany) to allow derivatization of the amino acids. Derivatization was achieved by shaking the mixture in a water bath at 40 °C for 1 h. To halt the derivatization process, 10 μL of formic acid (98–100% purity, Avantor Performance Materials, Gliwice, Poland) was added to the mixture. Finally, the derivatized sample was filtered into a chromatographic vial using a syringe filter (pore size: 0.22 μm, Captiva Econofilters, Agilent, Santa Clara, CA, USA) in preparation for subsequent chromatographic analysis.

Ultra-high performance liquid chromatography coupled with a high-resolution mass spectrometer (UHPLC-MS/MS) analysis was conducted to determine the concentration of free amino acids in the samples. Chromatographic separation of the compounds was achieved using a Cortecs UPLC C18 column (2.1 × 100 mm, 1.6 μm particle size, Waters, Milford, MA, USA). The mobile phase consisted of two phases: Phase A comprised water with acetonitrile (90:10) containing 0.1% formic acid (FA) and 5 mM ammonium formate, while Phase B comprised acetonitrile with water (90:10) containing 0.1% FA and 5 mM ammonium formate. Both LC-MS-grade water and LC-MS-grade acetonitrile were sourced from Witko (Łódź, Poland), while formic acid (98–100%) and ammonium formate (≥97%) for LC-MS were obtained from Chem-Lab (Zedelgem, Belgium). A liquid phase gradient was applied at a flow rate of 0.3 mL/min, with the following composition over time: 0–2 min: 90% A, 10% B; 2–22 min: 0% A, 100% B; 22–25 min: 0% A, 100% B; 25–26 min: 90% A, 10% B; 26–28 min: 90% A, 10% B. The UHPLC-MS/MS system utilized a Q Exactive Orbitrap Focus MS (Thermo Fisher Scientific, Waltham, MA, USA) equipped with a heated electrospray ionization (HESI) source. The analysis was performed in positive polarization mode, with an injection volume of 2.5 μL. Specific MS parameters were set as follows: spray voltage: 3 kV, capillary temperature: 256 °C, sheath gas flow rate: 48, auxiliary gas flow rate: 11, sweep gas flow rate: 2, probe heater temperature: 413 °C, S-lens RF level: 50, resolution: 70,000 in simultaneous scan and 35,000 in all ion fragmentation. Data acquisition and analysis were conducted using Xcalibur 4.2.47 software, facilitating precise determination and quantification of free amino acids in the plant samples.

4.1.5. Assessment of Hormone-like Activities

Indoleacetic acid (IAA)-like activity was estimated by assessing the reduction of roots of watercress (*Lepidium sativum* L.) after treatment with IAA and different aqueous solutions of the products, while gibberellin (GA)-like activity was evaluated by measuring the increase in length of lettuce epicotyls after treatment with increasing doses of GA and the microalgae-based product. The detailed protocol is reported in Ertani et al. [53].

4.2. Experimental Design

Two-week-old plantlets of *Lactuca sativa* L. var. gentile rossa were transplanted from the potting soil into 6 L hydroponic pots (35.5 × 12.5 × 21.5 cm) filled with a Hoagland modified nutrient solution, at a density of 6 plants per pot, which was renewed every 48 h. The nutrient solution had the following composition: KH_2PO_4 (0.63), $Ca(NO_3)_2$ (2), KNO_3 (3), $MgSO_4$ (1.5), FeNaEDTA (0.040), plus micronutrients. Before the transplant, the roots of plants were thoroughly washed with deionized water to gently remove the soil particles. Lettuce plants were cultivated inside a growth chamber at 22 ± 2 °C, with a relative humidity ranging within 70%–85% and a 16 h photoperiod under natural light. Seven days after transplantation, the microalgae-based product was added to the nutrient solution of half of the plants, at a concentration of 1mL L^{-1}, according to literature tests. This addition was done to induce a priming effect in the plants before subjecting them to the salt stress treatment. Two days later, lettuce plants, primed or not with the microalgae-based product, were treated with the following NaCl concentrations: 0, 50, and 100 mM. Three pots were set up for the control plants (unprimed and salt untreated) and each treatment (18 pots in total). Plants were sampled 5 days after salt treatment, divided into roots and leaves, and carefully washed. Sub-samples from control and treated plants were frozen with liquid nitrogen and stored at −80 °C. The fresh weight (FW) was determined for leaves and roots of 9 plants. Roots and leaves samples were then dried at 70 °C for 2–3 days, and their dry weight (DW) was determined. The experimental design is represented in Figure S3.

4.3. Root Morphological Analysis

Root morphological analysis was performed by using the WinRHIZO software 2012 (Regent Instruments). Total root length, root surface area, mean root diameter, root volume, total number of root tips, lateral root (0.000 < L < 0.500 and 0.500 < L < 1.000) were determined by computerized scanning (STD 1600, Regent Instruments, Québec, QC, Canada) [54,55].

4.4. Elemental Analyses

Dried leaves and roots of lettuce plants were ground before elemental analysis. The digestions were carried out as described in Ertani et al. [54] inside closed Teflon vessels of 120 mL volume using approximately 500 mg dry leaf or root material and 10 mL of 30% (*v*/*v*) HCl. After digestion, the resulting solution was transferred and diluted with 10 mL ultrapure water. The mineralized samples were then diluted with ultrapure water, and each element was analyzed by Inductively Coupled Plasma–Optical Emission Spectroscopy (ICP-OES) (Spectro Amatek Arcos II ICP-OES, Kleve, Germany). The analyses were performed in triplicate.

4.5. Analysis of Soluble Proteins, Chlorophyll Content and Total Phenols

The total content of proteins was measured in leaves (1 g) from three individual plants (biological replicates) per treatment, being estimated via the Bradford method and expressed as mg protein g^{-1} FW [55]. Chlorophyll a and b were determined photometrically in leaves according to Ertani et al. [54]. Soluble phenolic acids were extracted by crushing leaves (1 g) in a mortar in the presence of 3 mL pure methanol. The extracts were maintained in an ice bath for 30 min and centrifuged at 5000× *g* for 30 min at 4 °C. The supernatants were stored at −20 °C until analysis. Total phenols were measured according to Arnaldos et al. [56].

4.6. Lipid Peroxidation and Antioxidant Enzyme Activities

Lipid peroxidation was carried out by detecting the malondialdehyde content (MDA) via the reaction of thiobarbituric acid (TBA) [57]. Frozen leaf and root tissue (100 mg) samples were ground using a tissue lyser (Biosigma), added with 1 mL of 10% (*w*/*v*) trichloroacetic acid (TCA), and then centrifuged at 13,000 rpm for 10 min at 4 °C. Sample extracts (0.5 mL) were added with 1.5 mL of 0.5% (*w*/*v*) thiobarbituric acid (TBA) in 10%

(*w*/*v*) TCA. The samples were incubated in a water bath at 95 °C for 25 min. After that, the reaction was stopped in ice, and samples were centrifuged at 13,000 rpm and 4 °C for 5 min. The absorbance of the samples was measured at wavelengths of 532 nm, 600 nm, and 440 nm, where 532 nm corresponds to the maximum absorbance of TBA-MDA complexes, 600 nm serves as a correction factor for non-specific turbidity, and 440 nm corrects for interference generated by sucrose. The molar extinction coefficient used in the calculations was 157 mM^{-1} cm^{-1}.

Guaiacol peroxidase (GPX) and ascorbate peroxidase (APX) activities were measured in root tissues. Enzymes were extracted by homogenizing 200 mg of frozen tissue in 1.5 mL 0.1 M phosphate buffer (pH 7). The homogenate was centrifuged at 15,000× *g* for 5 min at 4 °C, and then the supernatant was collected. GPX activity was determined following guaiacol oxidation with H_2O_2 [58]. Shortly, 3 mL guaiacol was added with 20 µL extracts, in the presence of 10 µL H_2O_2 (30 vol.). The guaiacol oxidation (ε = 26.6 mM^{-1} cm^{-1}) reaction was followed spectrophotometrically as the increase of absorption at 470 nm.

For APX, the activity was determined following the decrease in ascorbate (extinction coefficient 2.8 mM cm^{-1}) and measuring the change in absorbance at λ = 290 nm over a 3 min interval. The reaction mixture contained 50 mM phosphate buffer (pH 7.0), 1 mM ethylenediaminetetraacetic acid disodium salt (EDTA)-Na, 0.5 mM ascorbic acid, 0.1 mM H_2O_2, and 50 µL of enzyme extract [59]. Three biological replicates from roots of plants of each experimental condition were analyzed.

4.7. Statistical Analysis

For all determinations, the analysis of variance (ANOVA) was performed using the R software (4.3.0) and was followed by pair-wise post hoc analyses (Student–Newman–Keuls test) to determine which means differed significantly at $p < 0.05$ (±SD). The number of biological replicates varied depending on the analysis performed and is indicated in the figures and tables' legends [60].

5. Conclusions

The current scarcity of farm strategies designed to mitigate abiotic stresses provides an opportunity for the adoption of biostimulants, or microorganisms, with the potential to enhance plant growth and/or health.

The findings of this study demonstrate that the use of the microalgae-based product was effective in promoting the growth and enhancing its nutritional quality, particularly by increasing the protein content and antioxidant levels. The abundant presence of certain compounds in the MR formulation was likely a key factor contributing to its positive effects on lettuce growth, both under normal conditions and in the presence of salinity.

When plants were subjected to NaCl, the microalgae-based product reduced stress indirectly by restricting Na^+ accumulation in plant tissues. Perhaps, the content in certain lipid fractions, known for their role in mitigating oxidative stress, may have directly contributed to the observed reduction in lipid peroxidation in NaCl-treated plants primed with the microalgae-based product. We can thus support the use of microalgae as a robust tool for improving the productivity of lettuce under both non-stress and salinity stress conditions through their biostimulant effects. The integration of microalgae-based biostimulants into agricultural practices may offer a promising strategy for sustainable crop production, contributing to improved plant health and resilience in challenging growing environments.

Supplementary Materials: The following supporting information can be downloaded at https://www.mdpi.com/article/10.3390/plants13233311/s1.

Author Contributions: Conceptualization, O.F., A.E. and S.N.; methodology, O.F., A.E., E.F. and S.N.; software, O.F., A.E., V.Z. and M.S.; validation, O.F., A.E., E.F., M.S. and S.N.; formal analysis, O.F., D.C., A.E., E.F., M.T.R.E. and M.C.d.L.; investigation, O.F., A.E., E.F., M.S. and S.N.; resources, O.F., S.N. and A.E.; data curation, A.E., O.F., S.N., E.F. and D.C.; writing—original draft preparation, A.E.;

writing—review and editing, A.E., O.F., S.N., M.S. and E.F.; visualization, A.E., O.F., S.N., M.S. and E.F.; supervision, O.F., S.N. and A.E.; project administration, S.N., O.F. and A.E.; funding acquisition, S.N. and O.F. All authors have read and agreed to the published version of the manuscript.

Funding: This research received no external funding.

Data Availability Statement: Data are contained within the article and Supplementary Materials.

Conflicts of Interest: The authors declare no conflicts of interest.

References

1. Gonçalves, A.L. The Use of Microalgae and Cyanobacteria in the Improvement of Agricultural Practices: A Review on Their Biofertilising, Biostimulating and Biopesticide Roles. *Appl. Sci.* **2021**, *11*, 871. [CrossRef]
2. Abinandan, S.; Subashchandrabose, S.R.; Panneerselvan, L.; Venkateswarlu, K.; Megharaj, M. Potential of Acid-Tolerant Microalgae, *Desmodesmus* sp. MAS1 and *Heterochlorella* Sp. MAS3, in Heavy Metal Removal and Biodiesel Production at Acidic pH. *Bioresour. Technol.* **2019**, *278*, 9–16. [CrossRef] [PubMed]
3. Kapoor, S.; Singh, M.; Srivastava, A.; Chavali, M.; Chandrasekhar, K.; Verma, P. Extraction and Characterization of Microalgae-derived Phenolics for Pharmaceutical Applications: A Systematic Review. *J. Basic. Microbiol.* **2022**, *62*, 1044–1063. [CrossRef] [PubMed]
4. Kapoore, R.V.; Wood, E.E.; Llewellyn, C.A. Algae Biostimulants: A Critical Look at Microalgal Biostimulants for Sustainable Agricultural Practices. *Biotechnol. Adv.* **2021**, *49*, 107754. [CrossRef] [PubMed]
5. Garg, M.; Sharma, N.; Sharma, S.; Kapoor, P.; Kumar, A.; Chunduri, V.; Arora, P. Biofortified Crops Generated by Breeding, Agronomy, and Transgenic Approaches Are Improving Lives of Millions of People around the World. *Front. Nutr.* **2018**, *5*, 12. [CrossRef] [PubMed]
6. Rady, M.M.; Elrys, A.S.; Selem, E.; Mohsen, A.A.A.; Arnaout, S.M.A.I.; El-Sappah, A.H.; El-Tarabily, K.A.; Desoky, E.-S.M. *Spirulina platensis* Extract Improves the Production and Defenses of the Common Bean Grown in a Heavy Metals-Contaminated Saline Soil. *J. Environ. Sci.* **2023**, *129*, 240–257. [CrossRef]
7. Ahmed, S.F.; Mofijur, M.; Parisa, T.A.; Islam, N.; Kusumo, F.; Inayat, A.; Le, V.G.; Badruddin, I.A.; Khan, T.M.Y.; Ong, H.C. Progress and Challenges of Contaminate Removal from Wastewater Using Microalgae Biomass. *Chemosphere* **2022**, *286*, 131656. [CrossRef]
8. Wang, C.; Qi, M.; Guo, J.; Zhou, C.; Yan, X.; Ruan, R.; Cheng, P. The Active Phytohormone in Microalgae: The Characteristics, Efficient Detection, and Their Adversity Resistance Applications. *Molecules* **2021**, *27*, 46. [CrossRef] [PubMed]
9. Zhou, J.-L.; Yang, L.; Huang, K.-X.; Chen, D.-Z.; Gao, F. Mechanisms and Application of Microalgae on Removing Emerging Contaminants from Wastewater: A Review. *Bioresour. Technol.* **2022**, *364*, 128049. [CrossRef] [PubMed]
10. Cordeiro, E.C.N.; Mógor, Á.F.; Amatussi, J.O.; Mógor, G.; Marques, H.M.C.; De Lara, G.B. Microalga Biofertilizer Improves Potato Growth and Yield, Stimulating Amino Acid Metabolism. *J. Appl. Phycol.* **2022**, *34*, 385–394. [CrossRef]
11. Berthon, J.-Y.; Michel, T.; Wauquier, A.; Joly, P.; Gerbore, J.; Filaire, E. Seaweed and Microalgae as Major Actors of Blue Biotechnology to Achieve Plant Stimulation and Pest and Pathogen Biocontrol—A Review of the Latest Advances and Future Prospects. *J. Agric. Sci.* **2021**, *159*, 523–534. [CrossRef]
12. Yang, T.; Samarakoon, U.; Altland, J.; Ling, P. Photosynthesis, Biomass Production, Nutritional Quality, and Flavor-Related Phytochemical Properties of Hydroponic-Grown Arugula (*Eruca sativa* Mill.) 'Standard' under Different Electrical Conductivities of Nutrient Solution. *Agronomy* **2021**, *11*, 1340. [CrossRef]
13. Puglisi, I.; La Bella, E.; Rovetto, E.I.; Stevanato, P.; Fascella, G.; Baglieri, A. Morpho-Biometric and Biochemical Responses in Lettuce Seedlings Treated by Different Application Methods of *Chlorella vulgaris* Extract: Foliar Spray or Root Drench? *J. Appl. Phycol.* **2022**, *34*, 889–901. [CrossRef]
14. Grzesik, M.; Romanowska-Duda, Z. Improvements in Germination, Growth, and Metabolic Activity of Corn Seedlings by Grain Conditioning and Root Application with Cyanobacteria and Microalgae. *Pol. J. Environ. Stud.* **2014**, *23*, 1147–1153.
15. Guzmán-Murillo, M.A.; Ascencio, F.; Larrinaga-Mayoral, J.A. Germination and ROS Detoxification in Bell Pepper (*Capsicum annuum* L.) under NaCl Stress and Treatment with Microalgae Extracts. *Protoplasma* **2013**, *250*, 33–42. [CrossRef]
16. Parmar, P.; Kumar, R.; Neha, Y.; Srivatsan, V. Microalgae as next Generation Plant Growth Additives: Functions, Applications, Challenges and Circular Bioeconomy Based Solutions. *Front. Plant Sci.* **2023**, *14*, 1073546. [CrossRef] [PubMed]
17. Rachidi, F.; Benhima, R.; Sbabou, L.; El Arroussi, H. Microalgae Polysaccharides Bio-Stimulating Effect on Tomato Plants: Growth and Metabolic Distribution. *Biotechnol. Rep.* **2020**, *25*, e00426. [CrossRef]
18. Mutale-joan, C.; Rachidi, F.; Mohamed, H.A.; Mernissi, N.E.; Aasfar, A.; Barakate, M.; Mohammed, D.; Sbabou, L.; Arroussi, H.E. Microalgae-Cyanobacteria–Based Biostimulant Effect on Salinity Tolerance Mechanisms, Nutrient Uptake, and Tomato Plant Growth under Salt Stress. *J. Appl. Phycol.* **2021**, *33*, 3779–3795. [CrossRef]
19. Hassani, A.; Azapagic, A.; Shokri, N. Global Predictions of Primary Soil Salinization under Changing Climate in the 21st Century. *Nat. Commun.* **2021**, *12*, 6663. [CrossRef]
20. Van Zelm, E.; Zhang, Y.; Testerink, C. Salt Tolerance Mechanisms of Plants. *Annu. Rev. Plant Biol.* **2020**, *71*, 403–433. [CrossRef] [PubMed]

21. Zhao, S.; Zhang, Q.; Liu, M.; Zhou, H.; Ma, C.; Wang, P. Regulation of Plant Responses to Salt Stress. *Int. J. Mol. Sci.* **2021**, *22*, 4609. [CrossRef] [PubMed]
22. Zhou, H.; Shi, H.; Yang, Y.; Feng, X.; Chen, X.; Xiao, F.; Lin, H.; Guo, Y. Insights into Plant Salt Stress Signaling and Tolerance. *J. Genet. Genom.* **2024**, *51*, 16–34. [CrossRef] [PubMed]
23. Seifikalhor, M.; Aliniaeifard, S.; Shomali, A.; Azad, N.; Hassani, B.; Lastochkina, O.; Li, T. Calcium Signaling and Salt Tolerance Are Diversely Entwined in Plants. *Plant Signal. Behav.* **2019**, *14*, 1665455. [CrossRef]
24. Gupta, B.; Shrestha, J. Editorial: Abiotic Stress Adaptation and Tolerance Mechanisms in Crop Plants. *Front. Plant Sci.* **2023**, *14*, 1278895. [CrossRef] [PubMed]
25. Rouphael, Y.; Kyriacou, M.C. Enhancing Quality of Fresh Vegetables Through Salinity Eustress and Biofortification Applications Facilitated by Soilless Cultivation. *Front. Plant Sci.* **2018**, *9*, 1254. [CrossRef]
26. Petropoulos, S.A.; Levizou, E.; Ntatsi, G.; Fernandes, Â.; Petrotos, K.; Akoumianakis, K.; Barros, L.; Ferreira, I.C.F.R. Salinity Effect on Nutritional Value, Chemical Composition and Bioactive Compounds Content of *Cichorium spinosum* L. *Food Chem.* **2017**, *214*, 129–136. [CrossRef] [PubMed]
27. El-Nakhel, C.; Cozzolino, E.; Ottaiano, L.; Petropoulos, S.A.; Nocerino, S.; Pelosi, M.E.; Rouphael, Y.; Mori, M.; Di Mola, I. Effect of Biostimulant Application on Plant Growth, Chlorophylls and Hydrophilic Antioxidant Activity of Spinach (*Spinacia oleracea* L.) Grown under Saline Stress. *Horticulturae* **2022**, *8*, 971. [CrossRef]
28. Acosta-Motos, J.; Ortuño, M.; Bernal-Vicente, A.; Diaz-Vivancos, P.; Sanchez-Blanco, M.; Hernandez, J. Plant Responses to Salt Stress: Adaptive Mechanisms. *Agronomy* **2017**, *7*, 18. [CrossRef]
29. Hassan, M.N.; Mekkawy, S.A.; Mahdy, M.; Salem, K.F.M.; Tawfik, E. Recent Molecular and Breeding Strategies in Lettuce (*Lactuca* spp.). *Genet. Resour. Crop Evol.* **2021**, *68*, 3055–3079. [CrossRef]
30. Grace, C.E.E.; Lakshmi, P.K.; Meenakshi, S.; Vaidyanathan, S.; Srisudha, S.; Mary, M.B. Biomolecular Transitions and Lipid Accumulation in Green Microalgae Monitored by FTIR and Raman Analysis. *Spectrochim. Acta Part A Mol. Biomol. Spectrosc.* **2020**, *224*, 117382. [CrossRef]
31. Bellamy, L.J. *The Infrared Spectra of Complex Molecules: Volume Two Advances in Infrared Group Frequencies*; Springer: Dordrecht, The Netherlands, 1980; ISBN 978-0-412-22350-1.
32. Pereira, L.; Gheda, S.F.; Ribeiro-Claro, P.J.A. Analysis by Vibrational Spectroscopy of Seaweed Polysaccharides with Potential Use in Food, Pharmaceutical, and Cosmetic Industries. *Int. J. Carbohydr. Chem.* **2013**, *2013*, 537202. [CrossRef]
33. Akhter, M.; Dutta Majumdar, R.; Fortier-McGill, B.; Soong, R.; Liaghati-Mobarhan, Y.; Simpson, M.; Arhonditsis, G.; Schmidt, S.; Heumann, H.; Simpson, A.J. Identification of Aquatically Available Carbon from Algae through Solution-State NMR of Whole ^{13}C-Labelled Cells. *Anal. Bioanal. Chem.* **2016**, *408*, 4357–4370. [CrossRef]
34. Hatzakis, E.; Agiomyrgianaki, A.; Kostidis, S.; Dais, P. High-Resolution NMR Spectroscopy: An Alternative Fast Tool for Qualitative and Quantitative Analysis of Diacylglycerol (DAG) Oil. *J. Am. Oil Chem. Soc.* **2011**, *88*, 1695–1708. [CrossRef]
35. Paes De Melo, B.; Carpinetti, P.D.A.; Fraga, O.T.; Rodrigues-Silva, P.L.; Fioresi, V.S.; De Camargos, L.F.; Ferreira, M.F.D.S. Abiotic Stresses in Plants and Their Markers: A Practice View of Plant Stress Responses and Programmed Cell Death Mechanisms. *Plants* **2022**, *11*, 1100. [CrossRef]
36. Zinn, K.E.; Tunc-Ozdemir, M.; Harper, J.F. Temperature Stress and Plant Sexual Reproduction: Uncovering the Weakest Links. *J. Exp. Bot.* **2010**, *61*, 1959–1968. [CrossRef]
37. Kusvuran, S. Microalgae (*Chlorella vulgaris* Beijerinck) Alleviates Drought Stress of Broccoli Plants by Improving Nutrient Uptake, Secondary Metabolites, and Antioxidative Defense System. *Hortic. Plant J.* **2021**, *7*, 221–231. [CrossRef]
38. Stirk, W.A.; Bálint, P.; Tarkowská, D.; Novák, O.; Strnad, M.; Ördög, V.; Van Staden, J. Hormone Profiles in Microalgae: Gibberellins and Brassinosteroids. *Plant Physiol. Biochem.* **2013**, *70*, 348–353. [CrossRef]
39. Siddhnath; Surasani, V.K.R.; Singh, A.; Singh, S.M.; Hauzoukim; Murthy, L.N.; Baraiya, K.G. Bioactive Compounds from Micro-Algae and Its Application in Foods: A Review. *Discov. Food* **2024**, *4*, 27. [CrossRef]
40. Ertani, A.; Nardi, S.; Francioso, O.; Sanchez-Cortes, S.; Foggia, M.D.; Schiavon, M. Effects of Two Protein Hydrolysates Obtained From Chickpea (*Cicer arietinum* L.) and *Spirulina platensis* on *Zea mays* (L.) Plants. *Front. Plant Sci.* **2019**, *10*, 954. [CrossRef]
41. Mazepa, E.; Malburg, B.V.; Mógor, G.; de Oliveira, A.C.; Amatussi, J.O.; Corrêa, D.O.; Lemos, J.S.; Ducatti, D.R.; Duarte, M.E.R.; Mogor, A.F.; et al. Plant Growth Biostimulant Activity of the Green Microalga *Desmodesmus subspicatus*. *Algal Res.* **2021**, *59*, 102434. [CrossRef]
42. Mishra, N.; Jiang, C.; Chen, L.; Paul, A.; Chatterjee, A.; Shen, G. Achieving Abiotic Stress Tolerance in Plants through Antioxidative Defense Mechanisms. *Front. Plant Sci.* **2023**, *14*, 1110622. [CrossRef]
43. Srivastava, R. Physicochemical, Antioxidant Properties of Carotenoids and Its Optoelectronic and Interaction Studies with Chlorophyll Pigments. *Sci. Rep.* **2021**, *11*, 18365. [CrossRef]
44. Carrasco-Ríos, L.; Pinto, M. Effect of Salt Stress on Antioxidant Enzymes and Lipid Peroxidation in Leaves in Two Contrasting Corn, "Lluteño" and "Jubilee". *Chil. J. Agric. Res.* **2014**, *74*, 89–95. [CrossRef]
45. Van Oosten, M.J.; Pepe, O.; De Pascale, S.; Silletti, S.; Maggio, A. The Role of Biostimulants and Bioeffectors as Alleviators of Abiotic Stress in Crop Plants. *Chem. Biol. Technol. Agric.* **2017**, *4*, 5. [CrossRef]
46. İkiz, B.; Dasgan, H.Y.; Balik, S.; Kusvuran, S.; Gruda, N.S. The Use of Biostimulants as a Key to Sustainable Hydroponic Lettuce Farming under Saline Water Stress. *BMC Plant Biol.* **2024**, *24*, 808. [CrossRef]
47. Gaschler, M.M.; Stockwell, B.R. Lipid Peroxidation in Cell Death. *Biochem. Biophys. Res. Commun.* **2017**, *482*, 419–425. [CrossRef]

48. Chokshi, K.; Pancha, I.; Ghosh, A.; Mishra, S. Salinity Induced Oxidative Stress Alters the Physiological Responses and Improves the Biofuel Potential of Green Microalgae *Acutodesmus dimorphus*. *Bioresour. Technol.* **2017**, *244*, 1376–1383. [CrossRef]
49. Boselli, E.; Velazco, V.; Fiorenza Caboni, M.; Lercker, G. Pressurized Liquid Extraction of Lipids for the Determination of Oxysterols in Egg-Containing Food. *J. Chromatogr. A* **2001**, *917*, 239–244. [CrossRef]
50. Folch, J.; Lees, M.; Stanley, G.H.S. A Simple Method for the Isolation and Purification of total Lipides from Animal Tissues. *J. Biol. Chem.* **1957**, *226*, 497–509. [CrossRef]
51. Toschi, T.G.; Cardenia, V.; Bonaga, G.; Mandrioli, M.; Rodriguez-Estrada, M.T. Coffee Silverskin: Characterization, Possible Uses, and Safety Aspects. *J. Agric. Food Chem.* **2014**, *62*, 10836–10844. [CrossRef]
52. Świder, O.; Roszko, M.Ł.; Wójcicki, M.; Szymczyk, K. Biogenic Amines and Free Amino Acids in Traditional Fermented Vegetables—Dietary Risk Evaluation. *J. Agric. Food Chem.* **2020**, *68*, 856–868. [CrossRef] [PubMed]
53. Ertani, A.; Francioso, O.; Tinti, A.; Schiavon, M.; Pizzeghello, D.; Nardi, S. Evaluation of Seaweed Extracts From *Laminaria* and *Ascophyllum nodosum* spp. as Biostimulants in *Zea mays* L. Using a Combination of Chemical, Biochemical and Morphological Approaches. *Front. Plant Sci.* **2018**, *9*, 428. [CrossRef] [PubMed]
54. Ertani, A.; Francioso, O.; Tugnoli, V.; Righi, V.; Nardi, S. Effect of Commercial Lignosulfonate-Humate on *Zea Mays* L. Metabolism. *J. Agric. Food Chem.* **2011**, *59*, 11940–11948. [CrossRef] [PubMed]
55. Bradford, M.M. A Rapid and Sensitive Method for the Quantitation of Microgram Quantities of Protein Utilizing the Principle of Protein-Dye Binding. *Anal. Biochem.* **1976**, *72*, 248–254. [CrossRef] [PubMed]
56. López Arnaldos, T.; Muñoz, R.; Ferrer, M.A.; Calderón, A.A. Changes in Phenol Content during Strawberry (*Fragaria* × *Ananassa*, Cv. Chandler) Callus Culture. *Physiol. Plant.* **2001**, *113*, 315–322. [CrossRef]
57. Heath, R.L.; Packer, L. Photoperoxidation in Isolated Chloroplasts. *Arch. Biochem. Biophys.* **1968**, *125*, 189–198. [CrossRef]
58. Tijssen, P. Practice and Theory of Enzyme Immunoassays. In *Laboratory Techniques in Biochemistry and Molecular Biology*; Elsevier: Amsterdam, The Netherlands, 1985; ISBN 978-0-08-085884-5.
59. Nakano, Y.; Asada, K. Hydrogen Peroxide Is Scavenged by Ascorbate-Specific Peroxidase in Spinach Chloroplasts. *Plant Cell Physiol.* **1981**, *22*, 867–880. [CrossRef]
60. Sokal, R.R.; Rohlf, F.J. *Biometry: The Principles and Practice of Statistics in Biological Research*, 4th ed.; [Extensively Rev.]; W.H. Freeman: New York, NY, USA, 2012; ISBN 978-0-7167-8604-7.
61. Carr, H.Y.; Purcell, E.M. Effects of Diffusion on Free Precession in Nuclear Magnetic Resonance Experiments. *Phys. Rev.* **1954**, *94*, 630. [CrossRef]
62. Meiboom, S.; Gill, D. Modified Spin-Echo Method for Measuring Nuclear Relaxation Times. *Rev. Sci. Instrum.* **1958**, *29*, 688–691. [CrossRef]

plants

MDPI

Article

Exogenous 2,4-Epibrassinolide Alleviates Alkaline Stress in Cucumber by Modulating Photosynthetic Performance

Wenjing Nie [1], Qinghai He [1,2], Jinzhao Ma [1], Hongen Guo [1,2,*] and Qinghua Shi [3,*]

1 Shandong Engineering Research Center of Functional Crop Germplasm Innovation and Cultivation Utilization, Yantai Engineering Research Center of Plant Stem Cell Targeted Breeding, Shandong Institute of Sericulture, Yantai 264001, China
2 Shandong Academy of Agricultural Machinery Science, Jinan 250100, China
3 Stage Key Laboratory of Crop Biology, College of Horticulture Science and Engineering, Shandong Agricultural University, Tai'an 271018, China
* Correspondence: cyswztd@163.com (H.G.); qhshi@sdau.edu.cn (Q.S.)

Abstract: Brassinosteroids (BRs) are recognized for their ability to enhance plant salt tolerance. While considerable research has focused on their effects under neutral salt conditions, the mechanisms through which BRs regulate photosynthesis under alkaline salt stress are less well understood. This study investigates these mechanisms, examining plant growth, photosynthetic electron transport, gas exchange parameters, Calvin cycle dynamics, and the expression of key antioxidant and Calvin cycle genes under alkaline stress conditions induced by $NaHCO_3$. The findings indicate that $NaHCO_3$ stress substantially impairs cucumber growth and photosynthesis, significantly reducing chlorophyll content, net photosynthetic rate (Pn), stomatal conductance (Gs), transpiration rate (E), maximum photochemical efficiency (Fv/Fm), actual photochemical efficiency (ΦPSII), antenna conversion efficiency (Fv′/Fm′), and photochemical quenching coefficient (qP). This disruption suggests a severe dysregulation of the photosynthetic electron transport system, impairing electron transfer from photosystem II (PSII) to photosystem I (PSI) and subsequently the Calvin cycle. Application of exogenous 24-epibrassinolide (EBR) alleviated these effects, reducing leaf chlorosis and growth inhibition and significantly enhancing the expression of key genes within the antioxidant system (AsA-GSH cycle) and the Calvin cycle. This intervention also led to a reduction in reactive oxygen species (ROS) accumulation and improved photosynthetic performance, as evidenced by enhancements in Pn, Gs, E, Fv/Fm, ΦPSII, Fv′/Fm′, and qP. Moreover, $NaHCO_3$ stress hindered chlorophyll synthesis, primarily by blocking the conversion from porphobilinogen (PBG) to uroporphyrinogen III (UroIII) and by increasing chlorophyllase (Chlase) and decreasing porphobilinogen deaminase (PBGD) activity. Exogenous EBR countered these effects by enhancing PBGD activity and reducing Chlase activity, thereby increasing chlorophyll content under stress conditions. In summary, EBR markedly mitigated the adverse effects of alkaline stress on cucumber leaf photosynthesis by stabilizing the photosynthetic electron transport system, accelerating photosynthetic electron transport, and promoting the Calvin cycle. This study provides valuable insights into the regulatory roles of BRs in enhancing plant resilience to alkaline stress.

Keywords: cucumber; alkaline stress; 2,4-epibrassinolide; photosynthesis; Calvin cycle

1. Introduction

Photosynthesis stands as the fundamental cornerstone of plant growth and development, driving the conversion of solar energy into chemical energy [1]. This energy conversion fuels the biosynthetic pathways that lead to biomass accumulation. The intricate process includes both the light-dependent reactions and the Calvin cycle, orchestrating the carbon assimilation from CO_2 into carbohydrates, which are essential for plant sustenance and productivity [2]. However, environmental stresses such as salt stress, drought, and extreme temperatures pose significant threats to photosynthetic efficiency [3,4]. These stressors initiate a cascade of physiological and biochemical disruptions, including stomatal closure, impaired electron transport, and oxidative damage, which collectively compromise the photosynthetic machinery [4,5]. The alteration in photosynthetic parameters under stress conditions, such as reduced chlorophyll content and disrupted photosystem activity, ultimately curtails plant growth and crop yield [6,7]. Thus, understanding the impact of environmental stress on photosynthesis is pivotal for developing strategies to enhance plant resilience and ensure food security in the face of climatic challenges.

Salt and alkaline stress significantly impairs plant physiological processes, especially photosynthesis, which is critical for growth and productivity. This stress leads to ionic imbalances and hyperosmotic stress, disrupting cellular homeostasis and nutrient assimilation, thereby reducing plant viability [8–10]. Elevated concentrations of Na^+, typical of salt conditions, and HCO_3^- and CO_3^{2-} under alkaline conditions result in cellular toxicity that manifests as chlorosis and necrosis, further impairing the photosynthetic apparatus [11]. Alkaline stress, characterized by elevated pH levels, poses a more severe threat than neutral salt stress. It not only exacerbates the toxic effects by inhibiting key enzymatic activities crucial for photosynthesis but also alters the structural integrity of chloroplasts, severely diminishing photosynthetic capacity [12]. These alkaline conditions, with higher pH, precipitate essential nutrients such as calcium and magnesium, further disrupting nutrient uptake and leading to more pronounced damage compared to neutral salt stress [13].

Salt and alkaline conditions also reduce water uptake due to osmotic stress, decreasing cell turgor and inducing stomatal closure, which limits CO_2 availability for the Calvin cycle and further diminishes photosynthetic capacity [14,15]. Additionally, oxidative stress from salt and alkaline conditions leads to an overproduction of ROS, damaging cellular structures and overwhelming plant antioxidant defenses, thus exacerbating reductions in photosynthetic efficiency [16–18]. Understanding the mechanisms by which plants respond to salt and alkaline stresses is crucial for developing crops with improved resilience, which is vital for maintaining agricultural productivity in affected regions.

Brassinosteroids (BRs), first discovered in the 1970s from rape pollen, are "the sixth major plant hormone". These compounds play a pivotal role in enhancing plant growth and resilience to environmental stresses, including salinity and alkalinity [19,20]. Their impact on plant biology extends to significant improvements in photosynthetic efficiency under stress conditions, making them an essential hormone in the adaptation of plants to challenging environments.

The role of BRs in photosynthesis, particularly under stress, is mediated through their regulation of cellular processes and gene expression that are directly linked to photosynthetic machinery [21–24]. BRs enhance the stability and functionality of PSII, crucial for maintaining optimal photosynthesis during salt and alkaline stress. This includes regulating genes that encode for proteins involved in the Calvin cycle and protecting these proteins from degradation induced by stress [25,26]. At the molecular level, BRs interact with specific receptors like BRI1 (Brassinosteroid Insensitive 1), initiating signaling cascades that ultimately lead to the activation of transcription factors. These transcription factors upregulate genes responsible for stress responses, thereby enhancing cellular homeostasis and the plant's ability to optimize nutrient and water uptake under stressful conditions [27–29]. Through these mechanisms, BRs not only support basic plant development but also empower plants to withstand adverse conditions by bolstering the photosynthetic apparatus.

BRs promoting growth and enhancing stress tolerance through improved photosynthesis underscores their potential in agricultural facing variable environmental stresses.

Exploring the effects of BRs on photosynthesis under salt and alkaline stress conditions is critical for enhancing plant resilience and agricultural productivity [30–32]. While there has been significant advancement in understanding the role of BRs in neutral salt stress mitigation, the specific photosynthetic responses to alkaline stress remain less defined, particularly concerning the complex mechanisms of photosynthesis that are affected [31,33–35]. This study aims to bridge this gap by investigating the impact of exogenous 24-epibrassinolide (EBR) application on the photosynthetic efficiency and stress tolerance of cucumber plants subjected to alkaline stress. It focuses on elucidating the physiological and molecular mechanisms underlying these effects. The findings from this research could contribute to the development of effective strategies for enhancing the resilience and productivity of crops grown in salt- and alkaline-affected soils, thus promoting sustainable agricultural practices in regions increasingly challenged by soil salinization [22].

2. Results

2.1. Effects of EBR on the Growth of Cucumber Under $NaHCO_3$ Stress

Exposure to $NaHCO_3$ significantly inhibited the growth of cucumber seedlings, as evidenced by the yellowing and wilting of leaves (Figure 1). This treatment also resulted in a reduction in the height and both above-ground and below-ground biomass of the cucumber seedlings ($p < 0.05$) (Table 1). The application of 0.2 μM exogenous EBR markedly mitigated these inhibitory effects under $NaHCO_3$ stress, enhancing the biomass of the cucumber seedlings (Table 1).

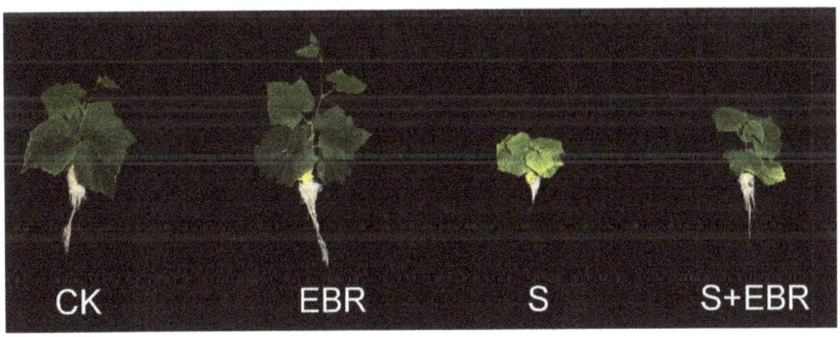

Figure 1. Effects of EBR on plant phenotypes under $NaHCO_3$ stress. (CK denotes plants grown under normal Hoagland nutrient solution; CK + EBR denotes the addition of EBR under normal growth conditions; S denotes $NaHCO_3$ stress treatment; S + EBR denotes the addition of EBR under $NaHCO_3$ stress).

Table 1. Effects of exogenous EBR on the growth of cucumber seedlings under $NaHCO_3$ stress.

Treatment	Plant Height (cm)	Shoot Fresh Weight (g·Plant^{-1})	Root Fresh Weight (g·Plant^{-1})	Total Fresh Weight (g·Plant^{-1})
CK	50.21 ± 0.73 a	32.74 ± 2.56 b	13.34 ± 1.03 a	46.08 ± 2.81 a
EBR	51.40 ± 0.67 a	34.26 ± 2.92 a	13.61 ± 1.80 a	47.87 ± 4.71 a
S	25.27 ± 4.12 c	11.67 ± 1.99 d	3.78 ± 0.43 c	15.45 ± 2.26 c
S + EBR	39.65 ± 4.68 b	15.51 ± 1.93 c	6.03 ± 0.40 b	21.54 ± 2.32 b

Note: different lowercase letters in the same column indicate significant differences at the 0.05 level among treatments. (CK denotes control plants grown under normal Hoagland nutrient solution; EBR denotes the addition of EBR under normal growth conditions; S denotes $NaHCO_3$ stress treatment; S + EBR denotes the addition of EBR under $NaHCO_3$ stress).

2.2. *Effects of EBR on Reactive Oxygen Species Accumulation and Antioxidant Gene Expression in Cucumber Leaves Under $NaHCO_3$ Stress*

As shown in Figure 2, histochemical staining of cucumber leaves with Nitroblue Tetrazolium (NBT) for $O_2^{\cdot -}$ and Diaminobenzidine (DAB) for H_2O_2 reveals that $NaHCO_3$ stress significantly elevated the levels of both reactive oxygen species. The darker staining indicates higher concentrations of these compounds. Conversely, the application of exogenous EBR markedly reduced their accumulation under $NaHCO_3$ stress, evidenced by lighter staining which corresponds to lower concentrations. These observations suggest that EBR treatment effectively mitigates oxidative damage in cucumber seedlings induced by $NaHCO_3$ stress.

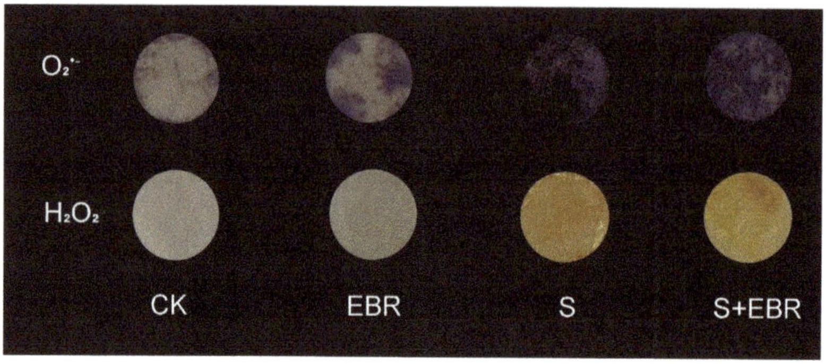

Figure 2. Effects of EBR on superoxide anion (O_2^-) and hydrogen peroxide (H_2O_2) staining in cucumber leaves under $NaHCO_3$ stress. (CK represents control plants grown in normal Hoagland nutrient solution; EBR denotes normal growth conditions with the addition of EBR; S indicates $NaHCO_3$ stress treatment; S + EBR indicates the addition of EBR under $NaHCO_3$ stress). Histochemical observations utilized Nitroblue Tetrazolium (NBT) for staining superoxide anion and Diaminobenzidine (DAB) for hydrogen peroxide.

To elucidate the effects of EBR on the expression of gene coding for antioxidant enzymes within the AsA-GSH cycle, qRT-PCR analysis was performed. As illustrated in Figure 3, transcript levels of *CsCu-ZnSOD, CsFeSOD, CsAPX, CsMDHAR, CsDHAR,* and *CsGR* were significantly enhanced by 48 h $NaHCO_3$ treatment, relative to the control. The application of EBR further upregulated the expression of these genes under $NaHCO_3$ treatment. These results confirm that EBR amplifies the expression of key genes responsible for scavenging reactive oxygen species and strengthening the AsA-GSH system.

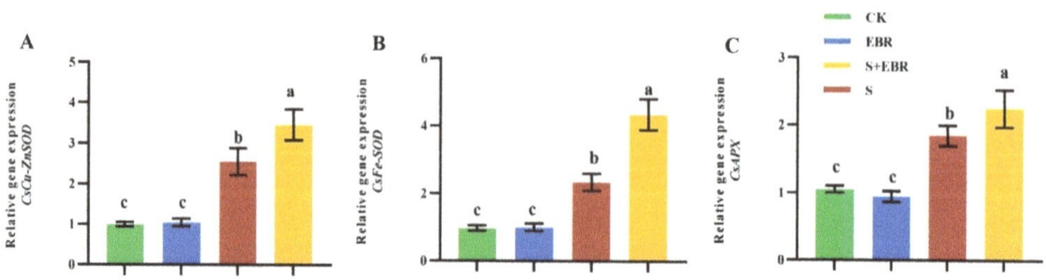

Figure 3. *Cont.*

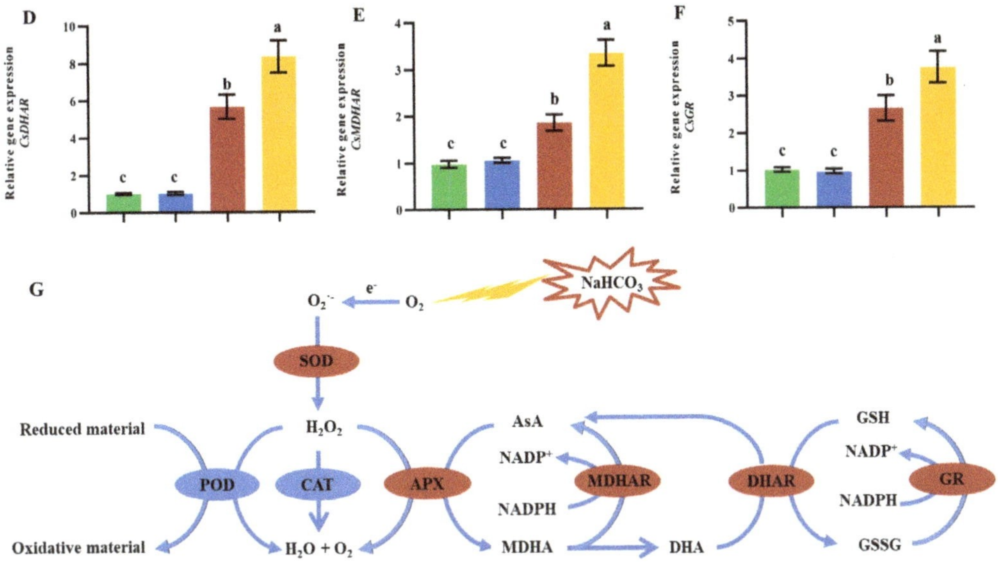

Figure 3. Influence of exogenous EBR on the transcript levels of key genes involved in the antioxidant and ascorbate–glutathione (ASA-GSH) cycles under $NaHCO_3$ stress in cucumber leaves. Panels represent transcript levels of (**A**) *CsCu-ZnSOD*, (**B**) *CsFeSOD*, (**C**) *CsAPX*, (**D**) *CsDHAR*, (**E**) *CsMDHAR*, (**F**) *CsGR*overall, (**G**) EBR modulation of the ASA-GSH cycle. Red symbols denote upregulated genes under stress conditions. Different lowercase letters indicate statistically significant differences among treatments at the 0.05 level ($p < 0.05$, n = 3). Key abbreviations: SOD, superoxide dismutase; APX, ascorbate peroxidase; MDHAR, monodehydroascorbate reductase; DHAR, dehydroascorbate reductase; GR, glutathione reductase; AsA, ascorbic acid; MDHA, monodehydroascorbate; DHA, dehydroascorbic acid; GSH, glutathione; GSSG, oxidized glutathione. (CK represents control plants grown in normal Hoagland nutrient solution; CK + EBR denotes normal growth conditions with the addition of EBR; S indicates $NaHCO_3$ stress treatment; S + EBR indicates the addition of EBR under $NaHCO_3$ stress).

2.3. Effects of EBR on Photosynthetic Parameters and Chlorophyll Synthesis in Cucumber Leaves Under $NaHCO_3$ Stress

Figure 4 shows that $NaHCO_3$ stress significantly reduces photosynthetic parameters in cucumber seedling leaves, including net photosynthetic rate (Pn), transpiration rate (Tr), stomatal conductance (Gs), and chlorophyll a, chlorophyll b, and total chlorophyll contents. The application of exogenous EBR improved these indices and alleviated the inhibition of photosynthesis in cucumber seedlings under $NaHCO_3$ stress, also enhancing the contents of chlorophyll a and b in the leaves.

As depicted in Figure 5, $NaHCO_3$ stress led to an increase in the contents of 5-aminolevulinic acid (ALA) and porphobilinogen (PBG), while the levels of uroporphyrinogen (UroIII), protoporphyrin (ProtoIX), magnesium protoporphyrin (Mg-ProtoIX), and protochlorophyllin acid ester (Pchl) decreased as the stress duration extended. This indicates that $NaHCO_3$ stress inhibited the synthesis pathway of chlorophyll, particularly at the UroIII synthesis stage from PBG. The application of exogenous EBR resulted in increased levels of chlorophyll precursors such as UroIII, ProtoIX, Mg-ProtoIX, and Pchl and reduced the accumulation of ALA and PBG under $NaHCO_3$ stress, thereby relieving the inhibition of chlorophyll synthesis and boosting chlorophyll content (Figure 4D–F).

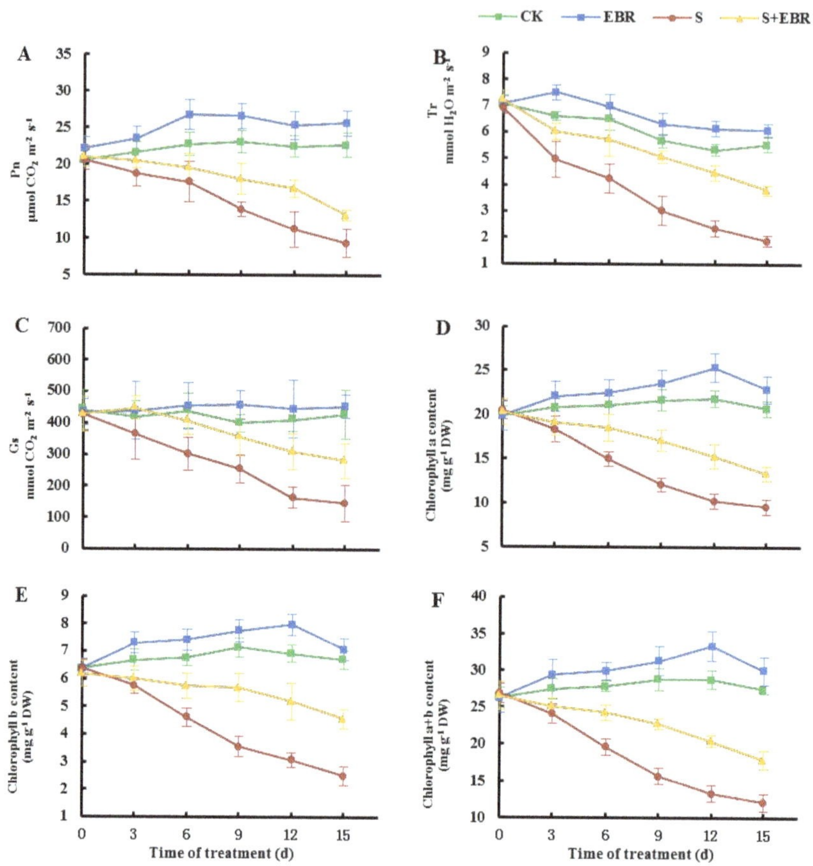

Figure 4. Effects of EBR on net photosynthetic rate, Pn (**A**), transpiration rate, Tr (**B**), stomatal conductance, Cs (**C**), chlorophyll a content (**D**), chlorophyll b content (**E**), and chlorophyll a + b content (**F**) of cucumber seedlings under $NaHCO_3$ stress. (CK represents normal Hoagland nutrient solution culture; EBR represents the addition of EBR under normal growth conditions; S represents $NaHCO_3$ stress treatment; S + EBR represents the addition of EBR under $NaHCO_3$ stress).

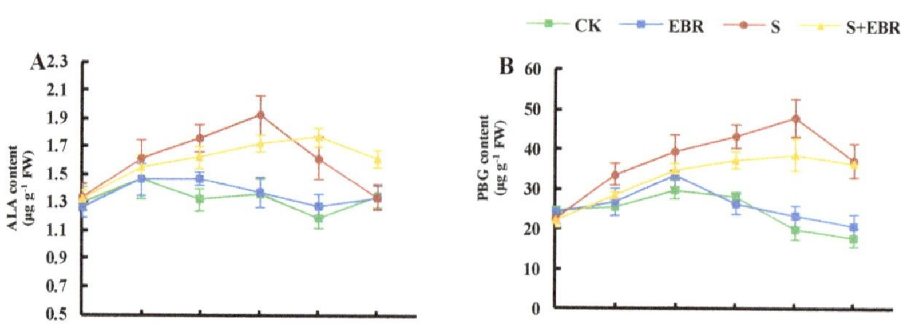

Figure 5. *Cont.*

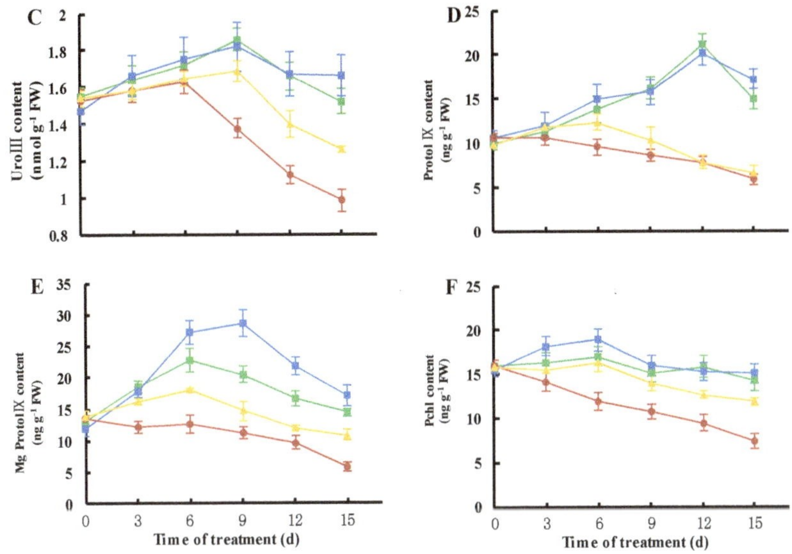

Figure 5. Effects of EBR on the contents of ALA (**A**), PBG (**B**), UroIII (**C**), ProtoIX (**D**), Mg-ProtoIX (**E**), and Pchl (**F**) in cucumber leaves under $NaHCO_3$ stress. (CK represents normal Hoagland nutrient solution culture; EBR represents the addition of EBR under normal growth conditions; S represents $NaHCO_3$ stress treatment; S + EBR represents the addition of EBR under $NaHCO_3$ stress).

Figure 6 illustrates that $NaHCO_3$ stress significantly enhanced chlorophyllase (Chlase) activity, which could be a major factor in the reduction in chlorophyll content under alkaline stress. It also notably inhibited porphobilinogen deaminase (PBGD) activity, blocking the conversion of PBG to UROIII. Exogenous EBR application alleviated the inhibition of Chlase activity by $NaHCO_3$ stress and increased PBGD activity under stress conditions.

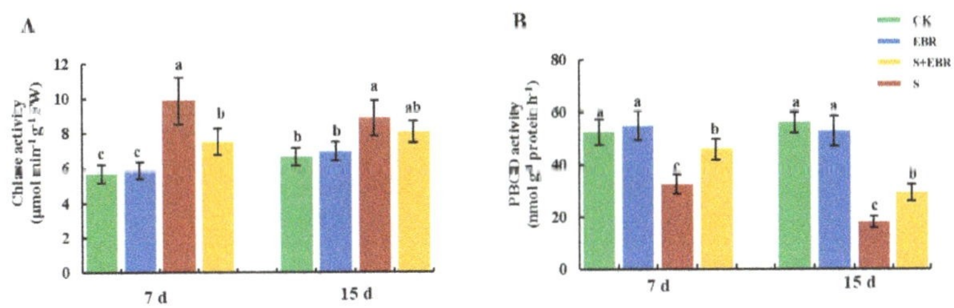

Figure 6. Effects of EBR on the activities of Chlase (**A**) and PBGD (**B**) in cucumber leaves under $NaHCO_3$ stress. Different lowercase letters indicate statistically significant differences among treatments at the 0.05 level ($p < 0.05$, n = 3). (CK represents normal Hoagland nutrient solution culture; EBR represents the addition of EBR under normal growth conditions; S represents $NaHCO_3$ stress treatment; S + EBR represents the addition of EBR under $NaHCO_3$ stress).

2.4. Effects of EBR on the Chlorophyll Fluorescence Characteristics of Cucumber Leaves Under $NaHCO_3$ Stress

To further investigate the regulatory mechanisms of brassinosteroids (BRs) on photosynthesis, we analyzed chlorophyll fluorescence to assess the functional state of photosystem II (PSII). The results, presented in Figure 7, show a time-dependent decrease in key photosynthetic parameters under $NaHCO_3$ stress: maximum photochemical efficiency (Fv/Fm), antenna conversion efficiency (Fv′/Fm′), photochemical quenching coefficient (qP), and actual photochemical efficiency (ΦPSII) of cucumber leaves. The application of exogenous EBR significantly mitigated the stress-induced decline in these indices, suggesting a disruption in the photochemical system of PSII under $NaHCO_3$ stress that impairs the absorption and utilization of light energy. Furthermore, exogenous EBR was observed to partially restore the physiological functions of the PSII photochemical system under these stress conditions.

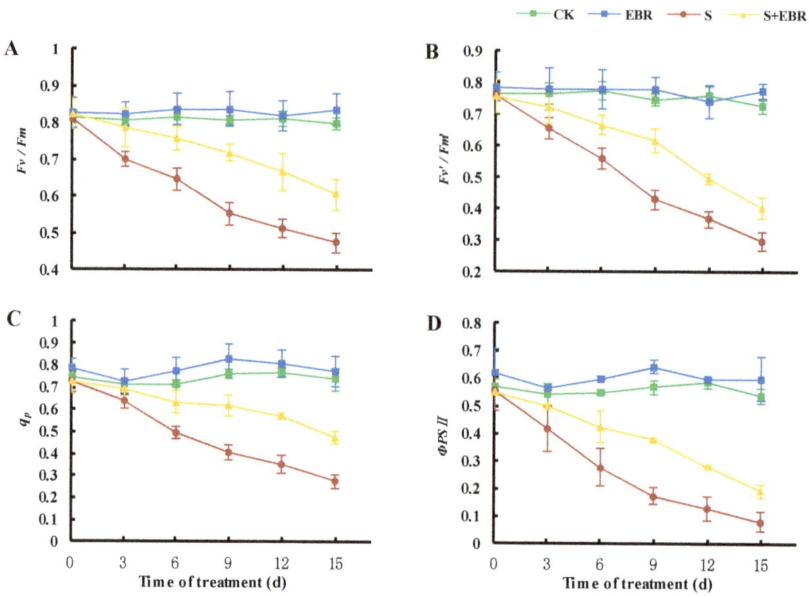

Figure 7. Effects of EBR on maximal photochemical efficiency of PSII, Fv/Fm (**A**); photochemical efficiency of PSII in the light, Fv′/Fm′ (**B**); photochemical quenching, qP (**C**); actual photochemical efficiency of PSII in the light, ΦPSII (**D**) in cucumber seedlings under $NaHCO_3$ stress. (CK represents normal Hoagland nutrient solution culture; EBR represents the addition of EBR under normal growth conditions; S represents $NaHCO_3$ stress treatment; S + EBR represents the addition of EBR under $NaHCO_3$ stress).

2.5. *Effects of EBR on the Expression of Photosynthesis-Related Genes in Cucumber Leaves Under $NaHCO_3$ Stress*

We assessed the expression levels of photosynthesis-related genes in cucumber leaves 48 h post-treatment using quantitative real-time PCR (qRT-PCR), as depicted in Figure 8. Under normal growth conditions, the application of exogenous EBR resulted in a slight enhancement of most photosynthesis-related genes, although these changes were not statistically significant. In contrast, under $NaHCO_3$ stress, there was a significant suppression of genes associated with PSII (*psbA* and *psbB*) and photosystem I (PSI) (*psaA* and *psaB*), along with genes involved in the Calvin cycle, such as *rbcL*, *rbcS*, *FBPase*, *SBPase*, *Rupe*, *PRK*, *Rca*, and *TPI*. However, the application of exogenous EBR under these stress conditions significantly counteracted this suppression, markedly enhancing the expression of these crucial photosynthetic genes. This enhancement included notable upregulation of PSII-related genes (*psbA* and *psbB*), PSI-related genes (*psaA* and *psaB*), and Calvin cycle-related genes. Specifically, *rbcL* and *rbcS*, encoding the large and small subunits of ribulose-1,5-bisphosphate carboxylase/oxygenase (Rubisco), respectively, exhibited significant increases. Additionally, substantial enhancements were also observed in the expression levels of *FBPase* (fructose-1,6-bisphosphatase), *SBPase* (sedoheptulose-1,7-bisphosphatase), *Rupe* (ribulose-5-phosphate isomerase), *PRK* (phosphoribulokinase), *TPI* (triose phosphate isomerase), and *Rca*, which encode Rubisco activase, underlining the profound regulatory effects of EBR on photosynthesis under $NaHCO_3$-induced stress conditions.

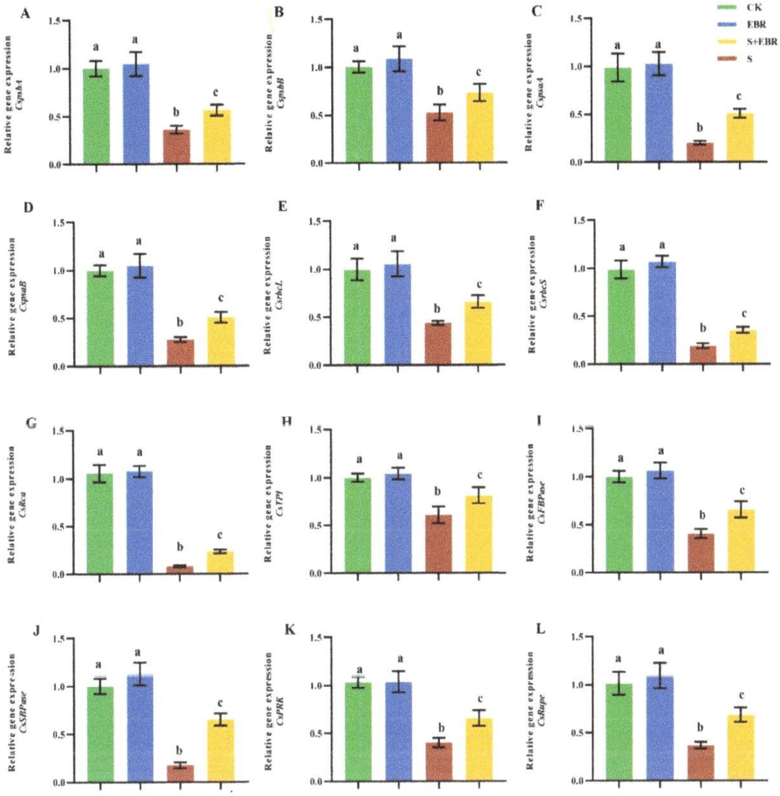

Figure 8. Effects of EBR on the expression of genes related to photosynthesis in cucumber leaves under $NaHCO_3$ stress. The relative expression levels of *psbA* (**A**), *psbB* (**B**), *psaA* (**C**), *psaB* (**D**), *rbcL* (**E**), *rbcS* (**F**), *Rca* (**G**), *TPI* (**H**), *FBPase* (**I**), *SBPase* (**J**), *PRK* (**K**), and *Rupe* (**L**) were measured. Abbreviations. *rbcL*, ribulose-1,5-bisphosphate carboxylase/oxygenase large subunit; *rbcS*, ribulose-1,5-bisphosphate carboxylase/oxygenase small subunit; *Rca*, Rubisco activase; *TPI*, triose phosphate isomerase; *FBPase*, fructose-1,6-bisphosphatase; *SBPase*, sedoheptulose-1,7-bisphosphatase; *Rupe*, ribulose-5-phosphate epimerase; *PRK*, phosphoribulokinase. Different letters indicate significant differences ($p < 0.05$; n = 3).

As shown in Figure 9, the application of exogenous EBR significantly alleviated the negative impact of $NaHCO_3$-induced alkaline stress on cucumber leaf photosynthesis. Gene expression analysis demonstrated a notable upregulation of key genes associated with PSII and PSI, including *psbA*, *psbB*, *psaA*, and *psaB*, indicating that EBR enhances electron transport efficiency from the primary electron acceptor in PSII to the terminal acceptors at the PSI side (Figure 9). Additionally, EBR treatment significantly increased the expression of essential Calvin cycle genes, such as *rbcL* and *rbcS* (which encode the large and small subunits of Rubisco), as well as other key enzymes including *Rca*, *TPI*, *PRK*, *Rupe*, *FBPase*, and *SBPase*, suggesting an accelerated Calvin cycle and improved carbon fixation efficiency. In contrast, $NaHCO_3$ treatment alone resulted in significant suppression of gene expression related to both photosystems and the Calvin cycle, as well as decreased Rubisco activity, leading to reduced electron transport efficiency and lower carbohydrate synthesis. These findings suggest that EBR plays a crucial role in mitigating the inhibitory effects of alkaline stress by enhancing photosynthetic electron transport.

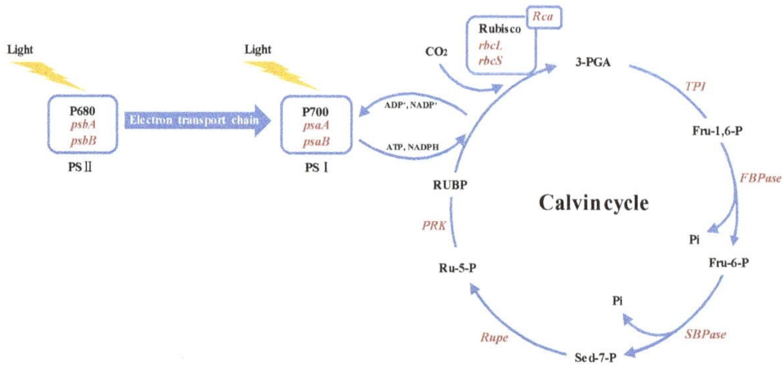

Figure 9. Impact of EBR on photosynthetic electron transport and Calvin cycle in cucumber leaves under $NaHCO_3$ stress. Compared to the $NaHCO_3$ treatment, the application of exogenous EBR upregulated the expression of crucial photosynthesis-related genes. These include *psbA* and *psbB* associated with PSII; *psaA* and *psaB* integral to photosystem I (PSI); and key genes involved in the Calvin cycle such as *rbcL* and *rbcS*, which encode the large and small subunits of ribulose-1,5-bisphosphate carboxylase/oxygenase (Rubisco), respectively. Additionally, genes like *Rca* (Rubisco activase), *TPI* (triose phosphate isomerase), *FBPase* (fructose-1,6-bisphosphatase), *SBPase* (sedoheptulose-1,7-bisphosphatase), *Rupe* (ribulose-5-phosphate isomerase), and *PRK* (phosphoribulokinase) also showed significant upregulation. The red highlights in the figure indicate areas where the expression levels of these genes were significantly enhanced due to EBR treatment under $NaHCO_3$ stress. Key abbreviations: PSII (photosystem II); P680 (primary electron donor of PSII); PSI (photosystem I); P700 (primary electron donor of PSI); rbcL, *rbcS* (genes encoding the large and small subunits of Rubisco, respectively); 3-PGA (3-phosphoglyceric acid); Fru-1,6-P (fructose-1,6-bisphosphate); Fru-6-P (fructose-6-phosphate); Sed-7-P (sedoheptulose-7-phosphate); Ru-5-P (ribulose-5-phosphate); RuBP (ribulose-1,5-bisphosphate); *Rubisco* (ribulose-1,5-bisphosphate carboxylase/oxygenase).

3. Discussion

Biomass accumulation serves as an essential indicator for a comprehensive assessment of plant stress tolerance [36]. In our investigation, we found that treatment with exogenous EBR significantly alleviated the growth suppression observed in cucumber seedlings under $NaHCO_3$-induced stress. This improvement was attributed to enhanced photosynthetic efficiency and increased reactive oxygen species scavenging capacity, as documented in Figure 1 and Table 1. Our findings are consistent with recent advancements reported in the literature. For instance, Zeng et al. (2024) demonstrated that EBR augments salt stress tolerance in *Quisqualis indica*, a mangrove species, by upregulating S-nitrosoglutathione reductase activity and modulating the nitric oxide signaling pathway [37]. In a similar vein, Chen et al. (2023) observed that EBR application in kiwifruit not only enhanced photosynthetic activity but also reduced hydrogen peroxide concentrations in the foliage, thereby effectively mitigating the detrimental effects of salt stress [38]. In alignment with these findings, both El-Azm and Elhady (2023) and Yu et al. (2023) have confirmed the significant role of EBR in reducing biomass decline due to salt stress in green beans and apple seedlings, respectively [39,40].

The reduction in plant biomass under stress conditions is primarily attributed to the inhibition of photosynthesis. Salt stress induces osmotic stress, leading to a decline in stomatal conductance and transpiration rates, which in turn diminishes photosynthetic activity [41,42]. In salt environments, the imbalance in intracellular Na^+/K^+ ratios contributes to stomatal closure, chlorophyll degradation, and photosystem damage [43]. Our findings indicate that $NaHCO_3$ stress substantially decreases Pn in cucumber seedlings (Figure 4), primarily due to the inhibition of chlorophyll synthesis and damage to the photosynthetic apparatus, resulting in impaired photosynthetic performance [16,17]. Chlorophyll, the predominant photosynthetic pigment in higher plants, is critical for the absorption, transmission, and distribution of light energy. It transforms light energy into chemical energy, essential for the functionality of the plant's photosystem and reaction centers [44]. Chlorophyll biosynthesis proceeds through a sequence of enzymatic reactions, delineated as follows: glutamic acid (Glu)→ALA→PBG→UroIII→ProtoIX→Mg-ProtoIX→Pchl→chlorophyll a→chlorophyll b [45]. Environmental and nutritional factors influence the dynamic equilibrium between chlorophyll synthesis and degradation [46,47]. Stress conditions such as salinity often result in reduced chlorophyll content, adversely affecting photosynthesis [48,49]. As shown in Figures 4 and 5, under $NaHCO_3$ stress, the decrease in chlorophyll content corresponds with reductions in Pchl, Mg-ProtoIX, UroIII, and ProtoIX, alongside increases in PBG and ALA, indicating that $NaHCO_3$ stress blocks the conversion of PBG to UroIII and inhibits its synthesis. The administration of EBR under $NaHCO_3$ stress led to increased levels of chlorophyll a, chlorophyll b, Pchl, Mg-ProtoIX, UroIII, and ProtoIX, while levels of PBG and ALA decreased, suggesting that EBR alleviates the inhibition of the chlorophyll synthesis pathway by $NaHCO_3$, thereby enhancing chlorophyll content and sustaining normal photosynthesis in cucumber leaves under stress [50,51].

Salt stress causes damage to the plant thylakoid membrane and inhibits enzymes crucial for chlorophyll synthesis, a primary factor behind reduced chlorophyll content under salt stress [50,51]. PBGD and uroporphyrin III synthase (UROS) are pivotal in the conversion of PBG to UroIII [47]. PBGD catalyzes the polymerization of four pheophorbide molecules, followed by UROS catalyzing the formation of cyclic UroIII [52]. Chlorophyll degradation is intricately linked to Chlase activity, typically low due to the spatial separation from chlorophyll [46]. Aging plants or those under adverse stress conditions experience chloroplast structural damage, facilitating chlorophyll contact with chloroplast enzymes and initiating degradation [53]. Salt stress is known to elevate the activity of enzymes associated with chlorophyll degradation [54,55], with post-translational modification of chloroplast enzymes being the rate-limiting step [56]. In our study, $NaHCO_3$ stress reduced PBGD activity (Figure 6B), obstructing chlorophyll synthesis at the PBG to UroIII step. Concurrently, $NaHCO_3$ stress increased Chlase activity (Figure 6A), hastening chlorophyll degradation, thereby leading to a decrease in chlorophyll content and inhibited photosynthesis (Figure 4). $NaHCO_3$ stress not only impedes PBGD activity, resulting in blocked PBG to UroIII conversion, but also exacerbates Na^+ and ROS toxicity [43], damaging the photosynthetic apparatus and enhancing Chlase activity, thus accelerating chlorophyll degradation. Introduction of exogenous EBR upregulated PBGD activity in cucumber leaves under $NaHCO_3$ stress, simultaneously inhibiting Chlase activity. This maintained chlorophyll content stability by promoting synthesis and reducing degradation, thereby protecting the photosynthetic apparatus, enhancing photosynthetic efficiency, and alleviating growth inhibition in cucumbers caused by $NaHCO_3$ stress.

Salt and alkaline stresses significantly compromise chloroplast structure, leading to chlorophyll degradation and the inhibition of photosynthesis, ultimately affecting plant growth adversely [57,58]. Our study demonstrates that $NaHCO_3$ stress induces a surge in ROS in cucumber plants, as depicted in Figure 2. This stress notably reduces photochemical efficiency (Fv/Fm, ΦPSII, Fv′/Fm′) (Figure 7) and impairs the functionality of the electron transport chain (qP), culminating in oxidative damage to the photosynthetic apparatus [59]. Furthermore, $NaHCO_3$ stress disrupts the functional balance between PSI and PSII in cucumber leaves. This is evidenced by the significant reduction in the expression of genes associated with PSII (psbA and psbB) and PSI (psaA and psaB), indicating that the alkaline environment impedes the synthesis of photosynthetic pigments and destabilizes the electron transport chain, further damaging vital photosynthetic structures, including chloroplasts [60].

The application of BRs has been shown to effectively protect the structural integrity of chloroplasts and enhance the efficiency of CO_2 absorption and utilization by boosting the activities of the cytochrome b6/f complex, D1 and D2 proteins, the large subunit of Rubisco, and carbonic anhydrase, thereby regulating the net photosynthetic rate [28,61]. Additionally, BRs significantly increase the net photosynthetic rate by upregulating the expression of related genes, which in turn improves the photochemical efficiency of PSII, chlorophyll content, and the activity of various key enzymes in the Calvin cycle [61]. The Calvin cycle is a crucial pathway for photosynthetic carbon assimilation and is highly dependent on Rubisco, a key rate-limiting enzyme [62]. The increase in Rubisco and other Calvin cycle enzyme activities is thought to be closely linked to the upregulation of related genes by BRs. These genes include those encoding *rbcL*, *rbcS*, and multiple enzymes associated with the Calvin cycle, such as *TPI*, *FBPase*, and *PRK* [61,63,64]. In our study, as shown in Figure 8, $NaHCO_3$ stress downregulated Rubisco-related genes *rbcL*, *rbcS*, and *Rca*, as well as Calvin cycle genes (e.g., *TPI*, *FBPase*, *SBPase*, *Rupe*, and *PRK*), indicating that the carbon fixation process was severely hindered, plant photosynthesis decreased overall, and the carbon assimilation capacity of cucumbers was limited, resulting in inhibited growth [65,66]. These results are consistent with previous findings on the effects of salt stress [67,68]. Under $NaHCO_3$ stress, exogenous EBR treatment significantly upregulated the expression of these photosynthesis-related genes, indicating that EBR alleviated the inhibition of the Calvin cycle by $NaHCO_3$ stress and enhanced carbon assimilation and photosynthetic capacity. Previous studies have shown that BRs can maintain cellular redox balance by activating the MAPK and NO signaling pathways involved in plant hormone regulation and increasing the activity of antioxidant enzymes, thereby protecting photosynthetic organs from free radical damage and regulating downstream transcription factors that are directly involved in the transcriptional regulation of photosynthetic genes, thus enhancing plant stress resistance [69–71]. This EBR-induced hormone regulation and increased antioxidant capacity may represent an important mechanism by which it improves the photosynthetic performance of cucumbers in this study.

Salt and alkaline stress severely inhibits plant photosynthesis and respiration, resulting in impaired electron transport and substantial accumulation of ROS [43]. The primary components of ROS include superoxide anion ($O_2^{\cdot-}$), hydrogen peroxide (H_2O_2), and hydroxyl radicals (OH). While low concentrations of ROS can trigger defense responses and enhance plant resistance, high concentrations may oxidize intracellular macromolecules and plasma membranes, leading to significant cellular damage and structural impairment [72]. ROS accumulation is a critical indicator of plant injury under salt and alkaline conditions [49]. In this study, $NaHCO_3$ stress led to a significant increase in ROS levels in cucumber, as illustrated in Figure 2. Exogenous EBR mitigates ROS accumulation induced by abiotic stresses, thereby enhancing plant stress tolerance [19,73]. In this experiment, exogenous EBR upregulated the expression of *SOD* and genes involved in the AsA-GSH cycle (*APX*, *MDHAR*, *DHAR*, *GR*) (Figure 3), enhancing antioxidant capacity and maintaining the stability of the AsA-GSH cycle, thus effectively reducing the ROS levels under $NaHCO_3$ stress and mitigating oxidative damage. Similar effects of exogenous BR have been observed in drought-treated barley [74] and salt-stressed perennial ryegrass [75].

Exogenous BR not only alleviates the inhibition of photosynthesis in cucumber seedlings under $NaHCO_3$ stress but also enhances their capacity to scavenge ROS, thereby reducing the growth inhibition experienced by cucumber seedlings under these stress conditions.

4. Materials and Methods

4.1. Experimental Materials and Design

This study was conducted in greenhouses at Shandong Agricultural University and the Shandong Institute of Sericulture using the cucumber cultivar Jinyan No. 4. Seeds were pre-soaked in warm water at 28 °C for 4 h, disinfected with a 0.3% sodium hypochlorite solution for 30 s, and thoroughly rinsed. After disinfection, the seeds were placed between two moistened layers of filter paper and incubated in darkness at 28 °C for germination. Once the cotyledons had fully expanded, uniformly germinated seedlings were selected, carefully washed to remove residual root substrate, and transplanted into hydroponic pots containing 5 L of Hoagland nutrient solution, with 4 seedlings per pot.

Experimental treatments began 8 days after pre-treatment and included the following groups:

CK: control plants grown in complete Hoagland nutrient solution;

EBR: Hoagland nutrient solution supplemented with 0.2 μmol/L EBR;

S: Hoagland nutrient solution supplemented with 30 mmol/L $NaHCO_3$;

S + EBR: Hoagland nutrient solution supplemented with 30 mmol/L $NaHCO_3$ and 0.2 μmol/L exogenous EBR.

The experiment followed a randomized block design with three replicates per treatment. Each replicate consisted of 15 pots, resulting in a total of 180 plants per treatment group (4 plants per pot × 15 pots × 3 replicates). EBR was applied daily, and nutrient solutions (including $NaHCO_3$, where applicable) were replaced every 2 days, with intermittent aeration provided. Sampling for photosynthetic parameters, chlorophyll fluorescence characteristics, and chlorophyll and chlorophyll precursor content was conducted on days 0, 3, 6, 9, 12, and 15 of the treatment period. Chlase and PBGD activities were measured on days 7 and 15. On day 10 of treatment, ROS (H_2O_2 and $O_2^{\cdot-}$) staining was performed. On day 15 of treatment, plant phenotypic images were captured, and biomass was determined.

4.2. Plant Biomass Measurement

Fifteen days after treatment, the plants were rinsed with deionized water, blotted dry with absorbent paper, and their biomass was assessed following the separation of aboveground and belowground components.

4.3. Photosynthetic Parameters Measurement

Photosynthetic rate (Pn), transpiration rate (Tr), and stomatal conductance (Gs) were measured using a Li-6400 photosynthesis system (Li-COR Biosciences, Lincoln, NE, USA) on the second functional leaf below the growing point.

4.4. Chlorophyll Content Determination

To determine the chlorophyll content, weigh 0.5 g of cucumber leaves, and cut them into small pieces in a mortar. Add a small amount of fine quartz sand, and grind the mixture into a paste. Extract the chlorophyll gradually with an 80% acetone aqueous solution until the residue becomes colorless. Filter the extract, and dilute it to a final volume of 15 mL. Measure the absorbance at 663 nm and 645 nm using the GENESYS 150 UV-Visible Spectrophotometer (Thermo Fisher Scientific, Waltham, MA, USA) and the UV-751 spectrophotometer.. The resulting optical density (OD) values are used in the following formulas to calculate the concentrations of chlorophyll a and chlorophyll b, as well as the chlorophyll a + b:

$$\text{chlorophyll a content} = 12.7 \times OD663 - 2.69 \times OD645$$

$$\text{chlorophyll b content} = 22.9 \times \text{OD645} - 4.86 \times \text{OD663}$$

$$\text{chlorophyll a + b content} = 20.2 \times \text{OD663} + 8.02 \times \text{OD645}$$

4.5. Chlorophyll Fluorescence Parameter Measurement

Chlorophyll fluorescence parameters Fo, Fm, Fs, Fo′, and Fm′ were assessed using an FMS-2 portable modulation fluorometer (FluorCam 7; Photon Systems Instruments, Brno, Czech Republic). Calculations were based on methods from Gong (2014) [17] and others, including maximum photochemical efficiency Fv/Fm = (Fm − Fo)/Fm; actual photochemical efficiency ΦPSII = (Fm′ − Fs)/Fm′; antenna conversion efficiency Fv′/Fm′ = (Fm′ − Fo′)/Fm′; and photochemical quenching coefficient qP = (Fm′ − Fs)/(Fm′ − Fo′).

4.6. Determination of the Content of Chlorophyll Precursor Substances

The determination of ALA was performed according to the method of Morton [76], with slight modifications. Specifically, 2 g of fresh sample was added to 6 mL of sodium acetate buffer (pH 4.6) and thoroughly ground while kept in an ice bath. After boiling the mixture in a water bath for 15 min, it was centrifuged at 10,000× *g* for 20 min. The precipitate was washed twice with 4 mL of extraction solution, and the supernatant was combined. Subsequently, 1 mL of the extract was taken, and 4 drops of ethyl acetoacetate were added. The mixture was then boiled in a water bath for 15 min before adding an equal volume of Alder's reagent. The absorbance was measured at 553 nm after 15 min. A standard curve was prepared using ALA-HCl (Sigma-Aldrich, St. Louis, MO, USA) as a standard sample to calculate the ALA content (nmol/g FW).

The content of UroIII and PBG was determined by following the method of Bogorad (1962) [77]. For the determination of UroIII, 1 g of fresh sample was weighed, and 5 mL of Tris-HCl (pH 7.2) buffer was added. The mixture was ground and extracted in an ice bath. After centrifugation at 5000× *g* for 15 min at 4 °C, the supernatant was taken, and 1 drop of glacial acetic acid was added to adjust the pH to 4.0. The mixture was centrifuged again at 5000× *g* for 15 min at 4 °C, and the supernatant was discarded. The precipitate was washed twice with 5 mL of water and centrifuged again to obtain a uracil-based precipitate. To extract uracil, 4 mL of pre-cooled concentrated ammonia water was added twice, and the supernatant was combined and transferred to an evaporating dish. The solution was evaporated to dryness at 55 °C, followed by the addition of 4 mL of 5% methanol sulfamate and incubation for 48 h. Afterward, 20 mL of water was added, and the pH was adjusted to 4.0 with saturated sodium acetate. The mixture was extracted twice with 4 mL of chloroform, evaporated to dryness at 55 °C, dissolved in 4 mL of chloroform, centrifuged at 5000× *g* for 10 min, and the absorbance was measured at 405.5 nm. Each treatment was repeated three times, and the content (nmol/g FW) was calculated using the following formula:

$$\text{UroIII} = 5.48\text{A}405.5 \times 10 \times 0.004 \times 3 \times 1000/\text{FW}$$

For the determination of PBG, 1 g of fresh sample was weighed, and 5 mL of Tris-HCl (pH 8.2) buffer was added, followed by grinding and extraction in an ice bath. The mixture was then centrifuged at 12,000× *g* and 4 °C for 10 min. The supernatant was taken, and 1 drop of glacial acetic acid was added to adjust the pH to 4.0 before centrifugation again. After discarding the supernatant, the precipitate was washed to obtain a PBG precipitate. To extract PBG, 4 mL of pre-cooled concentrated ammonia water was added twice, and the supernatant was combined and evaporated to dryness. The residue was treated with 4 mL of 5% sulfuric acid–methanol for esterification for 48 h. After adding 20 mL of water and adjusting the pH, the mixture was extracted with 4 mL of chloroform, and the colorimetric absorbance was finally measured at a wavelength of 553 nm, each treatment was repeated three times, and the PBG content was calculated using the following formula (nmol/g FW):

$$PBG = A553 \times 1.383 \times 10 \times 0.004 \times 3 \times 1000/FW$$

ProtoIX, Mg-ProtoIX, and Pchl were determined using the method of Hodgins and VanHuystee (1986) [45]. A fresh leaf sample without main veins weighing 0.5 g was taken, and 80% alkaline acetone (volume ratio 80:20 of acetone to 1% ammonia solution) was added. After grinding and extracting in an ice bath, the mixture was centrifuged at 15,000× *g* for 15 min at 4 °C. The supernatant was transferred to a 10 mL volumetric flask, made up to the mark with 80% alkaline acetone, shaken well, and the absorbance values were measured at wavelengths of 575 nm, 590 nm, and 628 nm.

The following equations were used to calculate the contents of UroIII, PBG, ProtoIX, Mg-ProtoIX, and Pchl, with three replicates for each treatment. Here, 'A' represents the absorbance measured at the specified wavelength (e.g., 405.5 nm, 553 nm, etc.); 'V' is the final volume (mL) of the extracted solution; and 'FW' is the fresh weight of the sample (g).

$$\text{Mg-ProtoIX} = (0.06077 \times A590 - 0.01937 \times A575 - 0.003423 \times A628) \times V/FW$$

$$\text{ProtoIX} = (0.18016 \times A575 - 0.04036 \times A628 - 0.04515 \times A590) \times V/FW$$

$$\text{Pchl} = (0.03563 \times A628 + 0.007225 \times A590 - 0.02955 \times A575) \times V/FW$$

4.7. Chlorase and Porphobilinogen Deaminase Activity

Chlase activity was determined using the method established by Holden (1961) [78] and further adapted by Zhou et al. (2016) [79]. A sample of 0.5 g of fresh cucumber leaves was weighed and ground in 20 mL of 50% ethanol. This mixture was then combined with 8 mL of a chlorophyll solution and 4 mL of a sodium citrate solution (0.04 M), and the pH was adjusted to approximately 8. The sample was incubated in the dark for 18 h. Following incubation, the mixture was filtered through a funnel to remove pigments. The filtrate was subsequently mixed with 10 mL of 2% sodium chloride and diluted to a final volume of 50 mL with 80% ethanol. The decomposition activity was assessed by measuring the absorbance at 607 nm using a spectrophotometer. Enzyme activity was expressed in μmol chlorophyllamide mg^{-1} FW min^{-1}, with one unit defined as the amount of enzyme that catalyzes the decomposition of 10% of the substrate to chlorophyllamide within one hour.

PBGD activity was quantified following the protocols of Rimington (1960) [80], Frydman and Frydman (1979) [81], and Zhou et al. (2016) [79]. A sample of 4.0 g of cucumber leaves was extracted using 20 mL of phosphate buffer (pH 6.8) to obtain the substrate (PBG). The material was ground in a mortar and pestle, and the supernatant was collected after centrifugation at 12,000× *g* for 10 min for subsequent analysis. Additionally, 1.0 g of cucumber material was ground in a mortar and pestle, and 10 mM Tris-HCl (pH 7.4, 4 °C) was added. This mixture was then centrifuged at 20,000× *g* for 15 min, and the supernatant was retained. In a test tube, 1 mL of PBG was combined with 1 mL of PBGD and 2 mL of phosphate buffer (pH 7.4). The reaction was conducted at 4 °C for 60 min. After adding Ehrlich's reagent (Frydman and Frydman, 1979), the absorbance was measured at 553 nm. Enzyme activity was expressed in nmol g^{-1} protein h^{-1} based on the absorbance changes induced by the reaction.

4.8. $O_2^{\cdot-}$ and H_2O_2 Staining

ROS staining, specifically for $O_2^{\cdot-}$ and H_2O_2, was conducted using an improved method based on Xia et al. (2009) [28].

O_2^{-} was detected in situ utilizing the NBT staining method. Leaf discs were vacuum-impregnated with 0.1 mg/mL NBT in 25 mM potassium phosphate buffer (pH 7.8) and incubated at room temperature in the dark for 4 h. Following staining, the leaf discs were washed with 80% ethanol at 70 °C for 10 min and subsequently mounted for photography.

H_2O_2 was detected in situ using the DAB staining method. Leaf discs were immersed in a 1 mg/mL DAB solution in 50 mM Tris-acetate buffer (pH 3.8) under vacuum and incubated at room temperature in the dark for 24 h. After incubation, the leaf discs were removed and washed with 80% ethanol, then observed under a microscope.

4.9. Fluorescence Quantification and Gene Expression Analysis Based on qRT-PCR

Total RNA was extracted from the plant samples using the TRIzol method, and complementary DNA (cDNA) was synthesized with the RevertAid First Strand cDNA Synthesis Kit (Thermo Fisher Scientific, Waltham, MA, USA). Quantitative real-time polymerase chain reaction (qRT-PCR) analysis was performed using the Power SYBR Green PCR Master Mix (Applied Biosystems, Foster City, CA, USA). The primers used in this study are listed in Table 2.

Table 2. Primer sequences.

Gene Name	Primer Sequences
actin	F: ATGGCCGATGCCGAGGATATR R: TAGGAGCATCATCACCAGCAAAAC
CsCu-ZnSOD	F: CTCCATTTTCAATCTCTCATTATCC R: ATAGAAGTGATTGTGCGGCCATAG
CsFe-SOD	F: TACACTGAACCTGAGTTCAACAAC R: ACGTTCTCTGAGAAACCAAACAGA
CsAPX	F: CTGCTACTGTTTTTGGAACCGCCG R: GCGGAGGAGAGGAAACGAGTAGTT
CsMDHAR	F: ACAGCCTTCTTCTGTTGCCTTCAG R: CTCTATTGTCGTTGGCGAAATCCG
CsDHAR	F: ATGTCGGGCTCCAGAATCCAACCA R: AAAGCGAGGAATTGGAAGGAAGGT
CsGR	F: GTCCGATAGTGCTGGAGGTGTTGG R: CCATCCAAAGCCATGACTCTCTTC
CspsbA	F: GTATTCCAGGCTGAGCACAACATC R: TACCTAAAGCGGTGGACCAGATAC
CspsbB	F: GGTATTTGGAGTTACGAAGGTGTG R: CCCAACCCTGAGAGAAATAAATGA
CspsaA	F: GATTTCTCATAGTTGGTGCTGCTG R: TACAAACCAAAACTGTGAAAGCCT
CspsaB	F: ATTTGGACATCTTGTTTGGGCTAC R: TGATGTAGAGGCAATCAAGAAAGC
CsrbcL	F: TACTGATATCTTGGCAGCATTCCG R: AAGATTCAGCAGCTACAGCGGC
CsrbcS	F: ATGGCTTCATCCATTCTCTCATCC R: CCAGTGAATGGTGCTACCATGCTA
CsRca	F: GAATATGGCAACATGCTCGTCATG R: TCCAAGAGCAGCTTCGTTCAGAT
CsTPI	F: CTCTCTTTCACAACGTCCACTCACA R: ACCAACGAAGAACTTGCCGGAG

Table 2. *Cont.*

Gene Name	Primer Sequences
CsFBPase	F: AATCTCTCGTTCTCTCTCTCGCCTC R: CGCTATGCCTTCTATTTCCACGG
CsSBPase	F: CAGTGTCCTCCTCATACTTGGGTTG R: CTGGGAAGAAAGATTGGGGAGAAA
CsRupe	F: TCCCAAGTCAGTGGGTTTATCGGAG R: AACCTTCTCCTCGAAACGGTAAGAG
CsPRK	F: ATCCACACCCTCATTCATTTCTCC R: GCAGTTGAGGGAGTGAAGAAGAAGA

4.10. Data Processing

In this study, a completely randomized experimental design was employed, with each treatment group consisting of three replicates. The data were expressed as mean values with the corresponding standard error. To identify significant differences among the means, Duncan's New Multiple Range Test was conducted using SPSS software version 22.0. Statistical significance was established with a p-value of less than 0.05. Furthermore, for additional data analysis and chart generation, Excel 2010 was utilized.

5. Conclusions

This study demonstrates that the exogenous application of EBR effectively mitigates the detrimental effects of $NaHCO_3$-induced alkaline stress in cucumber seedlings. EBR contributes to the restoration of growth and photosynthetic efficiency in stressed plants. Notably, it significantly enhances chlorophyll content, net photosynthetic rate (Pn), stomatal conductance (Gs), and crucial indicators of chlorophyll fluorescence, including maximum photochemical efficiency (Fv/Fm) and actual photochemical efficiency (ΦPSII). Additionally, EBR diminishes the accumulation of ROS and favorably regulates the expression of genes involved in the antioxidant ascorbate–glutathione (AsA-GSH) cycle, thereby mitigating oxidative stress. Moreover, EBR administration reverses the inhibition of chlorophyll biosynthesis pathways, primarily through elevated activity of PBGD and reduced chlorophyllase activity. This alteration leads to increased levels of chlorophyll precursors such as UroIII and ProtoIX, thereby enhancing chlorophyll accumulation under alkaline conditions.

Crucially, EBR stabilizes the photosynthetic electron transport system and enhances the expression of essential photosynthesis-related genes, improving light energy utilization and accelerating the Calvin cycle. Collectively, these findings underscore the potential of brassinosteroids to bolster plant resilience against harsh environmental conditions by reinforcing photosynthetic and antioxidant systems, thereby offering substantial agricultural benefits under alkaline stress.

Author Contributions: Conceptualization, Q.S. and W.N.; validation, J.M. and Q.H.; formal analysis, J.M. and Q.H; data curation, J.M. and Q.H; writing—original draft preparation, W.N.; writing—review and editing, Q.S. and H.G. All authors have read and agreed to the published version of the manuscript.

Funding: This research was funded by the Shandong Vegetable Industry Technology System (SDAIT-05); the Shandong Province Key R&D Program (2022CXGC020708 and 2022CXGC010610); the Agricultural Scientific and Technological Innovation Project of Shandong Academy of Agricultural Sciences (CXGC2024G05 and CXGC2023A16); the Yantai Science and Technology Program (2022XCZX087); and the Shandong Province "Bohai Grain Silo" Science and Technology Demonstration Project (2019BHLC005).

Data Availability Statement: The study's data can be obtained by contacting the corresponding author upon request.

Conflicts of Interest: The authors declare no conflict of interest.

References

1. Calzadilla, P.; Carvalho, F.; Gomez, R.; Neto, M.L.; Signorelli, S. Assessing photosynthesis in plant systems: A cornerstone to aid in the selection of resistant and productive crops. *Environ. Exp. Bot.* **2022**, *201*, 104950. [CrossRef]
2. Zhu, X.-G.; Long, S.P.; Ort, D.R. Improving photosynthetic efficiency for greater yield. *Annu. Rev. Plant Biol.* **2010**, *61*, 235–261. [CrossRef] [PubMed]
3. Muhammad, I.; Shalmani, A.; Ali, M.; Yang, Q.-H.; Ahmad, H.; Li, F.B. Mechanisms regulating the dynamics of photosynthesis under abiotic stresses. *Front. Plant Sci.* **2021**, *11*, 615942. [CrossRef] [PubMed]
4. Sharma, A.; Kumar, V.; Shahzad, B.; Ramakrishnan, M.; Singh Sidhu, G.P.; Bali, A.S.; Handa, N.; Kapoor, D.; Yadav, P.; Khanna, K. Photosynthetic response of plants under different abiotic stresses: A review. *J. Plant Growth Regul.* **2020**, *39*, 509–531. [CrossRef]
5. Yan, K.; Mei, H.; Dong, X.; Zhou, S.; Cui, J.; Sun, Y. Dissecting photosynthetic electron transport and photosystems performance in Jerusalem artichoke (*Helianthus tuberosus* L.) under salt stress. *Front. Plant Sci.* **2022**, *13*, 905100. [CrossRef]
6. Lotfi, R.; Ghassemi-Golezani, K.; Pessarakli, M. Salicylic acid regulates photosynthetic electron transfer and stomatal conductance of mung bean (*Vigna radiata* L.) under salinity stress. *Biocatal. Agric. Biotechnol.* **2020**, *26*, 101635. [CrossRef]
7. Wang, G.; Zeng, F.; Song, P.; Sun, B.; Wang, Q.; Wang, J. Effects of reduced chlorophyll content on photosystem functions and photosynthetic electron transport rate in rice leaves. *J. Plant Physiol.* **2022**, *272*, 153669. [CrossRef]
8. Hnilickova, H.; Kraus, K.; Vachova, P.; Hnilicka, F. Salinity stress affects photosynthesis, malondialdehyde formation, and proline content in *Portulaca oleracea* L. *Plants* **2021**, *10*, 845. [CrossRef]
9. Munns, R.; Tester, M. Mechanisms of salinity tolerance. *Annu. Rev. Plant Biol.* **2008**, *59*, 651–681. [CrossRef]
10. Shabala, S. *Plant Stress Physiology*; CABI: Wallingford, UK, 2017. [CrossRef]
11. Allakhverdiev, S.I.; Sakamoto, A.; Nishiyama, Y.; Inaba, M.; Murata, N. Ionic and osmotic effects of NaCl-induced inactivation of photosystems I and II in *Synechococcus* sp. *Plant Physiol.* **2000**, *123*, 1047–1056. [CrossRef]
12. Bose, J.; Rodrigo-Moreno, A.; Shabala, S. ROS homeostasis in halophytes in the context of salinity stress tolerance. *J. Exp. Bot.* **2014**, *65*, 1241–1257. [CrossRef] [PubMed]
13. Guo, S.-H.; Niu, Y.-J.; Zhai, H.; Han, N.; Du, Y.-P. Effects of alkaline stress on organic acid metabolism in roots of grape hybrid rootstocks. *Sci. Hortic.* **2018**, *227*, 255–260. [CrossRef]
14. Flexas, J.; Bota, J.; Loreto, F.; Cornic, G.; Sharkey, T. Diffusive and metabolic limitations to photosynthesis under drought and salinity in C3 plants. *Plant Biol.* **2004**, *6*, 269–279. [CrossRef] [PubMed]
15. Chaves, M.M.; Flexas, J.; Pinheiro, C. Photosynthesis under drought and salt stress: Regulation mechanisms from whole plant to cell. *Ann. Bot.* **2009**, *103*, 551–560. [CrossRef]
16. Abdel Latef, A.A.; Tran, L.-S.P. Impacts of priming with silicon on the growth and tolerance of maize plants to alkaline stress. *Front. Plant Sci.* **2016**, *7*, 243. [CrossRef]
17. Gong, B.; Li, X.; Bloszies, S.; Wen, D.; Sun, S.; Wei, M.; Li, Y.; Yang, F.; Shi, Q.; Wang, X. Sodic alkaline stress mitigation by interaction of nitric oxide and polyamines involves antioxidants and physiological strategies in Solanum lycopersicum. *Free Radic. Biol. Med.* **2014**, *71*, 36–48. [CrossRef]
18. Guo, R.; Zhou, J.; Hao, W.; Gong, D.; Zhong, X.; Gu, F.; Liu, Q.; Xia, X.; Tian, J.; Li, H. Germination, growth, photosynthesis and ionic balance in *Setaria viridis* seedlings subjected to saline and alkaline stress. *Can. J. Plant Sci.* **2011**, *91*, 1077–1088. [CrossRef]
19. Sharma, A.; Kumar, V.; Kanwar, M.; Thukral, A.; Bhardwaj, R. Ameliorating imidacloprid induced oxidative stress by 24-epibrassinolide in *Brassica juncea* L. *Russ. J. Plant Physiol.* **2017**, *64*, 509–517. [CrossRef]
20. Shahzad, B.; Tanveer, M.; Che, Z.; Rehman, A.; Cheema, S.A.; Sharma, A.; Song, H.; ur Rehman, S.; Zhaorong, D. Role of 24-epibrassinolide (EBL) in mediating heavy metal and pesticide induced oxidative stress in plants: A review. *Ecotoxicol. Environ. Saf.* **2018**, *147*, 935–944. [CrossRef]
21. Holá, D. Brassinosteroids and photosynthesis. In *Brassinosteroids: A Class of Plant Hormone*; Springer: Berlin, Germany, 2011; pp. 143–192. [CrossRef]
22. Fariduddin, Q.; Yusuf, M.; Ahmad, I.; Ahmad, A. Brassinosteroids and their role in response of plants to abiotic stresses. *Biol. Plant.* **2014**, *58*, 9–17. [CrossRef]
23. Li, S.; Zheng, H.; Lin, L.; Wang, F.; Sui, N. Roles of brassinosteroids in plant growth and abiotic stress response. *Plant Growth Regul.* **2021**, *93*, 29–38. [CrossRef]
24. Sadura, I.; Janeczko, A. Physiological and molecular mechanisms of brassinosteroid-induced tolerance to high and low temperature in plants. *Biol. Plant.* **2018**, *62*, 601–616. [CrossRef]

25. Wu, C.-y.; Trieu, A.; Radhakrishnan, P.; Kwok, S.F.; Harris, S.; Zhang, K.; Wang, J.; Wan, J.; Zhai, H.; Takatsuto, S. Brassinosteroids regulate grain filling in rice. *Plant Cell.* **2008**, *20*, 2130–2145. [CrossRef] [PubMed]
26. Vardhini, B.V.; Anjum, N.A. Brassinosteroids make plant life easier under abiotic stresses mainly by modulating major components of antioxidant defense system. *Front. Environ. Sci.* **2015**, *2*, 67. [CrossRef]
27. Cui, J.X.; Zhou, Y.H.; Ding, J.G.; Xia, X.J.; Shi, K.; Chen, S.C.; Asami, T.; Chen, Z.; Yu, J.Q. Role of nitric oxide in hydrogen peroxide-dependent induction of abiotic stress tolerance by brassinosteroids in cucumber. *Plant Cell Environ.* **2011**, *34*, 347–358. [CrossRef]
28. Xia, X.J.; Wang, Y.J.; Zhou, Y.H.; Tao, Y.; Mao, W.-H.; Shi, K.; Asami, T.; Chen, Z.; Yu, J.-Q. Reactive oxygen species are involved in brassinosteroid-induced stress tolerance in cucumber. *Plant Physiol.* **2009**, *150*, 801–814. [CrossRef]
29. Nie, W.F.; Wang, M.M.; Xia, X.J.; Zhou, Y.H.; Shi, K.; Chen, Z.; Yu, J.Q. Silencing of tomato RBOH1 and MPK2 abolishes brassinosteroid-induced H_2O_2 generation and stress tolerance. *Plant Cell Environ.* **2013**, *36*, 789–803. [CrossRef]
30. Sun, Z.; Zou, Y.; Xie, C.; Han, L.; Zheng, X.; Tian, Y.; Ma, C.; Liu, X.; Wang, C. Brassinolide improves the tolerance of *Malus hupehensis* to alkaline stress. *Front. Plant Sci.* **2022**, *13*, 1032646. [CrossRef]
31. Ma, Q.; Wu, E.; Wang, H.; Yuan, Y.; Feng, Y.; Liu, J.; Zhao, L.; Feng, B. Exogenous 24-epibrassinolide boosts plant growth under alkaline stress from physiological and transcriptomic perspectives: The case of broomcorn millet (*Panicum miliaceum* L.). *Ecotoxicol. Environ. Saf.* **2022**, *248*, 114298. [CrossRef]
32. Fang, S.; Hou, X.; Liang, X. Response mechanisms of plants under saline-alkali stress. *Front. Plant Sci.* **2021**, *12*, 667458. [CrossRef]
33. Nie, W.; Gong, B.; Geng, B.; Wen, D.; Qiao, P.; Guo, H.; Shi, Q. The effects of exogenous 2,4-epibrassinolide on the germination of cucumber seeds under $NaHCO_3$ stress. *Plants* **2024**, *13*, 394. [CrossRef] [PubMed]
34. Rattan, A.; Kapoor, D.; Kapoor, N.; Bhardwaj, R.; Sharma, A. Involvement of brassinosteroids in plant response to salt stress. In *Brassinosteroids in Plant Developmental Biology and Stress Tolerance*; Academic Press: New York, NY, USA, 2022; pp. 237–253. [CrossRef]
35. Soliman, M.; Elkelish, A.; Souad, T.; Alhaithloul, H.; Farooq, M. Brassinosteroid seed priming with nitrogen supplementation improves salt tolerance in soybean. *Physiol. Mol. Biol. Plants* **2020**, *26*, 501–511. [CrossRef] [PubMed]
36. Ondrasek, G.; Rathod, S.; Manohara, K.K.; Gireesh, C.; Anantha, M.S.; Sakhare, A.S.; Parmar, B.; Yadav, B.K.; Bandumula, N.; Raihan, F.; et al. Salt stress in plants and mitigation approaches. *Plants* **2022**, *11*, 717. [CrossRef] [PubMed]
37. Zeng, L.L.; Song, L.Y.; Wu, X.; Ma, D.N.; Song, S.W.; Wang, X.X.; Zheng, H.L. Brassinosteroid enhances salt tolerance via S-nitrosoglutathione reductase and nitric oxide signaling pathway in mangrove Kandelia obovata. *Plant Cell Environ.* **2024**, *47*, 511–526. [CrossRef]
38. Chen, C.; Cheng, D.; Li, L.; Sun, X.; He, S.; Li, M.; Chen, J. Physiological characteristics and transcriptome analysis of exogenous brassinosteroid-treated kiwifruit. *Int. J. Mol. Sci.* **2023**, *24*, 17252. [CrossRef]
39. El-Azm, A.; Elhady, A. Brassinosteroids or proline can alleviate yield inhibition under salt stress via modulating physio-biochemical activities and antioxidant systems in snap bean. *J. Hortic. Sci. Biotechnol.* **2023**, *98*, 526–539. [CrossRef]
40. Yu, B.; Wang, L.; Guan, Q.; Xue, X.; Gao, W.; Nie, P. Exogenous 24-epibrassinolide promoted growth and nitrogen absorption and assimilation efficiency of apple seedlings under salt stress. *Front. Plant Sci.* **2023**, *14*, 1178085. [CrossRef]
41. Zhang, H.; Liu, X.-L.; Zhang, R.-X.; Yuan, H.-Y.; Wang, M.-M.; Yang, H.-Y.; Ma, H.-Y.; Liu, D.; Jiang, C.-J.; Liang, Z.-W. Root damage under alkaline stress is associated with reactive oxygen species accumulation in rice (*Oryza sativa* L.). *Front. Plant Sci.* **2017**, *8*, 1580. [CrossRef]
42. Yang, C.; Jianaer, A.; Li, C.; Shi, D.; Wang, D. Comparison of the effects of salt-stress and alkali-stress on photosynthesis and energy storage of an alkali-resistant halophyte *Chloris virgata*. *Photosynthetica* **2008**, *46*, 273–278. [CrossRef]
43. Nie, W.; Gong, B.; Chen, Y.; Wang, J.; Wei, M.; Shi, Q. Photosynthetic capacity, ion homeostasis and reactive oxygen metabolism were involved in exogenous salicylic acid increasing cucumber seedlings tolerance to alkaline stress. *Sci. Hortic.* **2018**, *235*, 413–423. [CrossRef]
44. Taïbi, K.; Taïbi, F.; Abderrahim, L.A.; Ennajah, A.; Belkhodja, M.; Mulet, J.M. Effect of salt stress on growth, chlorophyll content, lipid peroxidation and antioxidant defence systems in *Phaseolus vulgaris* L. *S. Afr. J. Bot.* **2016**, *105*, 306–312. [CrossRef]
45. Hodgins, R.R.; Van Huystee, R.B. Rapid simultaneous estimation of protoporphyrin and Mg-porphyrins in higher plants. *J. Plant Physiol.* **1986**, *125*, 311–323. [CrossRef]
46. Jacob-Wilk, D.; Holland, D.; Goldschmidt, E.E.; Riov, J.; Eyal, Y. Chlorophyll breakdown by chlorophyllase: Isolation and functional expression of the Chlase1 gene from ethylene-treated Citrus fruit and its regulation during development. *Plant J.* **1999**, *20*, 653–661. [CrossRef]
47. Hu, L.; Xiang, L.; Li, S.; Zou, Z.; Hu, X.H. Beneficial role of spermidine in chlorophyll metabolism and D1 protein content in tomato seedlings under salinity–alkalinity stress. *Physiol. Plant.* **2016**, *156*, 468–477. [CrossRef]
48. Ping, W.S.; Rong, G.S.; Hui, H.X.; Jing, L.; Sheng, J.Y. Effects of NaCl stress on the content of photosynthetic pigments in the leaves of cucumber (*Cucumis sativus* L.) seedlings. *J. Jiangxi Agric. Univ.* **2006**. [CrossRef]

49. Liu, N.; Gong, B.; Jin, Z.; Wang, X.; Wei, M.; Yang, F.; Li, Y.; Shi, Q. Sodic alkaline stress mitigation by exogenous melatonin in tomato needs nitric oxide as a downstream signal. *J. Plant Physiol.* **2015**, *186*, 68–77. [CrossRef]
50. Qin, J.; Dong, W.; He, K.; Chen, J.; Yu, Y.; Wang, Z. Effects of salt stress on growth and photosynthetic physiological features of *Hippophae rhamnoides* seedlings. *Ecol. Environ. Sci.* **2009**, *18*, 1031–1036.
51. Lu, C.; Li, L.; Liu, X.; Chen, M.; Wan, S.; Li, G. Salt Stress Inhibits Photosynthesis and Destroys Chloroplast Structure by Downregulating Chloroplast Development–Related Genes in *Robinia pseudoacacia* Seedlings. *Plants* **2023**, *12*, 1283. [CrossRef]
52. Shoolingin-Jordan, P.M. Porphobilinogen deaminase and uroporphyrinogen III synthase: Structure, molecular biology, and mechanism. *J. Bioenerg. Biomembr.* **1995**, *27*, 181–195. [CrossRef]
53. Fang, Z.; Bouwkamp, J.C.; Solomos, T. Chlorophyllase activities and chlorophyll degradation during leaf senescence in non-yellowing mutant and wild type of *Phaseolus vulgaris* L. *J. Exp. Bot.* **1998**, *49*, 503–510. [CrossRef]
54. Yuan, R.; Shu, S.; Guo, S.; Sun, J.; Wu, J. The positive roles of exogenous putrescine on chlorophyll metabolism and xanthophyll cycle in salt-stressed cucumber seedlings. *Photosynthetica* **2018**, *56*, 557–566. [CrossRef]
55. Zhu, Y.-f.; Wu, Y.-x.; Hu, Y.; Jia, X.; Zhao, T.; Cheng, L.; Wang, Y. Tolerance of two apple rootstocks to short-term salt stress: Focus on chlorophyll degradation, photosynthesis, hormone and leaf ultrastructures. *Acta Physiol. Plant.* **2019**, *41*, 1–14. [CrossRef]
56. Matile, P.; Schellenberg, M. The cleavage of phaeophorbide a is located in the envelope of barley gerontoplasts. *Plant Physiol. Biochem.* **1996**, *34*, 55–59. [CrossRef]
57. Yamane, K.; Rahman, S.; Kawasaki, M.; Taniguchi, M.; Miyake, H. Pretreatment with Antioxidants Decreases the Effects of Salt Stress on Chloroplast Ultrastructure in Rice Leaf Segments (*Oryza sativa* L.). *Plant Prod. Sci.* **2004**, *7*, 292–300. [CrossRef]
58. Balasubramaniam, T.; Shen, G.; Esmaeili, N.; Zhang, H. Plants' Response Mechanisms to Salinity Stress. *Plants* **2023**, *12*, 2253. [CrossRef]
59. Gong, B.; Miao, L.; Kong, W.; Bai, J.-G.; Wang, X.; Wei, M.; Shi, Q. Nitric oxide, as a downstream signal, plays a vital role in auxin induced cucumber tolerance to sodic alkaline stress. *Plant Physiol. Biochem.* **2014**, *83*, 258–266. [CrossRef]
60. Jaspers, P.; Kangasjärvi, J. Reactive oxygen species in abiotic stress signaling. *Physiol. Plant.* **2010**, *138*, 405–413. [CrossRef]
61. Siddiqui, H.; Hayat, S.; Bajguz, A. Regulation of photosynthesis by brassinosteroids in plants. *Acta Physiol. Plant.* **2018**, *40*, 1–15. [CrossRef]
62. Raines, C.A. The Calvin cycle revisited. *Photosynth. Res.* **2003**, *75*, 1–10. [CrossRef]
63. Li, X.J.; Guo, X.; Zhou, Y.H.; Shi, K.; Zhou, J.; Yu, J.Q. Overexpression of a brassinosteroid biosynthetic gene dwarf enhances photosynthetic capacity through activation of Calvin cycle enzymes in tomato. *BMC Plant Biol.* **2016**, *16*, 33. [CrossRef]
64. Jiang, Y.-P.; Cheng, F.; Zhou, Y.-H.; Xia, X.-J.; Mao, W.-H.; Shi, K.; Chen, Z.-X.; Yu, J.-Q. Hydrogen peroxide functions as a secondary messenger for brassinosteroids-induced CO_2 assimilation and carbohydrate metabolism in *Cucumis sativus*. *J. Zhejiang Univ. Sci. B* **2012**, *13*, 811–823. [CrossRef] [PubMed]
65. Liu, Z.; Bie, Z.; Huang, Y.; Zhen, A.; Niu, M.; Lei, B. Rootstocks improve cucumber photosynthesis through nitrogen metabolism regulation under salt stress. *Acta Physiol. Plant.* **2013**, *35*, 2259–2267. [CrossRef]
66. El Sayed, A.; El-Hamahmy, M.; Rafudeen, M.; Ebrahim, M. Exogenous spermidine enhances expression of Calvin cycle genes and photosynthetic efficiency in sweet sorghum seedlings under salt stress. *Biol. Plant.* **2019**, *63*, 511–518. [CrossRef]
67. Yang, Y.; Yu, L.; Wang, L.; Guo, S. Bottle gourd rootstock-grafting promotes photosynthesis by regulating the stomata and non-stomata performances in leaves of watermelon seedlings under NaCl stress. *J. Plant Physiol.* **2015**, *186*, 50–58. [CrossRef] [PubMed]
68. Miao, Y.; Gao, X.; Li, B.; Wang, W.; Bai, L. Low red to far-red light ratio promotes salt tolerance by improving leaf photosynthetic capacity in cucumber. *Front. Plant Sci.* **2023**, *13*, 1053780. [CrossRef]
69. Kim, T.-W.; Michniewicz, M.; Bergmann, D.C.; Wang, Z.-Y. Brassinosteroid regulates stomatal development by GSK3-mediated inhibition of a MAPK pathway. *Nature* **2012**, *482*, 419–422. [CrossRef]
70. Gudesblat, G.E.; Schneider-Pizoń, J.; Betti, C.; Mayerhofer, J.; Vanhoutte, I.; Van Dongen, W.; Boeren, S.; Zhiponova, M.; De Vries, S.; Jonak, C. SPEECHLESS integrates brassinosteroid and stomata signalling pathways. *Nat. Cell Biol.* **2012**, *14*, 548–554. [CrossRef]
71. Li, X.; Zhang, L.; Ahammed, G.J.; Li, Z.-X.; Wei, J.-P.; Shen, C.; Yan, P.; Zhang, L.-P.; Han, W.-Y. Nitric oxide mediates brassinosteroid-induced flavonoid biosynthesis in *Camellia sinensis* L. *J. Plant Physiol.* **2017**, *214*, 145–151. [CrossRef]
72. You, J.; Chan, Z. ROS regulation during abiotic stress responses in crop plants. *Front. Plant Sci.* **2015**, *6*, 1091. [CrossRef]
73. Soares, C.; de Sousa, A.; Pinto, A.; Azenha, M.; Teixeira, J.; Azevedo, R.A.; Fidalgo, F. Effect of 24-epibrassinolide on ROS content, antioxidant system, lipid peroxidation and Ni uptake in *Solanum nigrum* L. under Ni stress. *Environ. Exp. Bot.* **2016**, *122*, 115–125. [CrossRef]
74. Gill, M.B.; Cai, K.; Zhang, G.; Zeng, F. Brassinolide alleviates the drought-induced adverse effects in barley by modulation of enzymatic antioxidants and ultrastructure. *Plant Growth Regul.* **2017**, *82*, 447–455. [CrossRef]
75. Wu, W.; Zhang, Q.; Ervin, E.H.; Yang, Z.; Zhang, X. Physiological mechanism of enhancing salt stress tolerance of perennial ryegrass by 24-epibrassinolide. *Front. Plant Sci.* **2017**, *8*, 1017. [CrossRef] [PubMed]

76. Morton, M. Biochemical Spectroscopy. 1975. Available online: https://core.ac.uk/download/pdf/82256489.pdf (accessed on 3 October 2024).
77. Bogorad, L. Porphyrin synthesis. *Methods Enzymol.* **1962**, *5*, 885–895. [CrossRef]
78. Holden, M. The breakdown of chlorophyll by chlorophyllase. *Biochem. J.* **1961**, *78*, 359–364. [CrossRef]
79. Zhou, H.; Guo, S.; An, Y.; Shan, X.; Wang, Y.; Shu, S.; Sun, J. Exogenous spermidine delays chlorophyll metabolism in cucumber leaves (*Cucumis sativus* L.) under high temperature stress. *Acta Physiol. Plant.* **2016**, *38*, 224. [CrossRef]
80. Rimington, C. Spectral-absorption coefficients of some porphyrins in the Soret-band region. *Biochem. J.* **1960**, *75*, 620–623. [CrossRef]
81. Frydman, R.B.; Frydman, B. Disappearance of porphobilinogen deaminase activity in leaves before the onset of senescence. *Plant Physiol.* **1979**, *63*, 1154–1157. [CrossRef]

plants

MDPI

Article

Enhanced Production by Terra-Sorb® Symbiotic Biostimulant in Two Model Species Under Nitrogen Stress

Laia Utgés-Minguell [1], Nuria Sierras-Serra [2], Cándido Marín [2] and Marta Pintó-Marijuan [1,3,*]

[1] Department of Evolutionary Biology, Ecology and Environmental Sciences, University of Barcelona, 08028 Barcelona, Spain; laiautgesminguell@gmail.com
[2] Plant Health R&D, Bioiberica, S.A.U, 08389 Palafolls, Spain; nsierras@bioiberica.com (N.S.-S.); cmarin@bioiberica.com (C.M.)
[3] Institute of Research in Biodiversity (IRBio), University of Barcelona, 08028 Barcelona, Spain
* Correspondence: martapinto@ub.edu

Abstract: The increasing soil pollution has accelerated the implementation of new agricultural regulations that significantly limit the use of synthetic nitrogen (N) fertilizers. Consequently, plants are likely to experience nutrient stress, leading to decreased productivity and potential threats to food security. To address these critical challenges, microbial-based biostimulant (BS) products, which utilize metabolites from microorganisms, offer a sustainable and eco-friendly solution to mitigate plant nutrient stress. This study evaluated the effects of the radicular application of a microbial-based BS containing L-α-amino acids on lettuce and pepper crops under two nitrogen regimes: optimal N availability and N stress (NS). Various parameters, including growth, production, soluble proteins, photosynthetic pigment content, and oxidative stress markers, were assessed. Under optimal N conditions, BS application enhanced commercial biomass in lettuce and vegetative biomass in pepper, indicating that BSs can reduce the need for nitrate uptake and endogenous amino acid synthesis, thereby conserving energy for other physiological processes. Despite BS application, NS conditions significantly reduced vegetative and reproductive growth in both species. However, BS treatment in pepper plants increased chloroplast pigments, improving light absorption and photosynthetic efficiency. The reduction in the carotenoid/chlorophyll ratio suggests efficient N allocation to growth and production. Thus, BS application proved effective in mitigating NS in pepper plants, enhancing pepper production, while under optimal conditions, it improved lettuce yield, particularly commercial biomass. These findings underscore the potential of symbiotic microbial-based BSs as a promising tool for sustainable agriculture under reduced N availability.

Keywords: *Capsicum annuum*; horticultural production; *Lactuca sativa*; microbial biostimulant; nitrogen stress; photosynthetic performance; plant growth-promoting bacteria

1. Introduction

In the coming decades, FAO studies forecast an increase in the global population exceeding 9.2 billion, requiring a 70% increase in food production to meet future demands [1]. Food security, a key focus of the Sustainable Development Goals, is becoming increasingly challenging due to the escalating impacts of climate change on agriculture. Modern agriculture has heavily relied on synthetic inputs, including pesticides and chemical fertilizers to maintain food security [2]. While these inputs have been crucial in stabilizing crop yields, the overuse of synthetic nitrogen fertilizers has raised relevant environmental concerns, including water contamination, soil degradation, and increased greenhouse gas emissions [3]. In response, the European Commission launched, in 2019, the European Green Deal which sets out policy initiatives to make Europe the first climate-neutral continent by 2050, including the Farm to Fork strategy, which aims to reduce the environmental impact of agriculture by decreasing chemical pesticide use and excessive synthetic nutrient inputs [4].

Nitrogen fertilization is crucial for maintaining global food production as it significantly enhances crop yields and quality. Nitrogen (N) is a plant macronutrient required for synthesizing proteins, nucleic acids, and chlorophylls, which are essential for growth and development [5]. Nitrogen deficiency stress (NS) impairs photosynthesis processes by reducing the biosynthesis of the photosynthetic enzymes involved in the Calvin–Benson Cycle, promotes stomatal limitation, and reduces light-harvesting protein availability as well as other proteins related to bioenergetics pathways such as Cytb6f or ATP synthase [6]. Therefore, the lack of nitrogen leads to lower energy production and reduced biomass, and crops fail to reach their full potential, resulting in lower yield and crop quality. On a global scale, this can threaten food security, which is exacerbated by climate change, and is expected to intensify hunger and malnutrition [7]. As a result, there is a growing need for sustainable practices in agriculture by adopting eco-friendly alternatives that optimize yields without compromising soil health or environmental sustainability [8]. Improved management practices that reduce dependency on synthetic inputs protect our natural resources while ensuring optimal productivity [9] such as biostimulants.

Biostimulants (BSs) are defined by the European Biostimulant Industry Council (EBIC) as substances or microorganisms that enhance plant nutrition efficiency, abiotic stress tolerance, and crop quality traits regardless of nutrient content. Due to their composition, they have been categorized and classified based on the substances and microorganisms that characterize them [10,11]. Among the various categories of biostimulants, protein hydrolysates (PHs), composed of amino acids and peptide mixtures, can be obtained from animal or plant by-products through chemical, enzymatic, or thermal hydrolysis, or a combination of these methods [11]. Enzymatic hydrolysis is considered a superior method for producing PHs because it preserves the biologically active form of amino acids which is the L-form, which is essential for crop growth and development [12]. This method of obtention avoids the conversion of the L-form to their less effective or even toxic D-forms, a drawback associated with chemical hydrolysis [13]. Additionally, the enzymatic mode of obtention is more sustainable and environmentally friendly compared to chemical methods, which can leave harmful residues such as sodium and sulfates [14]. PHs exhibit beneficial features in various physiological processes such as photosynthetic activity, nutrient assimilation, and translocation and positively impact quality parameters, enhancing fruit set and size in numerous crops [15,16]. Moreover, PHs contain bioactive peptides that can elicit hormone-like activities, influencing shoot and root growth and architecture, and contributing to the activation of both primary and secondary metabolism [17,18].

Another relevant BS group is the Plant Growth-Promoting Bacteria (PGPB), particularly the *Bacillus* genus, which promotes plant growth through nitrogen fixation [19] among other features and modes of action. They are also resistant to environmental stress conditions and are omnipresent in various types of soils [20,21]. *Bacillus* spp. can produce phytohormones such as abscisic acid, auxins, cytokinins, ethylene, jasmonic acid, and gibberellins [22] which are closely associated with the regulation of plant growth. Furthermore, these metabolites promote the induction of stress-related gene expression to maintain membrane stability and prevent the accumulation of reactive oxygen species (ROS) [19]. Another trait of *Bacillus* spp. is the facilitation of nutrient availability in the rhizosphere reducing the need for synthetic chemical fertilizers. Moreover, some studies have demonstrated increased uptake of N, P, and K when plants are under drought and salinity stress, showing that treatment with *Bacillus* increases the bioavailability of macronutrients [23,24]. One notable member of this genus is *B. velezensis*, which possesses strain-specific clusters of genes related to metabolites' biosynthesis, which play significant roles in both pathogen suppression and plant growth promotion [21]. *Bacillus velezensis* PH-023 CECT 30677 is a bacterial strain that, with its metabolites, is formulated combined with L-α-amino acids from enzymatic hydrolysis in a commercially available microbial biostimulant named Terra-Sorb® Symbiotic, (Bioiberica, S.A.U).

The objective of this study was to evaluate the effects of Terra-Sorb® Symbiotic on production and at a physiological level under N stress in two model horticultural species of agricultural economic interest: (1) a leafy species represented by lettuce (*Lactuca sativa* L.) and (2) a fruit producing species represented by pepper (*Capsicum annuum* L.). Moreover, we aimed to assess the potential of the stress mitigation by the PGPB in the microbial biostimulant based on L-α-amino acids application, as these crops require very high levels of fertilization to ensure adequate production. Therefore, this study considers the indirect provision of metabolites, L-α-amino acids, and symbiotic bacteria capable of interacting with the plant, with the hypothesis that the microbial BS will be able to compensate the provoked nitrogen stress with a more environmentally sustainable solution.

2. Results

To assess the effects of the nitrogen stress and the possible mitigation of the stress symptoms by BS application both, fruit horticultural species and leafy vegetables were included in the assay.

2.1. Nitrogen Stress Drastically Reduced Photosynthetic Activity, Pepper Production, and Plant Vegetative Biomass, but BS Application Promoted a Recovery of Both Vegetative and Reproductive Structures in Pepper Plants

The photosynthetic performance of pepper plants under NS conditions was affected by the N availability present in the substrate showing closed stomata (lower g_s), slower electron transport rate (ETR), and also lower maximum PSII efficiency (F_v/F_m) (Figure 1A–C). Accordingly, N stress symptoms were also observed by higher NPQ and Car/Chl (Figure 1D,F), although symbiotic BS application played a significant role in reducing the Car/Chl in the NS plants to the NC values. Another stress symptom was the lower content of Chls, both Chl*a* and Chl*b* (Figure 1E), and total soluble proteins (Figure 1H) compared to control plants. The non-stressed plants showed no effect of the BS application in any of the previously described parameters. In fact, the Chl*a*/*b* ratio (Figure 1G) was the only parameter particularly affected by the BS application in the NC plants (decreased a 3.12% after BS application). However, the BS application significantly enhanced chlorophyll biosynthesis in the plants under NS from the 3rd week (Table S2, Figure S2). Specifically, chlorophyll *a* content experienced a significant rise in comparison with the non-treated plants, higher than 70% (Figure 1E), and chlorophyll *b* content increased by more than 50%, and consequently, the Chl*a*/*b* ratio increased a 12.04% after the BS application in the NS plants. Therefore, pepper plant vegetative functions were significantly reduced by the N stress, but the BS application allowed chlorophyll biosynthesis and Car/Chl ratio recovery, alleviating stress symptoms, and might have positive consequences on pepper fruit production.

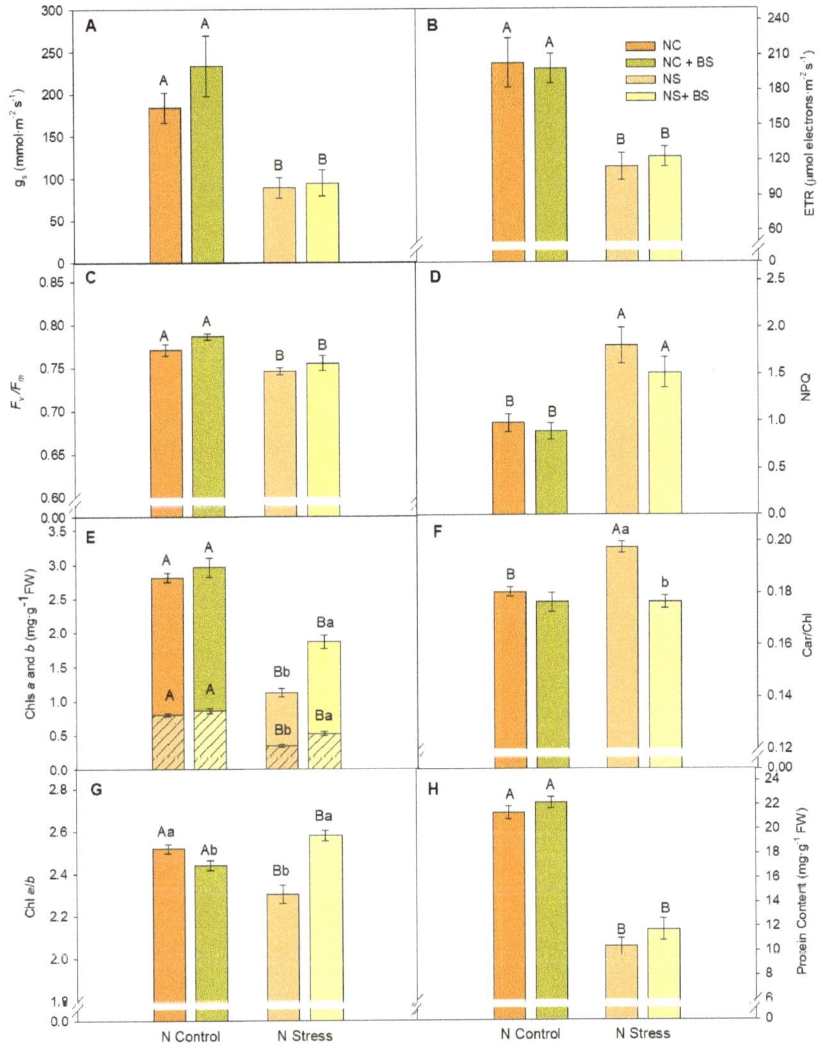

Figure 1. Effect of BS in photosynthetic performance under NC and NS treatments in pepper plants: stomatal conductance (g_S; **(A)**), electron transport rate (ETR; **(B)**), PSII maximum efficiency (F_v/F_m; **(C)**), non-photochemical quenching (NPQ; **(D)**), total chlorophyll content (Chl *b* (lower bars) Chl *a* (upper bars); **(E)**), carotenoids/chlorophylls (Car/Chl; **(F)**), chlorophyll *a*/*b* ratio (Chl *a*/*b*; **(G)**), and protein content **(H)**. Data show the mean ± SE (n = 8). Different letters indicate significant differences between nitrogen treatments (capital letters) or between BS applications (lower-case letters) with a $P_{value} < 0.05$.

The pepper plants were monitored weekly by quantifying multiple growth parameters (both vegetative and reproductive). Along the growing period, nitrogen stress provoked a strong shrinking on both the vegetative (Figures S3 and S4) and reproductive (Figures S5 and S6) growth parameters but the BS application had a significant mitigation of the stress symptoms from the 5th or 6th week (Tables S3–S5). Moreover, BS application also increased vegetative biomass in both nitrogen availabilities (Figure 2A). Nevertheless, the highest increase was observed in the N-stressed plants when BS was applied, showing a 110.5% vegetative biomass increase and around a 25.9% increase in pepper plant height (Figure 2B).

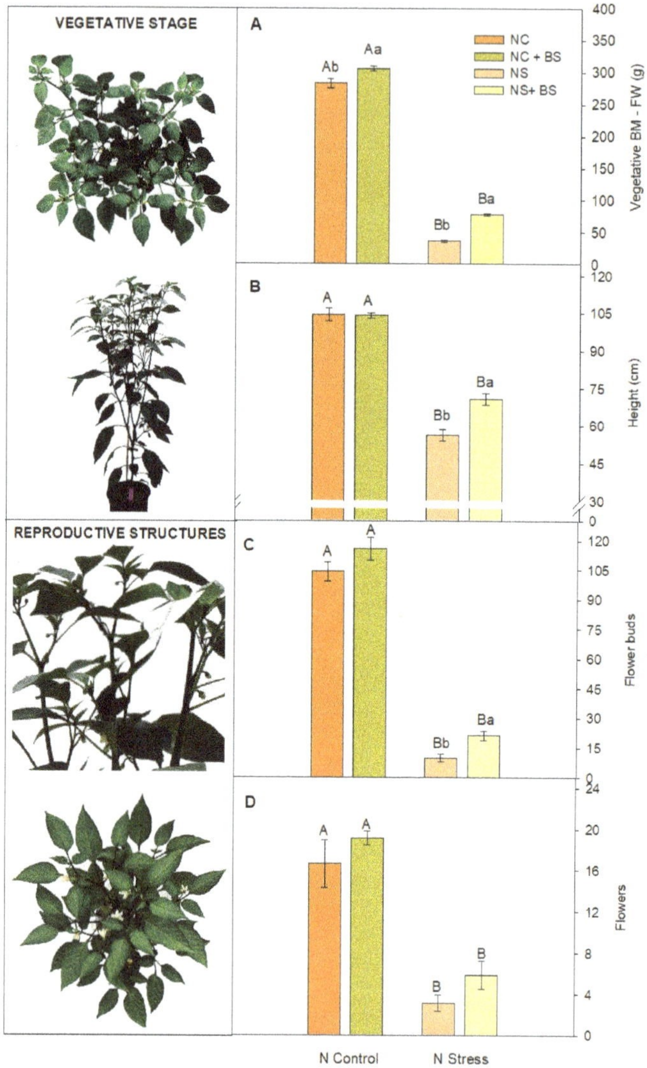

Figure 2. Vegetative and reproductive effects of BS under NC and NS treatments in pepper plants. Vegetative stage: vegetative BM-FW (**A**) and height (**B**); reproductive structures: flower buds (**C**) and flowers (**D**). Data show the mean ± SE (n = 8). Different letters indicate significant differences between nitrogen treatments (capital letters) or between BS applications (lower-case letters) with a $P_{value} < 0.05$.

Regarding the reproductive parameters, BS application effects were observed palliating the N stress consequences and enhancing the number of flower buds since the 6th week (Table S5 and Figure 2C), fruit set, and fruit yield at any of the development stages (Figure 3). Furthermore, the results show a tendency to develop more flowers (Figure 2D), being significant in the 10th week (Figure S4B). The stress palliation in terms of fruit pepper production rise a 300% after BS application (Figure 3D) in the NS plants. When optimal N concentration (NC) was available, only a trend to form more flowers and flower buds was observed (Figure 2C,D and Figure S5), but was not reflected in higher production (Figure 3). Regarding pepper harvest at three development stages (Figure 3), very similar results were shown: (1) NS significantly decreased both the number and weight in the three development stages, and (2) the BS application enhanced the number and weight of the three development stages under NS. In fact, from the 10th week, the 'just sprouted' number of peppers (Table S6) was significantly more abundant in the BS-treated plants. Surprisingly, though the number of 'just sprouted' and 'developed' fruits of the NS + BS-treated plants was lower than NC, the BS application allowed no differences between NC + BS and NS + BS in the number and the weight of the 'in development' peppers (Figure 3F,G), as well as in the weight of the 'developed stage' peppers (Figure 3E). Therefore, the BS treatment significantly enhanced pepper fruit production in the NS plants.

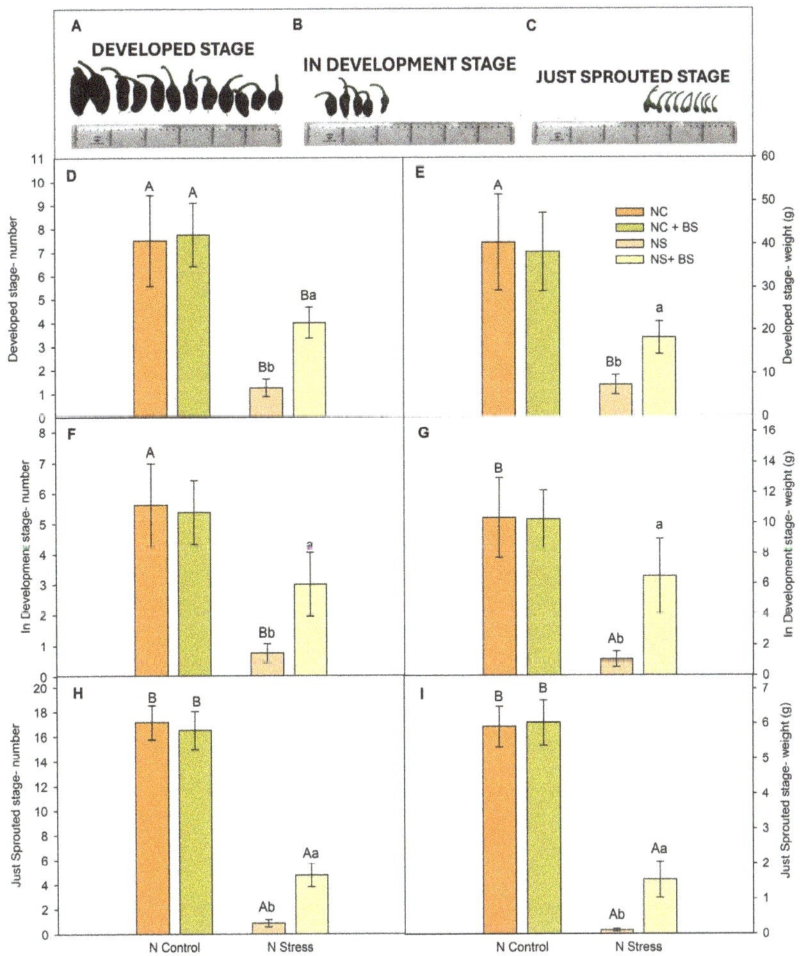

Figure 3. Images of the development stage in pepper plants (**A–C**) and the effect of BS under NC and NS treatments classified on the fruit set stage. Number of peppers: developed (**D**), in development (**F**), and just sprouted (**H**). Weight of peppers: developed (**E**), in development (**G**), and just sprouted (**I**). Data show the mean ± SE (n = 8). Different letters indicate significant differences between nitrogen treatments (capital letters) or between BS applications (lower-case letters) with a $P_{value} < 0.05$.

2.2. In Lettuce Plants, BS Application Promoted Lower Need for Photoprotection (In NC), Lower Need for Protein Content (In NS), and Significantly Enhanced Commercial Production in Lettuce Plants

Most of the photosynthesis-related parameters assessed in the study (g_s, ETR, and F_v/F_m) displayed no significant differences between the NC and NS treatments (Figure 4A–C) in lettuce plants. However, NS clearly affected energy dissipation by carotenoids in the PSII light-harvesting complex (LHC). For instance, NPQ and Car/Chl ratio (Figure 4D,E) were higher under NS (being significant differences between NC and NS in the BS-treated plants), indicating the need for enhanced photoprotective mechanisms under NS conditions. Another stress symptom was the significantly lower content of nitrogen-containing molecules in the NS plants compared to the NC plants: (1) chlorophylls, both Chl*a* and Chl*b* (Figure 4E), and (2) total soluble proteins (Figure 4H). Therefore, NS was affecting the lettuce plants' performance moderately.

Regarding the BS application, the results showed minimal impact on photosynthetic parameters under both the NC and NS conditions. However, a significant 11% reduction in the Car/Chl ratio was observed in the BS-treated lettuce plants under NC (NC + BS) compared to the untreated plants (NC) (Figure 4F), suggesting that BS may reduce the need for carotenoids under optimal nitrogen conditions. Moreover, the BS application under NS enhanced chlorophylls' biosynthesis (especially Chl*a*) that managed to have the same amount of Chl*a* as under control conditions (NC), promoting a significant rise in chlorophyll a/b ratio compared to the untreated plants (Figure 4F,G). SPAD weekly assessments demonstrated significant differences between both optimal N treatments (NC and NC + BS) compared to the N stress treatments (NS and NS + BS) from the 3rd week after transplant (Figure S7 and Table S7). Interestingly, protein content in both N availabilities experienced a decrease after BS application, being significant in the NS conditions with a 24.2% lower content in the NS + BS plants.

Lettuce growth development traits in both N availabilities (Figure 5) discovered a strong limitation in growth as a consequence of lower N conditions and NS significantly limited biomass, diameter, and height. For instance, height was significantly higher in NC than in NS since the 4th week in both BS treatments (Figure S8 and Table S8). BS application in the lettuces growing under NS did not lead to significant differences, although slight positive trends were observed in diameter and height. Under NC conditions, the BS-treated plants showed notable positive trends across all the measured parameters, including total BM, diameter, and height (Figure 5A,C,D and Figure S8), suggesting enhanced growth when BS was applied under optimal nitrogen availability. Moreover, under NC, commercial biomass (Figure 5B) was significantly higher after BS application, further boosting biomass compared to untreated plants.

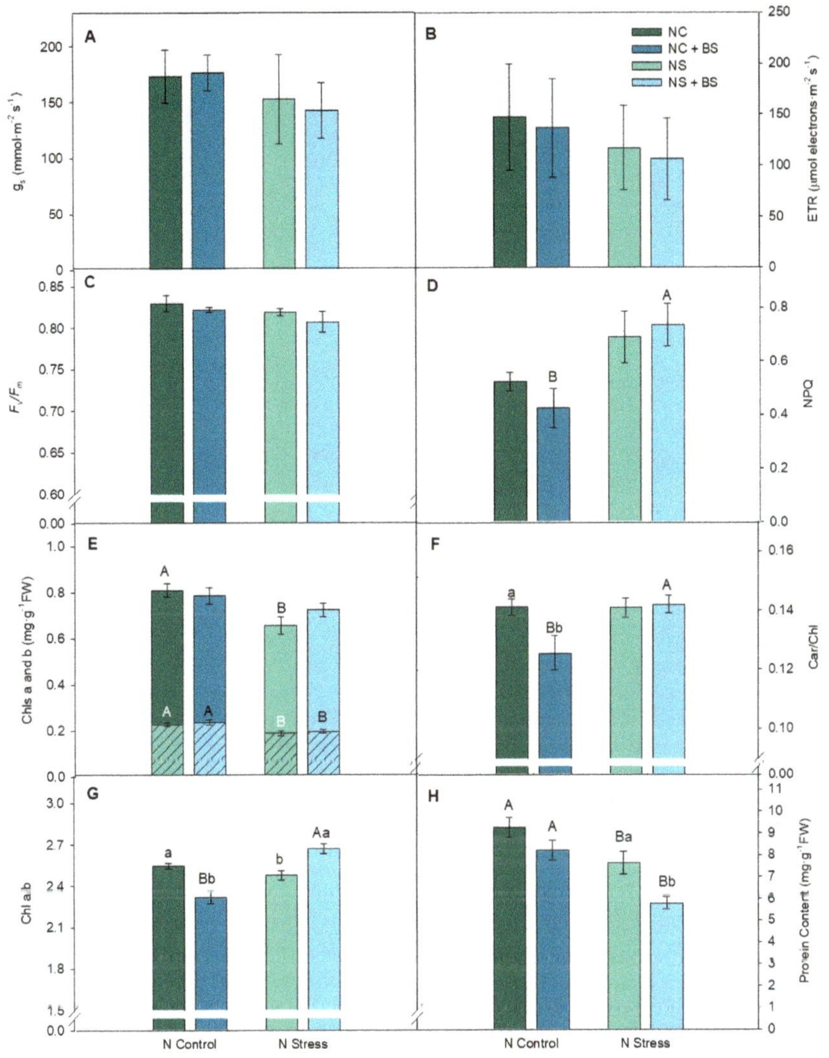

Figure 4. Effect of BS in photosynthetic performance under NC and NS treatments in lettuce: stomatal conductance (g_s; (**A**)), electron transport rate (ETR; (**B**)), PSII maximum efficiency (F_v/F_m; (**C**)), non-photochemical quenching (NPQ; (**D**)), total chlorophyll content (Chl *b* (lower bars) Chl *a* (upper bars); (**E**)), carotenoids/chlorophylls (Car/Chl; (**F**)), chlorophyll *a*/*b* ratio (Chl *a*/*b*; (**G**)), and protein content (**H**). Data show the mean ± SE (n = 8). Different letters indicate significant differences between nitrogen treatments (capital letters) or between BS applications (lower-case letters) with a $P_{value} < 0.05$.

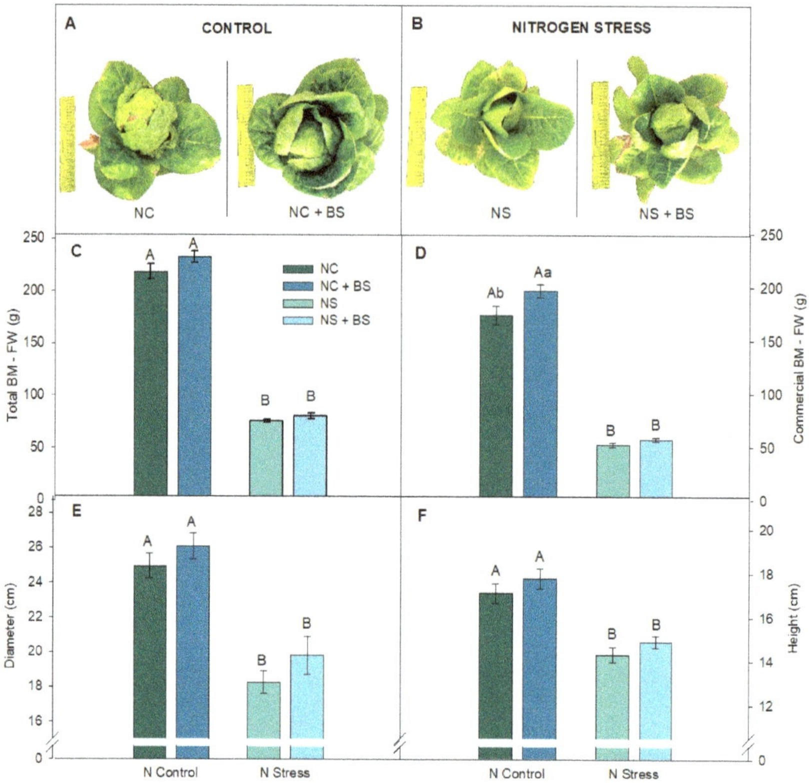

Figure 5. Images of the BS effects on harvested lettuce plants under NC (**A**) and NS (**B**) treatments and growth parameters. Production: total BM–FW (**C**), commercial BM–FW (**D**). Growth parameters: diameter (**E**), height (**F**). Data show the mean ± SE (n = 8). Different letters indicate significant differences between nitrogen treatments (capital letters) or between BS applications (lower-case letters) with a $P_{value} < 0.05$.

3. Discussion

Currently, the agricultural sector is grappling with several challenges: (1) ensuring food availability for a growing global population, (2) enhancing resource use efficiency, and (3) minimizing the ecological impact of crop production [25]. To avoid the damaging consequences brought by the abusive use of N synthetic fertilizers supplied to increase the crop yield, plant BSs are considered promising and eco-friendly tools to reduce the use of synthetic fertilizers by improving nutrient use efficiency and uptake from the soil [26,27], but the new policies implementation bring plant nutrient stress consequences. In this study, two horticultural (lettuce and pepper) species were tested under severe N reduction as recommended for future crop management [28] complemented with the use of a microbial BS, based on L-α-amino acids and containing a *B. velezensis* bacterial strain with its fermentation metabolites.

Nitrogen (N) plays a main role in plant nutrition as it is involved in most physiological reactions and is a well-known key element present in several important biomolecules such as proteins, nucleic acids, hormones, or chlorophylls [29,30]. Regarding the photosynthetic performance in the lettuce plants, neither stomatal conductance nor photo-oxidative stress markers ETR, F_v/F_m, and NPQ were affected by low N availability, suggesting that there were no stress-induced perturbations in the photosynthetic apparatus and plants were not suffering physiological stress under varying N regimes. Navarro-León et al. [31]) also found no significant differences in parameters such as analyzed chlorophyll *a* fluorescence in lettuce under the same N doses in lettuce. However, in pepper plants, N deficiency provoked significant stress with a marked stomatal closure and severely affected photosystem photochemical efficiency. Lower ETR and F_v/F_m with higher NPQ and Car/Chl content were clear stress markers of N availability stress suffered by the pepper plants. ETR quantifies the electron transport rate (depending on the Φ_{PSII} and the light incidence) that is used in photochemistry and F_v/F_m reflects the potential maximum PSII efficiency, being a sensitive indicator of plant photosynthetic performance, with optimal values typically ranging from 0.79 to 0.83 [32,33]. Our study recorded an average of 0.81 across all treatments in lettuce plants but N-stressed pepper plants without BS application presented values under 0.75. Several abiotic stresses require antioxidant protection mechanisms to safeguard the photosynthetic mechanisms such as thermal dissipation via NPQ is necessary to avoid photo-oxidation and protect PSII by preventing reactive oxygen species (ROS) formation [34]. Carotenoids play a crucial role in photoprotection, both in thermal dissipation (especially zeaxanthin and lutein) and in direct ROS scavenging [34]. In our study, N stress provoked, in the two different species, a reduction in the composition of the analyzed photosynthetic pigments: chlorophylls and carotenoids. The induction of stomatal closure and reduction in total chlorophyll and carotenoid content was also observed by Huang et al. [35] when N was limited in rice plants and by [25] in greenhouse-grown basil plants.

The BS application only reduced photoprotection needs by decreasing the Car/Chl ratio under NC conditions in lettuce plants and under NS in pepper plants, although the Chl*a* and Chl*b* content were also enhanced. Some studies proved that the application of amino acid-based BS boosted chlorophyll and carotenoid content in several species under nutrient reductions (Table S9) [36,37]. Our observations evidenced that the application of microbial BSs can manage lower N availability at the root level maintaining the biosynthesis of chlorophylls *a* and *b*. The presence of exogenous protein hydrolysates and amino acids such as glutamate, precursors of chlorophyll biosynthesis, restored chlorophyll levels to perform optimal photosynthesis rates under N deficiency [31]. It has been studied in tomatoes under reduced NPK nutrition that foliar applications of polysaccharides, protein hydrolysates, and humic acids-BS can also preserve chlorophyll content at optimal levels [36].

In both model species, N availability reduction showed a significant reduction in the measured N containing molecules: chlorophylls and soluble proteins independently of the BS treatment. With low N available in soil, once N is assimilated into the plant and reduced to NH_4^+, lower availability of NH_4^+ will restrict the biosynthesis of amino acids and future soluble proteins. Hence, the observed reduction in biomass can be suggested by a limited protein biosynthesis attributed to a decrease in N availability in the substrate. As has been demonstrated by Gao et al., [38] protein content shows a direct relationship with biomass production.

BS treatment to lettuce plants showed an unexpected decrease in soluble protein content, being significant under nitrogen stress. This observation strongly contrasts with the findings from other studies. For instance, Irani et al. [39] investigated the application of several BSs (amino acids, humic acids, fulvic acids, and seaweed extracts) under drought stress conditions in grapevine and showed that BS application significantly increased soluble protein content. Furthermore, Carillo et al. [17] confirmed that in spinach, under different N availabilities, the foliar application of protein hydrolysate-based BS promoted an increase in protein content. Therefore, our findings are not in accordance with these publications and the disruptive results obtained in our study must be understood.

Protein biosynthesis needs N to build the amino acid building blocks, whose biosynthesis uses energy (adenosine triphosphate, ATP) and reducing power (NADPH/NADH) both to reduce NO_3 from the soil to NH_4^+ and to synthesize the wide range of necessary enzymes involved in such reduction [40]. The required energy to assimilate N from the soil is provided by the glycolysis of sucrose at both roots and leaves. The triose phosphate resulting from the Calvin Cycle is the precursor of sucrose or other carbohydrates that might subsequently be converted into starch to store potential cell energy. RubisCO plays a crucial role in the carboxylation reaction to assimilate CO_2 molecules in the Calvin Cycle; this protein represents around 40% of total soluble proteins in most leaf plants. The following reduction phase of the Calvin Cycle expends three molecules of NADPH and two molecules of ATP, finally allowing the synthesis of the triose phosphate molecules. Briefly, the creation of amino acids and proteins is an expensive biosynthetic pathway that needs a huge amount of ATP, carbohydrates, and several enzymes. However, BS-treated plants may bypass these metabolic processes, as amino acids are already available for the plant with an income of the BS product. The available L-α-amino acids can be translocated and used directly, reducing the need for RubisCO and other soluble proteins involved in these physiological processes which provokes a plant's energy saving that can be used in other physiological processes. In our study, BS treatment in lettuce plants allowed the metabolic process to bypass the nitrogen stress, leading to lower protein content.

Analyzing the effects on growth, N stress decreased biomass in both the studied species as N soil availability is directly related to growth parameters and biomass production as is concluded in previous reports in cucumber [41], in tomato [42], and in canola [43]. Regarding BS application, on one side, the lettuce plants showed no effect on total biomass under any of the N availability conditions, indicating a non-limiting nitrogen concentration in this species not even under strong N reduction. However, commercial BM significantly increased due to BS application under NC. Nonetheless, BS application on pepper plants significantly increased the vegetative BM on both N regimes and fruit production under NS. This enhancement can be attributed to the exogenous L-α-amino acids pool provided by the BS which reduced the need for uptaking nitrate and the synthesis of endogenous amino acids and enzymes, conserving energy for other physiological processes. The higher amounts of early fruit set might be increased by the presence of metabolites from the microbial BS optimizing N use and allowing it to be spent to promote fruit development. Likewise, after spraying moringa leaf extract, seaweed extract, and fulvic acid BSs in apple trees, Mosa et al. [44] found improvement in fruit set and fruit yield due to better nutrient availability. Consequently, plants can use the available N more efficiently, promoting plant growth and increasing essential molecules such as chloroplast pigments. This increases the antenna pigments, allowing more light absorption by larger LHC used to enhance photosynthetic processes resulting in a final increase in pepper production. Therefore, the decrease in the Car/Chl ratio after the BS treatment showed that available N was used to increase vegetative growth and fruit production as LHC was fully equipped. Recently, several reviews have highlighted the combination of PGPB with non-PGPB biostimulant ingredients as a powerful tool to enhance growth and production under various abiotic stresses [45–49]. However, there are few studies focusing specifically on nutrient deficiencies (Table 1). Mainly, protein hydrolysates have been shown to increase yield and positively impact photosynthesis-related processes, including the electron transport chain and gas exchange. Additionally, some seaweed extracts or melatonin-based BS have promoted higher yields and contributed to reducing stress markers such as antioxidant enzymes. It is crucial to continue investigating the mechanisms of action for the available BSs on the market, as well as developing new products with the potential to overcome various stresses affecting crops globally.

In the current study, BS composition played a key role in promoting vegetative and reproductive growth due to the presence of metabolites generated by *B. velezensis* such as (1) gibberellins (GAs) which are involved in stem elongation [50]; (2) indole-3-acetamide (IAM), a precursor of indol-3-acetic acid (IAA) which is responsible for controlling several physiological processes such as cell elongation and division and play a pivotal role in plant development of shoots and roots [50,51]; and (3) pantothenic acid (vitamin B5) which promotes vegetative development, among others. When N was less available, the *B. velezensis* strain was performing mutualistic relations with the plant, enhancing nitrogen concentration in plant tissues by N_2 fixation from the atmosphere [20]. Regarding reproductive growth in pepper plants, both fruit yield and fruit setting precocity were enhanced only under N reduction stress. Our results are in accordance with Mesa et al. [37], whose application of a BS based on enzymatically hydrolyzed animal protein combined with low N priming, resulted in a significant increase in fruit production, effectively doubling the yield. Finally, the combined effect of metabolites, such as IAM and GA_4, acted as signaling molecules influencing various aspects of plant growth (in both root and shoot) allowing higher biomass rates [16,52].

Table 1. Effect of biostimulants under nutrient deficiency stress conditions.

Biostimulant		Crop	Application	Stress	Results	References
PHs	Amino acids, soluble sugars, and phenols	Jute	Foliar	Full Nutrient Stress	Increased yield, enhanced CO_2 assimilation, chlorophyll content, and nitrogen distribution	[53]
		Basil	Root	Full Nutrient Stress	Enhanced yield and functional quality attributes	[54]
		Lettuce	Foliar	Nitrogen Stress	Improved yield, chlorophyll content, N-uptake efficiency, and HAA	[55]
		Spinach	Foliar	Nitrogen Stress	Improved yield, chlorophyll content, N-uptake efficiency, and LAA	[55]
	Amino acids	Tomato	Root	Nutrient Stress	Enhanced primary and lateral root growth, by stimulating the biosynthesis of salicylic acid under stress	[56]
		Tomato	Root	Nitrogen Stress	Tendency to improve total fruit production	[37]
	Free L-α-amino acids and organic matter	Lettuce	Root	Nitrogen Stress	Enhanced photosynthesis, total N, and increased biomass production	[31]
Seaweed extract		Lettuce	Foliar	Nitrogen Stress	Increased photochemical efficiency and marketable fresh yield	[57]
		Tomato	Foliar	Nutrient Stress (NPK)	Decreased SOD and POD; maintained fruit quality and yield	[36]
Melatonin		Soybean	Root	Nitrogen Stress	Increased biomass, stem diameter, total leaf area, root nodule number, SOD, GPX and CAT	[58]
		Cucumber	Foliar	Nitrogen Stress	Enhanced activity of enzymes involved in N metabolism and N assimilation efficiency	[59]
Others	Betaines, alginic acid, and kaidrin	Lettuce	Foliar	Nitrogen Stress	Increased yield	[60]

Abbreviations: catalase (CAT); guaiacol peroxidase (GPX); hydrophilic antioxidant activity (HAA); lipophilic antioxidant activity (LAA); peroxidase (POD); protein hydrolysates (PHs); superoxide dismutase (SOD).

4. Materials and Methods

4.1. Plant Material and Experimental Design

Two different species were grown in the greenhouse of the Servei de Camps Experimentals (University of Barcelona, Spain) from the 27 February to the 8 May 2024. The selected model plant species were (1) a horticultural leafy species represented by the Little Gem lettuce (*Lactuca sativa* L. var. Little Gem Cerler) and (2) a fruit-producing species represented by Padrón pepper (*Capsicum annuum* L. var. Longum). Seedlings from both species were transplanted at the 4–5 true-leaf stage into a substrate composed of peat, perlite, and vermiculite in a 2:1:1 ratio. Both species seedlings were obtained from Semilleros Deitana S.L. (Murcia, Spain). During the experiment, temperature and relative humidity inside the greenhouse were recorded (Figure S1).

The trial was designed to assess the application or absence of a PGPB biostimulant: Terra-Sorb® Symbiotic (TSS), combined with two nitrogen regimes. The plants were irrigated with Hoagland Solution (HS) following the nutrient recipe developed by Hoagland and Arnon (1938) [61] adapted to our studied species and using different contents on nitrogen compounds (Table S1). The nitrogen contents in both nitrogen regimes were: (1) Nitrogen Control (NC) with 8 mmol N L^{-1}; (2) nitrogen stress (NS) with 2.4 mmol N L^{-1}. Moreover, half of each plant species, lettuces and peppers, received radicular Terra-Sorb® Symbiotic biostimulant (+BS) application (as indicated in Figure S1). This design resulted in a total of four treatment groups per species (NC, NC + BS, NS, and NS + BS). To ensure statistical robustness, each treatment group was replicated in 8 different plants randomly distributed throughout the greenhouse (n = 8).

The strain PH-023 has been chosen among extremophile bacteria for its great growth capacity and development in difficult soil conditions for agriculture. PH-023 strain can fix atmospheric nitrogen, produce siderophores, synthesize biosurfactants, and promote the production of phytohormones, especially auxins. PH-023 is produced by fermentation technology using partially as a nitrogen source the same carrier, a hydrolyzed protein obtained through enzymatic hydrolysis, where the bacteria are formulated. The product contains amino acids, the bacteria PH-023, as well as the metabolites synthesized through the fermentation process. Specifically, Terra-Sorb® Symbiotic contains 6% w/w of L-α-free amino acids, 10^7 cfu/mL PH-023, and 8% w/w of organic matter with a density of 1.12 kg L^{-1} and a pH of 5.0. Finally, the aminogram of the product (in g kg^{-1}) is as follows: Asp (4.3), Ser (3.2), Glu (5.9), Gly (5.8), His (1.7), Arg (3.2), Thr (3.6), Ala (3.9), Pro (2.8), Cis (0.2), Tyr (3.6), Val (4.0), Met (3.8), Lys (3.9), Ile (3.0), Leu (5.3), Phe (2.9), and Trp (0.5).

All the treatments were applied through root irrigation to the substrate at intervals of 10 to 12 days throughout the experimental period. Therefore, lettuce, with a growth duration of 45 days, received a total of 4 BS applications, while pepper, with a growth duration of 65 days, received 6 applications (Figure S1). The dose was adjusted based on the planting frame: lettuce plants received 0.13 mL BS/plant and pepper plants received 1 mL BS/plant. Therefore, as Terra-Sorb® Symbiotic provides 2.5 mmol N L^{-1}, the extra nitrogen supplied to each treatment due to BS application were as follows: (1) in lettuce plants 1.3 μmol N/plant; (2) in pepper plants 15 μmol N/plant.

4.2. Growth Quantification

Lettuce growth was quantified weekly (from 1st to 5th week) by number of mature leaves and height. Moreover, at harvest, diameter, total biomass (BM), and commercial biomass in fresh weight (FW) were assessed. The pepper plants were monitored weekly depending on their developmental stage. Along with the experiment, plant height was measured. During the first stage of plant development (from the 1st to 7th week), the number of mature (fully developed) and developing leaves was quantified. When the reproductive stage began (from the 4th week), the number of flower buds and open flowers was also counted, continuing until the end of the experiment. In the 8th week, fruit set began and the number of just sprouted peppers was counted and classified based on size, distinguishing between those larger and smaller than 1 cm. At the end of the trial, the total BM of the vegetative tissue of each plant was measured, and the harvested fruits were classified depending on their developmental stage: (1) 'just sprouted' peppers (length $\leq$ 1 cm); (2) 'in development' peppers (length $1 < x \geq 3$); (3) 'developed' peppers (length $\geq$ 3 cm and Ø $\geq$ 1.5 cm). To assess the pepper production across the four treatments, the weight for every group was measured, and all the fruits were counted.

4.3. Photosynthesis Related Parameters

At the end of the trial, a MINI-PAM-II modulated pulse fluorimeter combined with a porometer (Heinz Walz GmbH, Effeltrich, Germany) was used to measure stomatal conductance (g_s) and chlorophyll fluorescence. Measurements under light conditions were used to quantify g_s, the relative efficiency of PSII (Φ_{PSII}), and to obtain the F'_m for non-photochemical quenching (NPQ) calculation. Afterward, leaves were dark-adapted for more than 20 min, and measurements of maximum efficiency of PSII (F_v/F_m) were taken in the dark. Finally, NPQ was calculated as $(F_m - F'_m)/F'_m$ [62].

Photosynthetic pigments were measured at the end of the trial on fully developed lettuce and pepper leaves, which were immediately frozen into liquid nitrogen and stored at −80 °C until analysis. To determine the chlorophyll *a* (Chl*a*), chlorophyll *b* (Chl*b*), and total carotenoids (Car), 100 mg of lettuce frozen leaves and 50 mg of pepper frozen leaves were ground to powder. Extraction was performed by adding 1 mL of cold pure methanol. After the vortex, the extracts were ultrasonicated (Bransonic ultrasonic bath 2800, Emerson Industrial, Ferguson, MO, USA). The supernatant was collected, and the pellet was re-extracted once with the same procedure, pooling at the end both the supernatants to proceed with measurements. The absorbance of the diluted extracts (1:4 for lettuce and 1:5 for pepper) was measured at wavelengths of 750, 665.2, 652.4, and 470 nm using a UV-visible spectrophotometer (Cecil Aquarius CE7400, Cecil Instruments, Cambridge, UK) according to Lichtenthaler [63].

Moreover, total chlorophyll (Chl) content was estimated weekly with a SPAD-502 Plus chlorophyll meter (Konica Minolta Business Solutions Europe GmbH, Chicago, IL, USA). Measurements were made in three fully developed leaves per plant and the average was used as individual value.

4.4. Total Soluble Proteins

The total soluble protein content in fully developed leaves was determined with 100 mg of frozen lettuce leaves and 75 mg of frozen pepper leaves. Plant tissue was ground to powder, and the extraction of soluble proteins was performed using 1 mL of phosphate-buffered saline and vortexing. The extracts were centrifuged, and the supernatant was collected. The absorbance of 50 μL supernatant reacted with the Bradford solution was measured at 595 nm using the same UV-Vis spectrophotometer (UV/VIS3000, DINKO Instruments, Barcelona, Spain), and the protein content was determined using a standard curve performed with lyophilized bovine serum albumin (BSA).

4.5. Statistical Analysis

Program IMB SPSS Statistics 29 for Macintosh (SPSS Inc, Chicago, IL, USA) was used for statistical analysis. An analysis of the variance (ANOVA) was used to compare mean values and set differences between the nitrogen stress treatments and differences between the BS application treatments. The significance level was set at 95% ($p < 0.05$). Significantly different means were marked with different letters. The SigmaPlot 10.0 (Systat software, Palo Alto, CA, USA) software was used to create all the figures in this publication.

5. Conclusions

Worldwide food demand is increasing, while at the same time, high amounts of synthetic N fertilizers used in agriculture are polluting our ecosystems. Fertilizer reduction to avoid environmental drawbacks provokes plant stress with the consequent severe productivity drop. Thus, environment-friendly tools must be implemented in agriculture regulation policies. In this study, our results proved that microbial BS benefits help to maintain or even improve production. The two model horticultural species demonstrated that nitrogen stress reduced biomass and productivity, but different modes of action were used after BS application: (1) on one side, pepper plants growing under nitrogen stress, showed a significant increase in vegetative growth (BM and height) and reproductive parameters (production and fruit setting); (2) on the other side, lettuce commercial biomass was enhanced only when plants were under optimal N availability. Therefore, the plants treated with microbial BS showed higher commercial outputs, both commercial biomass in lettuce and enhanced production of all the different pepper development stages. As previously suggested, the metabolites of the microbial BS acted as growth promoters, enhancing plant development and the plant's metabolic efficiency. In conclusion, symbiotic BS will provide a sustainable solution to face food production in the near future and counteract the stress caused by reduced fertilization, specifically the reduction in synthetic nitrogen sources due to current regulations and trends promoted by society.

Supplementary Materials: The following supporting information can be downloaded at https://www.mdpi.com/article/10.3390/plants14071087/s1, Table S1: Macronutrient and micronutrient composition of fertigation solution; Figure S1: Temperature and relative humidity in the greenhouse during the experiment; Table S2 and Figure S2: Chlorophyll content of pepper plants assessed weekly; Table S3 and Figure S3: Height of pepper plants assessed weekly; Table S4 and Figure S4: Number of 'in development leaves' (A) and 'mature leaves' (B) of pepper plants assessed weekly; Table S5 and Figure S5: Number of flower buds (A) and developed flowers (B) of pepper plants assessed weekly; Table S6 and Figure S6: Number of just sprouted pepper fruits smaller or equal (A) to 1 cm or bigger (B) than 1 cm assessed weekly; Table S7 and Figure S7: Chlorophyll content of lettuce plants assessed weekly; Table S8 and Figure S8: Height (A) and number of Mature Leaves (B) of the lettuce plants assessed weekly.

Author Contributions: Conceptualization: L.U.-M., N.S.-S., C.M. and M.P.-M.; methodology: N.S.-S., C.M. and M.P.-M.; validation: L.U.-M. and M.P.-M.; formal analysis: L.U.-M. and M.P.-M.; investigation: L.U.-M. and M.P.-M.; resources: N.S.-S., C.M. and M.P.-M.; data curation: L.U.-M.; writing—original draft preparation: L.U.-M., N.S.-S. and M.P.-M.; writing—review and editing: L.U.-M., N.S.-S., C.M. and M.P.-M.; visualization: L.U.-M., N.S.-S., C.M. and M.P.-M.; supervision: N.S.-S., C.M. and M.P.-M.; project administration: L.U.-M., N.S.-S. and M.P.-M.; funding acquisition: N.S.-S. and M.P.-M. All authors have read and agreed to the published version of the manuscript.

Funding: This research received no external funding.

Data Availability Statement: Data will be made available upon request.

Acknowledgments: We want to acknowledge Josep Matas, Francesc Prenyanosa, and Xavier Aranda (Servei de Camps Experimentals, University of Barcelona) for their technical assistance during the plant growing period. We would also like to thank the Department of Evolutive Biology, Ecology and Environmental Sciences at the University of Barcelona for facilitating the use of the measuring equipment.

Conflicts of Interest: Authors Nuria Sierras Serra and Cándido Marín were employed by the company Bioiberica, S.A.U. The remaining authors declare that the research was conducted in the absence of any commercial or financial relationships that could be construed as a potential conflict of interest.

Abbreviations

The following abbreviations are used in this manuscript:

ATP	adenosine triphosphate
BM	total biomass
BS	biostimulants
Car	total carotenoids
Chl*a*	chlorophyll *a*
Chl*b*	chlorophyll *b*
ETR	electron transport rate
FW	fresh weight
F_v/F_m	maximum efficiency of PSII
g_s	stomatal conductance
LHC	light-harvesting complex
NC	Nitrogen Control Treatment
NPQ	non-photochemical quenching
NS	Nitrogen Deficiency Stress Treatment
PGPB	Plant Growth-Promoting Bacteria
PHs	protein hydrolysates
ROS	reactive oxygen species
Φ_{PSII}	relative efficiency of PSII

References

1. Larramendy, M.; Soloneski, S. *Organic Fertilizers History, Production and Applications*; IntechOpener: London, UK, 2019; pp. 23–25.
2. Dar, G.H.; Bhat, R.A.; Mehmood, M.A.; Hakeem, K.R. (Eds.) *Microbiota and Biofertilizers, Vol 2: Ecofriendly Tools for Reclamation of Degraded Soil Environs*; Springer International Publishing: Cham, Switzerland, 2021. [CrossRef]
3. European Environment Agency. Food in a Green Light—A Systems Approach to Sustainable Food. Available online: https://www.eea.europa.eu/publications/food-in-a-green-light (accessed on 15 April 2024).
4. European Commission. Farm to Fork Strategy. Available online: https://food.ec.europa.eu/horizontal-topics/farm-fork-strategy_en (accessed on 13 April 2024).
5. de Bang, T.C.; Husted, S.; Laursen, K.H.; Persson, D.P.; Schjoerring, J.K. The Molecular–Physiological Functions of Mineral Macronutrients and Their Consequences for Deficiency Symptoms in Plants. *New Phytol.* **2021**, *229*, 2446–2469. [CrossRef] [PubMed]
6. Mu, X.; Chen, Y. The Physiological Response of Photosynthesis to Nitrogen Deficiency. *Plant Physiol. Biochem.* **2021**, *158*, 76–82. [CrossRef]
7. Burgos, S. *Advancing Climate Action on Agriculture and Food Security*; FAO: Rome, Italy, 2024; pp. 1–5.
8. Hereu-Morales, J.; Vinardell, S.; Valderrama, C. Towards Climate Neutrality in the Spanish N-Fertilizer Sector: A Study Based on Radiative Forcing. *Sci. Total Environ.* **2024**, *946*, 174131. [CrossRef] [PubMed]
9. Bonini, P.; Rouphael, Y.; Miras-Moreno, B.; Lee, B.; Cardarelli, M.; Erice, G.; Cirino, V.; Lucini, L.; Colla, G. A Microbial-Based Biostimulant Enhances Sweet Pepper Performance by Metabolic Reprogramming of Phytohormone Profile and Secondary Metabolism. *Front. Plant Sci.* **2020**, *11*, 567388. [CrossRef]
10. Bulgari, R.; Franzoni, G.; Ferrante, A. Biostimulants Application in Horticultural Crops under Abiotic Stress Conditions. *Agronomy* **2019**, *9*, 306. [CrossRef]
11. Du Jardin, P. Plant Biostimulants: Definition, Concept, Main Categories and Regulation. *Sci. Hortic.* **2015**, *196*, 3–14. [CrossRef]
12. Sierras, N.; Botta, A.; Staasing, L.; Martinez, M.J.; Bru, R. Understanding the Effect of Amino Acids Based Biostimulant by an Enantiomeric Analysis of Their Active Principles and a Proteomic Profiling Approach. *Acta Hortic.* **2016**, *1148*, 93–100. [CrossRef]
13. Magnabosco, P.; Masi, A.; Shukla, R.; Bansal, V.; Carletti, P. Advancing the Impact of Plant Biostimulants to Sustainable Agriculture through Nanotechnologies. *Chem. Biol. Technol. Agric.* **2023**, *10*, 117. [CrossRef]
14. Colantoni, A.; Recchia, L.; Bernabei, G.; Cardarelli, M.; Rouphael, Y.; Colla, G. Analyzing the Environmental Impact of Chemically-Produced Protein Hydrolysate from Leather Waste vs. Enzymatically-Produced Protein Hydrolysate from Legume Grains. *Agriculture* **2017**, *7*, 62. [CrossRef]
15. Acin-Albiac, M.; García-Jiménez, B.; Marín Garrido, C.; Borda Casas, E.; Velasco-Alvarez, J.; Serra, N.S.; Acedo, A. Lettuce Soil Microbiome Modulated by an L-α-Amino Acid-Based Biostimulant. *Agriculture* **2023**, *13*, 344. [CrossRef]
16. Malécange, M.; Sergheraert, R.; Teulat, B.; Mounier, E.; Lothier, J.; Sakr, S. Biostimulant Properties of Protein Hydrolysates: Recent Advances and Future Challenges. *Int. J. Mol. Sci.* **2023**, *24*, 9714. [CrossRef] [PubMed]

17. Carillo, P.; Colla, G.; Fusco, G.M.; Dell'Aversana, E.; El-Nakhel, C.; Giordano, M.; Pannico, A.; Cozzolino, E.; Mori, M.; Reynaud, H.; et al. Morphological and Physiological Responses Induced by Protein Hydrolysate-Based Biostimulant and Nitrogen Rates in Greenhouse Spinach. *Agronomy* **2019**, *9*, 450. [CrossRef]
18. Colla, G.; Nardi, S.; Cardarelli, M.; Ertani, A.; Lucini, L.; Canaguier, R.; Rouphael, Y. Protein Hydrolysates as Biostimulants in Horticulture. *Sci. Hortic.* **2015**, *196*, 28–38. [CrossRef]
19. Miljaković, D.; Marinković, J.; Balešević-Tubić, S. The Significance of *Bacillus* spp. in Disease Suppression and Growth Promotion of Field and Vegetable Crops. *Microorganisms* **2020**, *8*, 1037. [CrossRef]
20. Etesami, H.; Jeong, B.R.; Glick, B.R. Potential Use of *Bacillus* spp. as an Effective Biostimulant against Abiotic Stresses in Crops—A Review. *Curr. Res. Biotechnol.* **2023**, *5*, 100128. [CrossRef]
21. Rabbee, M.F.; Ali, M.S.; Choi, J.; Hwang, B.S.; Jeong, S.C.; Baek, K. *Bacillus velezensis*: A Valuable Member of Bioactive Molecules within Plant Microbiomes. *Molecules* **2019**, *24*, 1046. [CrossRef]
22. Castillo, P.; Escalante, M.; Gallardo, M.; Alemano, S.; Abdala, G. Effects of Bacterial Single Inoculation and Co-Inoculation on Growth and Phytohormone Production of Sunflower Seedlings under Water Stress. *Acta Physiol. Plant.* **2013**, *35*, 2299–2309. [CrossRef]
23. da Silva, M.S.R.A.; Dos Santos, B.M.S.; da Silva, C.S.R.A.; da Silva, C.S.R.A.; Antunes, L.F.S.; Dos Santos, R.M.; Santos, C.H.B.; Rigobelo, E.C. Humic Substances in Combination With Plant Growth-Promoting Bacteria as an Alternative for Sustainable Agriculture. *Front Microbiol.* **2021**, *12*, 719653. [CrossRef]
24. Barnawal, D.; Maji, D.; Bharti, N.; Chanotiya, C.S.; Kalra, A. ACC Deaminase-Containing *Bacillus subtilis* Reduces Stress Ethylene-Induced Damage and Improves Mycorrhizal Colonization and Rhizobial Nodulation in Trigonella Foenum-Graecum Under Drought Stress. *J. Plant Growth Regul.* **2013**, *32*, 809–822. [CrossRef]
25. Consentino, B.B.; Vultaggio, L.; Sabatino, L.; Ntatsi, G.; Rouphael, Y.; Bondì, C.; De Pasquale, C.; Guarino, V.; Iacuzzi, N.; Capodici, G.; et al. Combined Effects of Biostimulants, N Level and Drought Stress on Yield, Quality and Physiology of Greenhouse-Grown Basil. *Plant Stress* **2023**, *10*, 100268. [CrossRef]
26. Consentino, B.B.; Aprile, S.; Rouphael, Y.; Ntatsi, G.; De Pasquale, C.; Iapichino, G.; Alibrandi, P.; Sabatino, L. Application of PGPB Combined with Variable N Doses Affects Growth, Yield-Related Traits, N-Fertilizer Efficiency and Nutritional Status of Lettuce Grown under Controlled Condition. *Agronomy* **2022**, *12*, 236. [CrossRef]
27. Martinez Esteso, M.J.; Vilella-Antón, M.; Sellés-Marchart, S.; Martínez-Márquez, A.; Botta-Catala, A.; Piñol-Dastis, R.; Bru-Martinez, R. A DIGE Proteomic Analysis of Wheat Flag Leaf Treated with TERRA-SORB® Foliar, a Free Amino Acid High Content Biostimulant. *J. Integr. OMICS* **2016**, *6*, 188. [CrossRef]
28. Beckman, J. Economic and Food Security Impacts of Agricultural Input Reduction Under the European Union Green Deal's Farm to Fork and Biodiversity Strategies. 2020. Available online: https://ageconsearch.umn.edu/record/307277/?v=pdf (accessed on 13 April 2024).
29. Kalaji, H.M.; Jajoo, A.; Oukarroum, A.; Brestic, M.; Zivcak, M.; Samborska, I.A.; Cetner, M.D.; Łukasik, I.; Goltsev, V.; Ladle, R.J. Chlorophyll a Fluorescence as a Tool to Monitor Physiological Status of Plants under Abiotic Stress Conditions. *Acta Physiol. Plant.* **2016**, *38*, 102. [CrossRef]
30. Shewangizaw, B.; Kassie, K.; Assefa, S.; Lemma, G.; Gete, Y.; Getu, D.; Getanh, L.; Shegaw, G.; Manaze, G. Tomato Yield, and Water Use Efficiency as Affected by Nitrogen Rate and Irrigation Regime in the Central Low Lands of Ethiopia. *Sci. Rep.* **2024**, *14*, 13307. [CrossRef]
31. Navarro-León, E.; López-Moreno, F.J.; Borda, E.; Marín, C.; Sierras, N.; Blasco, B.; Ruiz, J.M. Effect of L-amino Acid-based Biostimulants on Nitrogen Use Efficiency (NUE) in Lettuce Plants. *J. Sci. Food Agric.* **2022**, *102*, 7098–7106. [CrossRef]
32. Attar, A.Z.; Ahmed, T.; Kato, A.; Saadaoui, I.; Shabala, S. Understanding Impact of Heat, Drought, and Salinity Stresses on Growth and Physiological Attributes of Chenopodium Album under Field Conditions. *Plant Growth Regul.* **2023**, *100*, 107–118. [CrossRef]
33. Maxwell, K.; Johnson, G.N. Chlorophyll Fluorescence—A Practical Guide. *J. Exp. Bot.* **2000**, *51*, 659–668. [CrossRef]
34. Pintó-Marijuan, M.; Munné-Bosch, S. Photo-Oxidative Stress Markers as a Measure of Abiotic Stress-Induced Leaf Senescence: Advantages and Limitations. *J. Exp. Bot.* **2014**, *65*, 3845–3857. [CrossRef]
35. Huang, Z.-A.; Jiang, D.-A.; Yang, Y.; Sun, J.-W.; Jin, S.-H. Effects of Nitrogen Deficiency on Gas Exchange, Chlorophyll Fluorescence, and Antioxidant Enzymes in Leaves of Rice Plants. *Photosynthetica* **2004**, *42*, 357–364. [CrossRef]
36. Koleška, I.; Hasanagić, D.; Todorović, V.; Murtić, S.; Klokić, I.; Parađiković, N.; Kukavica, B. Biostimulant Prevents Yield Loss and Reduces Oxidative Damage in Tomato Plants Grown on Reduced NPK Nutrition. *J. Plant Interact.* **2017**, *12*, 209–218. [CrossRef]
37. Mesa, T.; Polo, J.; Casadesús, A.; Gómez, Í.; Munné-Bosch, S. Application of a Biostimulant (Pepton) Based in Enzymatic Hydrolyzed Animal Protein Combined with Low Nitrogen Priming Boosts Fruit Production Without Negatively Affecting Quality in Greenhouse-Grown Tomatoes. *Front. Plant Sci.* **2022**, *13*, 828267. [CrossRef] [PubMed]
38. Gao, F.; Guo, J.; Shen, Y. Advances from Chlorophyll Biosynthesis to Photosynthetic Adaptation, Evolution and Signaling. *Plant Stress* **2024**, *12*, 100470. [CrossRef]

39. Irani, H.; ValizadehKaji, B.; Naeini, M.R. Biostimulant-Induced Drought Tolerance in Grapevine Is Associated with Physiological and Biochemical Changes. *Chem. Biol. Technol. Agric.* **2021**, *8*, 5. [CrossRef]
40. Anas, M.; Liao, F.; Verma, K.K.; Sarwar, M.A.; Mahmood, A.; Chen, Z.-L.; Li, Q.; Zeng, X.-P.; Liu, Y.; Li, Y.-R. Fate of Nitrogen in Agriculture and Environment: Agronomic, Eco-Physiological and Molecular Approaches to Improve Nitrogen Use Efficiency. *Biol. Res.* **2020**, *53*, 47. [CrossRef]
41. Zargar Shooshtari, F.; Souri, M.K.; Hasandokht, M.R.; Jari, S.K. Glycine Mitigates Fertilizer Requirements of Agricultural Crops: Case Study with Cucumber as a High Fertilizer Demanding Crop. *Chem. Biol. Technol. Agric.* **2020**, *7*, 19. [CrossRef]
42. Ronga, D.; Parisi, M.; Pentangelo, A.; Mori, M.; Di Mola, I. Effects of Nitrogen Management on Biomass Production and Dry Matter Distribution of Processing Tomato Cropped in Southern Italy. *Agronomy* **2019**, *9*, 855. [CrossRef]
43. Evans, C.C.; Qaderi, M.M. Supplemental Nitrogen Alleviates the Negative Effects of Higher Temperature on the Vegetative Growth of Canola Regardless of Carbon Dioxide Concentration. *Plant Stress* **2024**, *13*, 100521. [CrossRef]
44. Mosa, W.F.A.; Sas-Paszt, L.; Głuszek, S.; Górnik, K.; Anjum, M.A.; Saleh, A.A.; Abada, H.S.; Awad, R.M. Effect of Some Biostimulants on the Vegetative Growth, Yield, Fruit Quality Attributes and Nutritional Status of Apple. *Horticulturae* **2023**, *9*, 32. [CrossRef]
45. Franzoni, G.; Cocetta, G.; Prinsi, B.; Ferrante, A.; Espen, L. Biostimulants on Crops: Their Impact under Abiotic Stress Conditions. *Horticulturae* **2022**, *8*, 189. [CrossRef]
46. Castiglione, A.M.; Mannino, G.; Contartese, V.; Bertea, C.M.; Ertani, A. Microbial Biostimulants as Response to Modern Agriculture Needs: Composition, Role and Application of These Innovative Products. *Plants* **2021**, *10*, 1533. [CrossRef]
47. Kaushal, P.; Ali, N.; Saini, S.; Pati, P.K.; Pati, A.M. Physiological and Molecular Insight of Microbial Biostimulants for Sustainable Agriculture. *Front. Plant Sci.* **2023**, *14*, 1041413. [CrossRef] [PubMed]
48. Van Oosten, M.J.; Pepe, O.; De Pascale, S.; Silletti, S.; Maggio, A. The Role of Biostimulants and Bioeffectors as Alleviators of Abiotic Stress in Crop Plants. *Chem. Biol. Technol. Agric.* **2017**, *4*, 5. [CrossRef]
49. Di Sario, L.; Boeri, P.; Matus, J.T.; Pizzio, G.A. Plant Biostimulants to Enhance Abiotic Stress Resilience in Crops. *Int. J. Mol. Sci.* **2025**, *26*, 1129. [CrossRef]
50. Vega-Celedón, P.; Canchignia Martínez, H.; González, M.; Seeger, M. Review Biosynthesis of indole-3-acetic acid and plant growth promoting by bacteria. *Cultiv. Trop* **2016**, *37*, 33–39. [CrossRef]
51. Ertani, A.; Sambo, P.; Nicoletto, C.; Santagata, S.; Schiavon, M.; Nardi, S. The Use of Organic Biostimulants in Hot Pepper Plants to Help Low Input Sustainable Agriculture. *Chem. Biol. Technol. Agric.* **2015**, *2*, 11. [CrossRef]
52. Colla, G.; Rouphael, Y.; Canaguier, R.; Svecova, E.; Cardarelli, M. Biostimulant Action of a Plant-Derived Protein Hydrolysate Produced through Enzymatic Hydrolysis. *Front. Plant Sci.* **2014**, *5*, 448. [CrossRef]
53. Carillo, P.; Colla, G.; El-Nakhel, C.; Bonini, P.; D'Amelia, L.; Dell'Aversana, E.; Pannico, A.; Giordano, M.; Sifola, M.I.; Kyriacou, M.C.; et al. Biostimulant Application with a Tropical Plant Extract Enhances Corchorus olitorius Adaptation to Sub-Optimal Nutrient Regimens by Improving Physiological Parameters. *Agronomy* **2019**, *9*, 249. [CrossRef]
54. Ciriello, M.; Formisano, L.; Kyriacou, M.C.; Colla, G.; Graziani, G.; Ritieni, A.; De Pascale, S.; Rouphael, Y. Biostimulatory Action of Vegetal Protein Hydrolysate Compensates for Reduced Strength Nutrient Supply in a Floating Raft System by Enhancing Performance and Qualitative Features of "Genovese" Basil. *Front. Plant Sci.* **2022**, *13*, 906686. [CrossRef]
55. Di Mola, I.; Cozzolino, E.; Ottaiano, L.; Nocerino, S.; Rouphael, Y.; Colla, G.; El-Nakhel, C.; Mori, M. Nitrogen Use and Uptake Efficiency and Crop Performance of Baby Spinach (*Spinacia oleracea* L.) and Lamb's Lettuce (*Valerianella locusta* L.) Grown under Variable Sub-Optimal N Regimes Combined with Plant-Based Biostimulant Application. *Agronomy* **2020**, *10*, 278. [CrossRef]
56. Casadesús, A.; Pérez-Llorca, M.; Munné-Bosch, S.; Polo, J. An Enzymatically Hydrolyzed Animal Protein-Based Biostimulant (Pepton) Increases Salicylic Acid and Promotes Growth of Tomato Roots Under Temperature and Nutrient Stress. *Front. Plant Sci.* **2020**, *11*, 953. [CrossRef]
57. Di Mola, I.; Cozzolino, E.; Ottaiano, L.; Giordano, M.; Rouphael, Y.; Colla, G.; Mori, M. Effect of Vegetal- and Seaweed Extract-Based Biostimulants on Agronomical and Leaf Quality Traits of Plastic Tunnel-Grown Baby Lettuce under Four Regimes of Nitrogen Fertilization. *Agronomy* **2019**, *9*, 571. [CrossRef]
58. Wang, H.; Ren, C.; Cao, L.; Jin, X.; Wang, M.; Zhang, M.; Zhao, Q.; Li, H.; Zhang, Y.; Yu, G. The Mechanisms Underlying Melatonin Improved Soybean Seedling Growth at Different Nitrogen Levels. *Funct. Plant Biol.* **2021**, *48*, 1225–1240. [CrossRef] [PubMed]
59. Xu, Y.; Xu, R.; Li, S.; Ran, S.; Wang, J.; Zhou, Y.; Gao, H.; Zhong, F. The Mechanism of Melatonin Promotion on Cucumber Seedling Growth at Different Nitrogen Levels. *Plant Physiol. Biochem.* **2024**, *206*, 108263. [CrossRef]
60. Amanda, A.; Ferrante, A.; Valagussa, M.; Piaggesi, A. Effect of Biostimulants on Quality of Baby Leaf Lettuce Grown Under Plastic Tunnel. *Acta Hortic.* **2009**, *807*, 407–412. [CrossRef]
61. Hoagland, D.R.; Arnon, D.I. The Water Culture Method for Growing Plants Without Soil. *Circular* **1938**, *347*, 32.

62. Ruban, A.V. Nonphotochemical Chlorophyll Fluorescence Quenching: Mechanism and Effectiveness in Protecting Plants from Photodamage. *Plant Physiol.* **2016**, *170*, 1903–1916. [CrossRef]
63. Lichtenthaler, H. Chlorophylls and Carotenoids: Pigments of Photosynthetic Biomembranes. *Methods Enzymol.* **1987**, *148C*, 350–382. [CrossRef]

plants

MDPI

Article

Transcriptional Profiling to Assess the Effects of Biological Stimulant Atlanticell Micomix on Tomato Seedlings Under Salt Stress

María Salud Justamante [1,†], Eduardo Larriba [1,†], Ernesto Alejandro Zavala-González [2], Almudena Aranda-Martínez [2] and José Manuel Pérez-Pérez [1,*]

[1] Instituto de Bioingeniería, Universidad Miguel Hernández, 03202 Elche, Spain; mjustamante@umh.es (M.S.J.); elarriba@umh.es (E.L.)
[2] Atlántica Agrícola S.A., 03400 Villena, Spain; ezavala@atlanticaagricola.com (E.A.Z.-G.); aaranda@atlanticaagricola.com (A.A.-M.)
* Correspondence: jmperez@umh.es
† These authors contributed equally to this work.

Abstract: Recent environmental changes in the Mediterranean region, attributable to anthropogenic climate change, present a substantial challenge to the adaptive evaluation of crops and the development of novel improvement strategies. In this study, we established a hydroponic tomato cultivation protocol under in vitro conditions to analyze the transcriptomic profile of seedlings exposed to salinity stress. The study also examined the impact of Atlanticell Micomix, a biological stimulant derived from a mixture of mycorrhizal microorganisms and rhizobacteria, on plant growth and development under standard conditions and in response to moderate salinity. Our transcriptomic analysis indicated a differential effect of biostimulant inoculation compared to the effect induced by salinity stress, involving genes such as *GOX3* or *DIR1*, which are associated with the plant's defense response to adverse conditions. In addition, the presence of a cross-regulatory module between jasmonic acid and auxin, involving potential orthologs of *IAA29* and *JAZ*, was proposed. The application of the biostimulant demonstrated a potential priming effect on the tomato seedlings, which might be useful in reversing the transcriptomic effects caused by salt stress. A comprehensive analysis of the pathways differentially affected by the treatments facilitates further investigation into the mechanisms underlying these effects.

Keywords: biostimulants; differentially expressed genes; moderate salinity; *Solanum lycopersicum*; transcriptome profiling

1. Introduction

Currently, the agricultural sector faces the challenge of increasing food production to meet the demands of a growing global population while minimizing its impact on ecosystems and human health, particularly in the context of an accelerated anthropogenic climate change scenario [1,2]. Climate change has a major impact on global precipitation patterns, which is expected to result in longer drought periods in some regions, such as the countries in the Mediterranean basin [3,4]. This region is especially affected by the salinization of agricultural soils [5,6]. The use of fertilizers and pesticides has been essential to increase yield and ensure crop production [7]. However, new methods and approaches have been proposed to enhance the sustainability of agricultural production by reducing synthetic chemical compounds and helping crops to adapt to challenging abiotic conditions [8].

The use of plant biostimulants is a promising and environmentally friendly innovation that enhances the various processes of plant development, including vegetative growth, flowering, and fruit production. Furthermore, plant biostimulants can enhance nutrient uptake efficiency and improve tolerance to biotic and abiotic stress factors, such as those associated with anthropogenic climate change [9,10]. The development of new biostimulants requires precise testing and confirmation of their effects on the morphological and physiological traits of the crop for which they have been designed. Additionally, a detailed understanding of the mechanism of action of the formulations on the host plant is necessary [11,12]. Several authors have employed evaluation approaches for new biostimulant products based on omics technologies [13–16], which are more efficient and cost-effective. Biostimulants based on microorganisms, mainly fungi and bacteria, are used in agriculture to improve crop quality, production, and stress tolerance. One promising type of biostimulants is formulated with arbuscular mycorrhizal fungi (AMF). These AMF are commonly found in soils and form symbiotic associations with 80% of terrestrial plant species, including most crops [17,18]. These fungi are useful as biofertilizers in organic agriculture due to their ability to improve water and nutrient absorption and increase resistance to abiotic and biotic stresses, particularly in resource-limited environments [19].

In 2023, Spain ranked 9th in tomato worldwide production, with a total volume of 3968 million tons of tomatoes produced on just over 50,090 hectares of land [20]. Recent studies have shown that AMF colonization has positive effects on important agronomic traits in this species, such as fruit quality [21,22] and resistance to aphid infection [23]. It also leads to increased production and better efficiency in responding to biological stress situations. Previous studies have shown that AMF can decrease the accumulation of certain viruses, such as tomato bushy stunt virus and tomato mosaic virus, by upregulating the genes of pathogenesis-related protein in colonized plant tissues [24]. Furthermore, AMF colonization has been observed to increase the endogenous production of compounds derived from alkaloids and fatty acids, resulting in higher mortality rates of herbivorous insects feeding on colonized tomato plants [25,26]. These and other results [27] demonstrate the enhanced disease resistance induced by AMF colonization in tomatoes. The impact of AMF symbiosis on the response to abiotic stresses such as drought, salinity, and high temperatures, has been also studied in this species [28–31]. Tomato plants colonized with AMF showed the differential regulation of some drought response genes in both the roots and leaves during water deficit conditions [30,32]. High salinity stress can reduce growth and water absorption in many crops, including tomatoes [33,34]. Inoculating tomato seedlings with AMF from species such as *Piriformospora indica* [35,36] or *Rhizophagus irregularis* [37] can enhance salt stress tolerance through various mechanisms. These mechanisms include improved nutrient uptake, increased fatty acid and phospholipid contents, and reduced sodium accumulation by regulating the expression of specific genes [35–37]. These findings indicate that inoculation with AMF can enhance plant tolerance and growth under drought and high salinity conditions, as well as activate natural plant defenses. This process is known as priming and is regulated by metabolic and hormonal pathways [38]. To comprehend the effect of AMF inoculation on plant metabolism, it is helpful to analyze changes in the transcriptomic profile of AMF-colonized and non-AMF-colonized plants when exposed to different stressors. Through this approach, several studies have identified the key regulatory genes that control endogenous responses to heavy metals [39], high salinity [40], and low phosphate availability [41]. Recent studies have been conducted to ascertain the impact of plant growth-promoting rhizobacteria (PGPR) with biostimulant properties on the germination and growth of tomato seedlings. These investigations aim to elucidate the role of PGPR in mitigating the deleterious effects of various stresses, including salinity, drought, and elevated temperatures [42–47].

Numerous studies have demonstrated the multiple advantages of AMF and/or PGPR inoculation in tomato plants [48–50], as well as other crops [51,52]. However, their practical implementation in field conditions as biostimulants requires a thorough comprehension of the mechanisms involved in the plant physiological responses during mycorrhization. Identifying key molecular factors is crucial to fully exploiting the potential of these microorganisms as potential biostimulants. In addition, to counteract the harsh environmental conditions driven by current climate change, it is crucial to conduct a comprehensive assessment of the mechanisms that mediate plant adaptive responses, including enhanced nutrient absorption, improved water efficiency, and resistance to biotic and other abiotic stresses. These mechanisms should be evaluated while considering the characteristics of the cultivars of interest and utilizing the most suitable symbiotic combinations. The application of assays using omics technologies, such as transcriptomics or metabolomics, will provide new insights into the mechanisms that mediate the interaction and response of plants to different biostimulants and under different stress conditions.

In this study, we propose the utilization of an in vitro hydroponic culture system for the implementation of transcriptomic studies in tomato seedlings. The experimental design is validated by studying young tomato seedlings subjected to salt stress and control conditions, in the presence and absence of a commercial biostimulant. A comprehensive analysis of the transcriptomic outcomes will ascertain whether the influence of the biostimulant treatment on salinity stress alleviation is attributable to a direct or indirect effect of the biostimulant on the stress response pathways under investigation.

2. Results

2.1. Growth Responses of Young Tomato Seedlings to Atlanticell Micomix and Salt Stress

After 2–3 days of transferring tomato seedlings to the hydroponic medium, hyphae were observed in the nutrient solution only in the treatments containing the Atlanticell Micomix solution, hereafter referred to as the AMM solution (Figure S1 and see Section 4). To confirm the effective establishment of mycorrhizae, the presence of intracellular arbuscules in the root cortex was analyzed in five roots of each of the four treatments after five days of growth in hydroponic culture. These structures were observed only in treatments containing AMM in all cases (Figure 1a). No significant differences were observed between the treatment with AMM and the treatment with AMM and 100 mM NaCl.

Experiments E1 and E2 included a morphological study to characterize the growth response of 8-day-old tomato seedlings grown in vitro under the tested conditions. The seedlings were at the same stage of development with fully expanded cotyledons at stage 10–100 [53] (Figure S1). After five days of hydroponic cultivation, some of the seedlings treated with NaCl appeared yellow and did not grow (Figure 1b). The mortality rate was 32% at 100 mM NaCl and 20% at 50 mM NaCl (Figure 1c). The control group did not experience any adverse effects, and the mortality rate was below 5% in the group treated with the AMM solution. Adding the AMM solution to the saline treatment decreased mortality rates to 16% in 100 mM NaCl and 4% in 50 mM NaCl (Figure 1c). The saline treatment significantly reduced the viability of the tomato seedlings (p-value = 0.000). The presence of AMM had a significant positive effect on viability under stress conditions at 100 mM NaCl (p-value = 0.0036). For experiments E3 and E4, we utilized 13-day-old seedlings that had the first pair of leaf primordia visible at stage 11–101 [53]. To quantitatively determine the impact of different treatments on seedling growth, we used new leaf formation under hydroponic conditions as a growth estimator (Figure 1d,e). A significant increase in the percentage of new leaf formation was observed after 5 days under hydroponic conditions in samples treated with AMM, in both 8-day-old and 13-day-old seedlings. Also, a significant increase in the leaf formation percentage was observed in AMM-treated plants at 100 mM NaCl compared with the same salt stress condition without treatment, while no significant effect of AMM was observed under the 50 mM NaCl condition (Figure 1d).

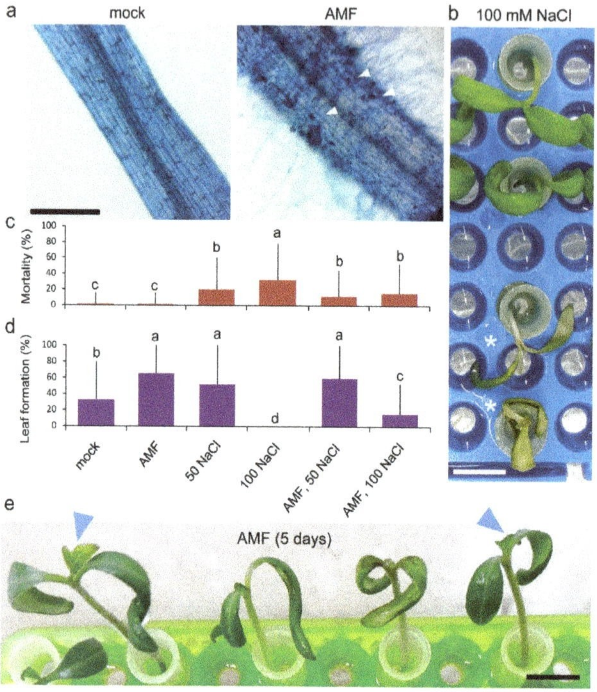

Figure 1. Colonization and shoot morphology of Atlanticell Micomix (AMM)-treated plants under salt stress. (**a**) Representative images of trypan blue-stained roots from *S. lycopersicum* var. 'Rosa de Altea' seedlings grown for 5 days in hydroponics; white arrowheads indicate some intracellular arbuscules in the cortical layer. (**b**) Detailed view of the shoot morphology of salt-treated seedlings; asterisks indicate yellow plants that did not develop further. (**c**) Percentage of seedling mortality in the different conditions assayed. (**d**) Percentage of new leaf formation in the different conditions assayed. Letters in (**c**,**d**) indicate statistically significant differences (p-value < 0.05; Chi-Squared test). (**e**) Representative images of tomato seedlings at hydroponic culture. Blue arrowheads indicate the new leaf formation. Scale bars: (**a**) 0.5 mm, and (**b**,**e**) 5 mm.

To evaluate the effect of the different treatments on root system growth, we measured several parameters, including primary root (PR) length, lateral root (LR) number, and adventitious root (AR) number, in 8-day-old seedlings grown for 5 days in hydroponic conditions (Figure 2a). The combined treatment of AMM solution with 50 or 100 mM NaCl significantly (p-value = 0.0087) reduced PR length, as well as with 50 mM salt stress compared to the other conditions (Figure 2b). A higher rate of LR formation was observed under the 50 or 100 NaCl salt stress conditions compared to the control conditions, while the AMM treatment did not significantly change the average LR number (Figure 2c). This higher rate of LR formation in both salt stress conditions also corresponded with an increased number of LRs produced compared with the control condition (Figure 2d). In the control conditions, we did not observe the appearance of ARs from the hypocotyl, while we found a positive and significant effect of either the AMM solution or salt stress (50 and 100 mM) on AR emergence (p-value = 0.0078). The percentage increase in AR capacity was statistically significant in both AMM solution and salinity conditions (Figure 2e). The AMM treatment promoted the formation of up to five ARs in comparison to the mock conditions, as well as in the 100 mM NaCl salt stress condition or the combination of AMM plus 100 mM NaCl. The number of ARs was highly variable between the samples and conditions, with the highest average value observed with 100 mM NaCl (average ARs = 1.0 ± 0.3) (Figure 2f), despite no individual showing more than six roots.

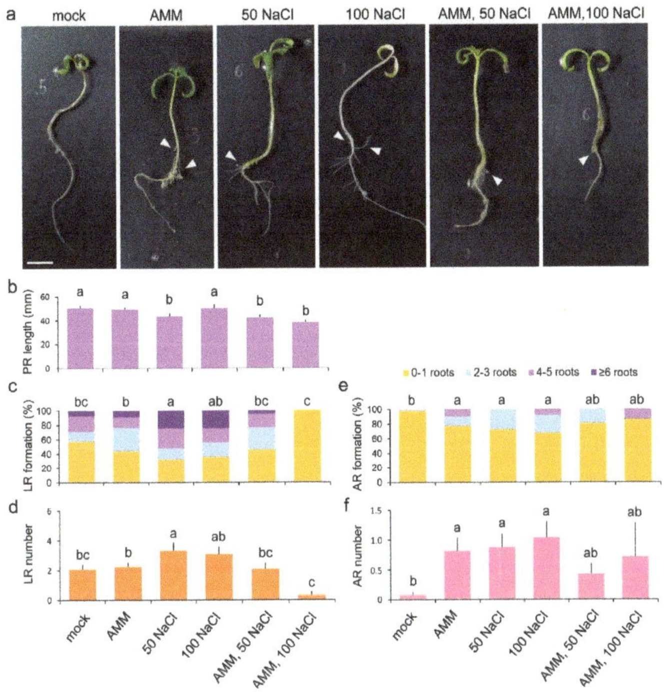

Figure 2. Root morphology of Atlanticell Micomix (AMM)-treated plants under salt stress. (**a**) Representative images of *S. lycopersicum* var. 'Rosa de Altea' seedlings grown for 5 days in hydroponics; white arrowheads indicate some adventitious roots (AR). (**b**) Average primary root (PR) length in the different conditions assayed. (**c**,**e**) Percentage of lateral roots (LR) (**c**) and AR (**e**) formation in the different conditions assayed. (**d**,**f**) Distribution of LR (**d**) and AR number (**f**) in the different conditions assayed. The lines in bar graphs indicate standard deviation. Letters in (**b**–**f**) indicate statistically significant differences (p-value < 0.05; Fisher's least significant difference). Scale bar: 10 mm.

2.2. Gene Expression Analysis in Response to AMM and Moderate Salinity

To elucidate the molecular pathways that are modulated in response to AMM and moderate salinity (50 mM NaCl) in our experimental setup, a transcriptome analysis of polyadenylated RNA in leaves was performed (see Section 4). The results obtained included the expression data of 21,275 protein-coding genes and 908 non-coding RNAs (ncRNAs) (Figure S2a–c and Table S1a,b). Principal component analyses indicated that most of the observed variability in gene expression data between samples was due to the treatment, validating our experimental design (Figure 3a). To further characterize the gene expression profiles, we performed a k-means analysis on the most variable genes (n = 6600), which yielded four distinct expression clusters (Figure S3a). The analysis revealed that genes with an upregulated expression in response to moderate salinity were predominantly distributed across clusters 1 and 3, with low effect of the AMM treatment (Figure S3a). Genes within cluster 2 were found upregulated in response to the AMM treatment and downregulated by moderate salinity, while those in cluster 4 were found downregulated in all three treatments under the mock conditions (Figure S3a). Gene Ontology (GO) enrichment was performed in genes from these four clusters, and the results are indicated in Figure S3b. We identified 3658 differential expressed genes (DEGs) in four different contrasts (Figure 3b and Figure S3c and Table S2), representing 17.2% of all expressed genes. The UpSet plots of DEGs in the different contrasts allowed us to identify either treatment-specific or shared genes between the treatments (Figure 3c).

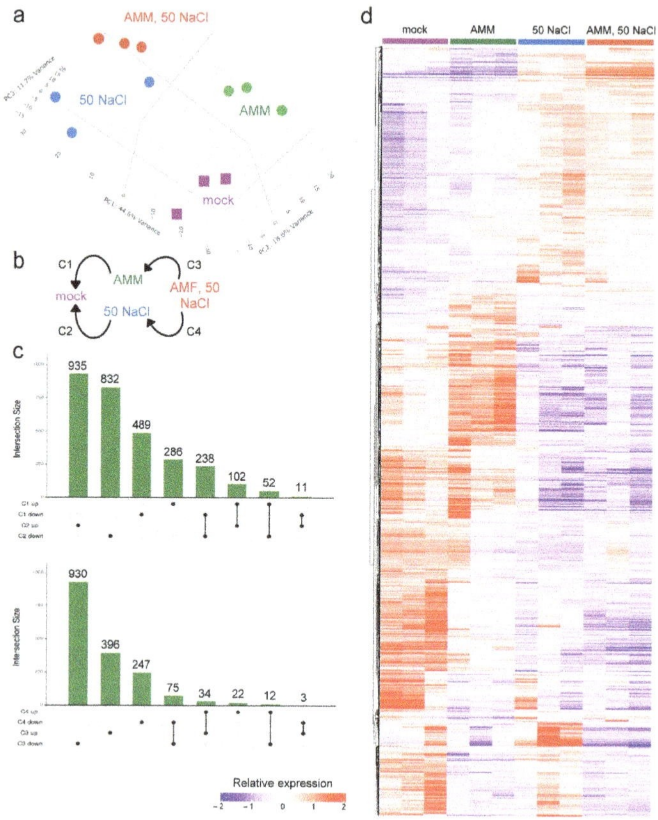

Figure 3. Expression analysis of RNA-Seq data. (**a**) Principal component analysis of the RNA-Seq results from protein coding genes. (**b**) Diagram of the four comparisons (C1 to C4) carried out for the analysis of DEGs. (**c**) UpSet plot of the number of DEGs from indicated contrasts. (**d**) K-means clustering of DEGs. Predicted clusters are grouped according to their expression profile. Expression values are relative and adjusted to −2 and +2.

Among the DEGs that were found upregulated in response to the AMM treatment (C1 contrast), we observed a significant enrichment of GO terms associated with a defense response (Figure 4a and Table S3). Among these genes, we found *Solyc06g070990.3* and *Solyc03g095770.3* (Figure 4b and Table S4a), which encode two transcription factors of the WRKY family [54], *Solyc03g121900.1* encoding the putative apoplastic lipid transfer protein DEFECTIVE IN INDUCED RESISTANCE 1 (Figure 4b and Table S4a) involved in systemic acquired resistance in *Arabidopsis thaliana* [55], and *Solyc10g007600.3* encoding a glycolate oxidase, GOX3 (Figure 4b and Table S4a), that is known to modulate reactive oxygen species (ROS) mediated signal transduction during non-host resistance [56]. It is hypothesized that these genes could regulate the systemic responses to root colonization by the AMF and PGPR present in the AMM solution. Conversely, downregulated genes in the leaves of the

AMM-treated seedlings were associated with a response to oxidative stress, particularly linked to hydrogen peroxide metabolism (Figure 4a and Table S3). It is notable that the expression levels of genes *SlPrx52* (*Solyc08g069040.3*) and *SlPrx102* (*Solyc12g005370.2*), which encode two peroxidases of the class III, were strongly downregulated by the AMM treatment, even in the presence of moderate salinity (Table S1a). These results are consistent with the observed establishment of an effective symbiosis between the AMM and the tomato roots, which can be detected transcriptionally in distant tissues, such as leaves, using some of these specifically deregulated genes as biomarkers.

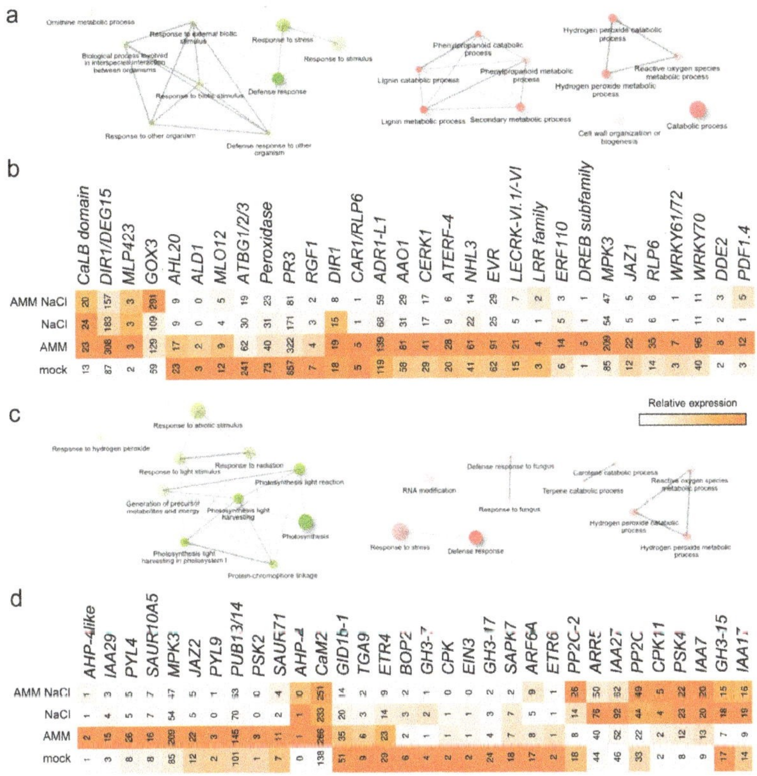

Figure 4. Functional annotation of differentially expressed genes (DEGs). (**a**,**c**) The interaction networks of the different termini in Gene Ontology Biological Process (GO BP) significantly enriched (False discovery rate = 0.05) in the upregulated (green) and downregulated (red) genes in C1 (**a**) and C2 (**c**) contrasts. (**b**,**d**) DEGs involved in defense response (**b**) and phytohormone signaling (**d**). Grey/orange in (**b**,**d**) indicate normalized counts per million (CPM) values in each row, where orange indicates the highest abundance of transcripts. Gene annotations are found in Table S4.

In response to moderate salinity (C2 contrast, cluster 1), upregulated DEGs were significantly enriched for GO terms related to photosynthesis (light-dependent reactions) and response to abiotic stimulus (Figure 4c and Table S3). We identified 82 expressed genes assigned to photosynthesis-related KEGG pathways, and the expression of 44 of these genes (53.7%) was deregulated upon moderate salinity (Table S4b). A significant increase in the expression of key genes in the different photosynthetic complexes was observed in response to the different conditions, which exhibit an additive effect in AMM treatment and salinity (Figure S4 and Table S4b), Interestingly, among the DEGs that were downregulated by moderate salinity, genes related to defense response were also found (Figure 4b). Several of these salt-downregulated genes, including *Solyc06g070990.3* and *Solyc03g095770.3*, encode WRKY transcription factors that were found highly upregulated by the AMM treatment (Figure 4b and Table S4a), suggesting some antagonism between AMM treatment and moderate salinity. Two genes encoding two 9-cis-epoxycarotenoid dioxygenase (NCED) enzymes involved in abscisic acid (ABA) biosynthesis were identified—*NOTABILIS (NOT; SlNCED1; Solyc07g056570.1)* and *SlNCED2 (Solyc08g016720.1)*. These genes showed a significant decrease in expression in response to moderate salinity in the tomato leaves (Figure S4d and Table S4c). In addition, *SlCCD8 (Solyc08g066650.3)*, a gene involved in strigolactone biosynthesis, also showed a significant decrease in expression under moderate salinity conditions in the tomato leaves (Figure S4d and Table S4c).

In this regard, we have annotated 117 genes related to the plant hormone transduction pathway in KEGG (Figure S5 and Table S4c). Of particular interest are *Solyc08g021820.3*, which encodes the IAA29 corepressor of auxin signaling [57], *Solyc10g076410.1*, which encodes the ABA receptor SlPYL10 [58], and *Solyc12g009220.2*, which encodes SlJAZ2 [59]. These genes exhibited a substantial increase in expression in response to AMM and were potently repressed by salt (Figure 4d and Table S4c).

Recent examples have demonstrated the effectiveness of AMF and/or PGPR inoculation of seeds/seedlings in enhancing salt stress tolerance in various crops [60–62]. To ascertain the combined effect of AMM treatment and moderate salinity in our experimental setup, the DEGs obtained in the C3 and C4 contrasts were analyzed in detail (Table S3). It was determined that, among the DEGs that were found to be expressed at higher levels in both contrasts, there was a highly significant enrichment of GO terms related to de novo post-translational protein folding and ROS response (Table S1a).

In order to validate the results obtained in our RNA-Seq, an additional experiment was carried out. This experiment entailed reverse transcription, followed by quantitative PCR (RT-qPCR), utilizing RNA obtained from plant samples that had been frozen (see Section 4). A selection of five DEGs that exhibited differential expression were chosen, including the upregulation in AMM treatment (*IAA29, WRKY53*) or in AMM treatment and moderate salinity (*GLO2*), as well as *Prx52* and *Prx102*, which were downregulated under AMM treatment and moderate salinity conditions (Table S2 and Figure S6a). The results of the RT-qPCR analysis demonstrate a congruent expression pattern with those obtained through RNA-Seq (Figure S6b,c). In addition, a high degree of correlation is observed between both datasets (Figure S6d).

Our transcriptomic analysis suggested that moderate salinity combined with AMM inoculation triggered the expression of several key genes related to the unfolded protein response, which is a cellular response that aims to restore protein homeostasis in the endoplasmic reticulum after prolonged stress [63]. It is noteworthy that this specific response does not appear to be highly activated by AMM treatment or moderate salinity alone. Conversely, a subset of downregulated DEGs in the combination of AMM treatment and moderate salinity were associated with water transport, including some aquaporin-encoding genes, such as *SlTIP2;3 (Solyc06g060760.3)*, *SlTIP1;2 (Solyc06g075650.3)*, and *SlPIP2;4 (Solyc06g011350.3)* (Figure S4d and Table S1a). These findings collectively imply that root inoculation with AMM solution could regulate water loss in a non-cell autonomous manner under moderate salinity conditions, as previously described with tomato plants treated with *Azotobacter chroococcum* [64].

2.3. Regulatory Responses to AMM and Moderate Salinity

In a previous study, we annotated 1940 tomato genes encoding transcription factors (TFs) that were classified into 48 families [65]. In the present study, we identified differential regulations of 254 of the 1326 TF-encoding genes expressed in our experimental setup

(Table S5). Hierarchical clustering and K-means analysis of these DEGs revealed significant disparities in their expression patterns, which were categorized into three clusters (Figure 5a). Cluster 1 comprised 96 transcription factor (TF) genes that were repressed in all conditions tested relative to control conditions. Cluster 2 included 77 TF-encoding genes that increased in expression in samples treated with AMM and decreased in the rest, including AMM and moderate salinity. Cluster 3 had 81 TF-encoding genes that increased under salinity conditions, both with and without AMM (Figure 5a). These findings confirm that AMM treatment does not reverse the transcriptional effect of moderate salinity. The enrichment of these DEGs has been estimated according to the TF family they encode and as a function of the contrast performed (Figure 5b). It is evident that certain TF-encoding families, including ERF, C2H2, MYB, NAC, and bHLH, exhibit members in numerous contrasts with comparable frequencies. Conversely, families such as Dof, WRKY, and bZIP exhibit a concentration of deregulated members in specific contrasts (Figure 5b). It is important to examine the case of TF-encoding DEGs belonging to the WRKY family, where a considerable proportion of these genes exhibit a substantial increase in expression in response to AMM treatment compared to the control conditions. However, the levels of these genes are found to be repressed by saline treatment, irrespective of the AMM treatment (Figure 5c). In contrast, the TF-encoding genes of the bZIP family show a significant increase in expression in response to salinity, except for bZIP62 which shows a pattern like WRKY (Figure 5c).

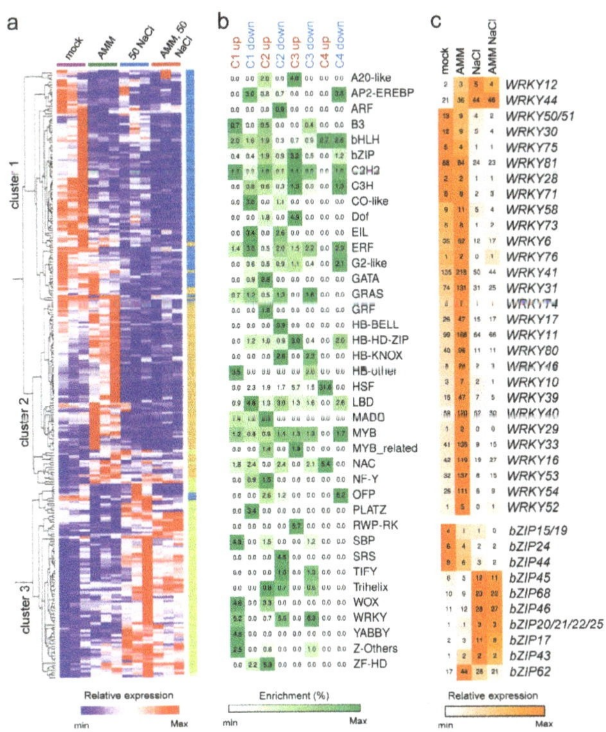

Figure 5. Expression analysis of transcription factor genes. (**a**) K-means clustering of differentially expressed genes (DEGs). Predicted clusters are grouped according to their expression profile. Expression values are relative and adjusted to −2 and +2. (**b**) Fold enrichment of the different transcription factors (TF) families expressed; the intensity of the green color indicates the strength of the fold enrichment (**c**) TF-encoding DEGs belonging to the WRKY and bZIP families. Grey/orange indicate normalized counts per million (CPM) values in each row, where orange indicates the highest abundance of transcripts.

In the present study, we identified 908 non-coding RNA sequences that showed consistent expression according to our experimental design (Figure S2e,f and Table S1b). We found that 78 ncRNAs (8.6% of expressed ncRNAs) exhibited differential expression in some of the treatments tested (Figure S7a), and hierarchical clustering revealed the following dynamic expression patterns: (cluster i) 19 ncRNAs appeared to be downregulated in all treatments (AMM; 50 mM NaCl; AMM and 50 mM NaCl) compared to the mock conditions; (cluster ii) 6 ncRNAs showed downregulation of their expression only under moderate salinity conditions; (cluster iii) 21 ncRNAs exhibited an increase in expression in response to the AMM treatment but not in moderate salinity or AMM plus salinity; and (cluster iv) 24 ncRNAs showed higher expression levels in moderate salinity and AMM plus salinity conditions. A further eight ncRNAs exhibited an atypical expression pattern (Figure S7b). Of the 78 degraded ncRNAs, 37 were classified as antisense, 31 as intergenic and the remaining 10 corresponded to exonic or intronic ncRNA. No significant enrichment in ncRNA type compared to the total expressed ncRNAs was observed. In addition, we did not find any association between specific ncRNA, and the protein coding genes located in the same genetic locus.

Despite the absence of evidence supporting the functionality of these ncRNAs, the stringent regulation of their expression under the conditions examined suggests a potential for involvement in the regulation of gene expression in response to stress. Further research is warranted to investigate this possibility.

3. Discussion

Salt stress represents one of the most pernicious abiotic stresses, exerting deleterious effects on plant growth and productivity. These conditions trigger a series of physiological responses, including stomatal closure and the generation of ROS. Other notable changes encompass metabolic imbalance and photosynthesis limitation, collectively impacting plant growth and development [66]. Salinity is known to limit plant growth and productivity; however, the effect of salinity on tomato growth can be highly variable. In a recent study [67], it was observed that different concentrations of NaCl had different effects on the *S. lycopersicum* cultivar 'Moneymaker' grown in vitro. Specifically, low-to-moderate salt concentrations (25–50 mM) enhanced lateral root development and modified other morphological characteristics, including shoot fresh weight. In contrast, higher salt concentrations (>100 mM) in the culture medium had been shown to exert a negative effect on seedling growth [67].

A hydroponic system was utilized to cultivate and assess the impacts of diverse compounds and stresses on juvenile tomato seedlings (Figure S1). This system facilitated the precise regulation of experimental parameters and provided sufficient biological material for omics studies. The present study investigated the effects of moderate salinity in the culture medium on various growth parameters, thereby enabling its application in the evaluation of other stresses, including nutrient deficiency, temperature fluctuations, or simultaneous exposure to multiple stressors. Additionally, the impact of incorporating biostimulants to foster effective symbiosis with the root system under these conditions was examined, with the objective of leveraging this experimental system for quantitative assessment in the identification of novel biostimulant compounds. The methodology established in this study, with appropriate modifications, has the potential to serve as the basis for the evaluation of other plant species, such as lettuce or pepper.

Observations were made regarding the impact of moderate salinity on the seedlings of the tomato cultivar 'Rosa de Altea', a traditional variety from La Marina Baixa (Alicante, Spain) with a certain tolerance to salinity due to its cultivation near the Mediterranean coast. Moderate salinity has been shown to exert a deleterious effect on the growth of the primary root and to induce the development of lateral and adventitious roots in juvenile tomato seedlings in our experimental conditions (Figures 1 and 2). The inoculation of these seedlings with the AMM biostimulant, containing a proprietary mixture of mycorrhizal microorganisms and rhizobacteria, facilitates the establishment of a robust root symbiosis. This results in an augmentation of the root system area, characterized by an increase in the number and length of lateral and adventitious roots (Figure 2). Moreover, the AMM treatment has been observed to mitigate the deleterious effects of moderate salinity conditions on seedling viability (Figure 1), a finding that is consistent with the results of other research groups who have demonstrated that inoculation with AMF or PGPR enhances the response of tomato plants to salinity [45,48,68,69].

To validate the use of our experimental approach for transcriptomic analysis, we performed an assay with samples in control conditions, treated with the biostimulant (AMM), in moderate salinity conditions (50 mM NaCl), and in the combination of both factors (AMM and 50 mM NaCl). The analysis revealed that approximately 74.9% of the observed variability in the expression of protein-coding genes and non-coding RNAs could be attributed to the treatment effect, with a mere 5.6% being attributable to experimental replicates (Figure 3a,d and Table S1).

In the present study, we identified differential expression in 3658 genes in the leaves of tomato seedlings grown in vitro under three different conditions (AMM, NaCl, and AMM-NaCl), as regards the control conditions (Figure 3b,c and Table S2). These genes were found to be enriched in processes related to photosynthesis, response to stress (both abiotic and biotic), hormone regulation, and response to ROS, among others (Figure 4 and Table S3). The moderate salinity treatment elicited the most substantial expression changes in the transcriptome, exerting a highly significant effect on photosynthesis-related and abiotic stress response genes (Table S4). The saline treatment increased the expression of these genes, both in the absence and presence of the biostimulant, which agrees with the mild improvement observed in on salinity-treated seedlings treated with AMM in our study. The upregulation of photosynthesis genes in salt-treated tomato leaves was counterintuitive, given the reduced growth observed due to salt stress (Figures 1 and 2; [70]), and the known effect of salt stress on lowering the photosynthetic induction rate and limiting carbon fixation [71]. We hypothesize that the upregulation of these photosynthesis-related genes by moderate salinity may imply a regulatory mechanism to compensate for the damage to chloroplasts caused by salt stress. This hypothesis is further supported by been recent findings in *Robinia pseudoacacia* seedlings [72]. Additionally, two genes required for ABA biosynthesis (*SlNCED1*) and *SlNCED2*—were found to be downregulated by moderate salinity in young tomato leaves. A recent study has demonstrated that *SlNCED1*, *SlNCED2*, and *SlNCED6* are strongly induced under severe salt stress (94 mM NaCl) in tomato cultivar 'Ailsa Craig', while this upregulation was not observed in young leaves at moderate salt levels (47 mM NaCl) [73]. Considering the observation that the tomato variety used in our study ('Rosa de Altea') displays tolerance to moderate salinity, the low expression levels of these *NCED* genes may be indicative of an adaptive response. However, further studies are needed to determine the endogenous levels of ABA in young tomato leaves in our experimental system to test this hypothesis.

The present study offers evidence that increased salinity levels could render plants more susceptible to pathogen attack or exert a negative influence on mycorrhization. This phenomenon is characterized by a decrease in the expression by moderate salinity of defense genes and the *SlCCD8*, which is involved in the biosynthesis of the strigolactone hormones required for effective mycorrhization [74]. The latter finding could offer a potential explanation for the increased incidence of adventitious roots observed in tomato plants cultivated under saline conditions [75]. The interplay between jasmonic acid (JA) and auxin signaling has been extensively studied in various developmental processes and in response to stress [76]. The ability to JA-induced leaf senescence to be antagonized by auxin via the transcription factor WRKY57, which has been shown to interact with JASMONATE ZIM-DOMAIN4/8 (JAZ4/8) and the AUX/IAA protein IAA29, repressors of the JA and auxin signaling pathways, respectively [57], is of particular interest. In addition, WRKY53 has been shown to negatively regulate the JA biosynthesis pathway, thereby contributing to plant defense mechanisms in Arabidopsis [77]. In the present study, we identified that multiple WRKY genes were found to be upregulated by the biostimulant treatment, with *Solyc08g008280.3*, a putative orthologue encoding WRKY53, being among the most expressed (Figure S6). Furthermore, *Solyc08g021820.3*, putatively encoding IAA29 (Figure S6), and *Solyc12g009220.2* which encodes SlJAZ2, exhibited increased expression in biostimulant-treated samples and were found to be inhibited by the saline treatment. These findings imply the presence of a JA and auxin cross-regulation module in tomato involving WRK53, IAA29, and JAZ2, which could be adversely affected by salinity. This pathway will be the subject of further research.

4. Materials and Methods

4.1. Plant Materials

The seeds used in this study, *Solanum lycopersicum* L. var. 'Rosa de Altea', were kindly provided by Dr. Santiago García Martínez (Universidad Miguel Hernández de Elche, Spain). The biostimulant solution used, Atlanticell Micomix (AMM), was obtained from Atlántica Agrícola S.A. (Villena, Spain). The AMM solution is a proprietary biological stimulant composed of mycorrhizal microorganisms (*Rhizoglomus irregulare, Funneliformis mosseae* and *Funneliformis caledonium*; 12,500 propagules/g) and rhizobacteria (*Bacillus* spp., equivalent to a concentration of 1×10^{10} UFC/g).

4.2. Culture of Tomato Seedlings

The seeds underwent surface sterilization using chlorine gas (100 mL NaOCl 10% and 4 mL HCl 37%) for 3.5 h in an airtight container. Next, the seeds were transferred to multi-well plates with 25 seeds per well and soaked in sterile water for 24 h in darkness inside a growth chamber at 25.4 ± 1.6 °C. The following day, the germinated seeds were transferred to square Petri dishes (120 mm × 120 mm × 10 mm). Each Petri dish contained 75 mL of 2.15 g/L of Murashige & Skoog (MS) salt medium, 2.5 g/L of gelling agent Gelrite (Duchefa Biochemie, Haarlem, The Netherlands), 0.5 g/L 2-(N-morpholino) ethanesulfonic acid (Duchefa Biochemie), and 2 mL/L Gamborg B5 vitamin solution (Duchefa Biochemie) at pH 5.8. The seedlings grown in vitro were incubated in a growth chamber under controlled conditions, with a photoperiod of 16 h light (day) and 8 h darkness (night) until the end of the experiment provided by white and red LED panels. The average temperature during the day was 25.4 ± 1.6 °C and during the night was 22.8 ± 1.3 °C. The average relative humidity during the day was 49.5 ± 9.1% and during the night was 64.6 ± 12.2%. A graphic visualization of the procedure is shown in Figure S1.

Four different experiments were performed, named E1 to E4. Eight (E1, E2) or thirteen (E3, E4) days after germination, the seedlings from the same developmental stage were transferred to perforated plastic floats positioned above 1 L of liquid culture medium in square plastic containers (165 × 165 × 135 mm), with enforced aeration using an aquarium air pump. The liquid medium was prepared as indicated above, without the addition of the gelling agent Gelrite (Duchefa Biochemie). The salt treatment consisted of 100 mM NaCl (E1 experiment) or 50 mM NaCl (E2 to E4 experiments). The AMM solution, containing a mixture of AMF and rhizobacteria (see above), was added to the liquid culture medium at a concentration of 1 g/L. The following four different conditions were tested: mock, salinity, AMM, and salinity plus AMM. The number of plant samples ranged from 17 to 25 for each condition. The nutrient solution was renewed every two days. All the four experiments were conducted between April and May 2021.

4.3. Sample Collection and RNA Extraction

For transcriptomic analysis, three experimental replicates were taken for each condition (mock, salinity, AMM, salinity plus AMM) from E3 and E4 experiments. Each replicate consisted of all aerial tissues, including hypocotyl, cotyledons, and leaves, from six different plants. The samples were collected five days after growing in hydroponic culture, immediately frozen in liquid nitrogen, and stored at −80 °C. Total RNA was extracted from approximately 100 mg of powdered tissue using TRIzol reagent (Invitrogen, Carlsbad, CA, USA), following the manufacturer's protocol in duplicate (RNA-Seq and RT-qPCR). The RNA was then stored at −80 °C. The integrity of the RNA was assessed using a 2100 Bioanalyzer (Agilent Technologies, Santa Clara, CA, USA). Directional RNA sequencing was performed in (BGI-Tech, Shenzhen, China) using DNB-Seq technology. Sequencing was carried out in paired-end mode with 150 sequencing cycles. The reads were processed using SOAPNuke to obtain the clean reads [78]. Three libraries were sequenced from E4 plants in each condition. The raw data will be provided upon request to the corresponding author.

4.4. RNA-Seq Analysis

The bioinformatics workflow utilized in this study was based on the previously described protocol by [79]. In brief, cleaned RNA-Seq reads were mapped to the *S. lycopersicum* genome build SL4.0 [80] using STAR 2.7 [81]. This approach resulted in 92% of average reads from different libraries mapped to the SL4 genome (Figure S2a). Gene count was performed using featureCounts from the Subread package [82] with the ITAG4 gene models from SolGenomics [83]. Raw counts were transformed to CPM for normalization (Figure S2b), and genes with less than 0.5 CPM in at least three libraries were excluded. A correlation matrix was used to analyze the relationship between the samples (Figure S2c). DESeq2 integrated into the Differential Expression and Pathway analysis (iDEP 2.0) web application was used for read count normalization and differential gene expression analysis [84]. All DEGs were obtained using an FDR threshold of >0.05 and a fold change threshold of $>|2|$. GO enrichment analysis was performed using the iDEP 2.0 [85]. Putative tomato orthologs from *A. thaliana* were identified using Proteinortho [86]. For the identification of expressed and deregulated transcription factors, we used those annotated in [65]. The fold enrichment of the different TF families was obtained by applying the following formula: (number of DEGs up or down of a TF family/total number of TF DEGs up or down)/(number of expressed genes of a TF family/total TF genes expressed). Plant hormone signal transduction was obtained from the KEGG database (January 2025). For heatmap and hierarchical clustering analyses, we used Morpheus web tool [87]. The ncRNAs annotation was primarily retrieved from the SolGenomics database. The JustRNA database was also consulted for ncRNA annotation (January 2025) [88].

4.5. Reverse Transcription-Quantitative PCR (RT-qPCR)

RNA samples were subjected to a retrotranscription using RevertAid (ThermoFisher Scientific, Waltham, MA, USA), in accordance with the manufacturer's instructions. The cDNA obtained was adjusted to 20 ng/µL, and 1.5 µL was used as a template for quantitative PCR, using the PowerTrack SYBR Green kit in a QuantStudio 3 Real-Time PCR System (ThermoFisher Scientific, Waltham, MA, USA). To calculate the PCR efficiency of each primer, a linear regression equation was calculated for each target using serial dilutions of a pool of cDNA. Subsequently, a melting curve analysis was performed for each reaction to verify the synthesis of a single PCR product. The analyses described above, in addition to the quality control analysis, the determination of the baseline, the interplate adjustment, and the calculation of the relative expression values (RQ) using *GADPH* as the endogenous reference gene, were performed with the Standard Curve and Relative Quantification apps in the Connect Data Analysis platform (ThermoFisher Scientific, Waltham, MA, USA). The primers used for RT-qPCR are indicated in Figure S6a.

4.6. Microscopy Observation

Basal root fragments (approximately 5 mm) were obtained using a razor blade from hydroponically grown seedlings either in mock condition or in the AMS treatment from E1 and E2 experiments (n = 5 each). The root fragments were then placed separately in multi-well plates and stained with trypan blue solution (ThermoFisher Scientific, Waltham, MA, USA) as follows: 2 mL of 10% KOH was added to each well, and the plate was incubated at 85 °C for 10 min. The KOH was then removed, and the roots were rinsed twice with 2 mL of distilled water. The distilled water was removed, and 2 mL of 2% HCl were added to each well. The plate was incubated at room temperature (RT) for 15 min. Next, the HCl solution was removed, and 2 mL of 0.05% trypan blue were added to each well and incubated at RT for 30 min. The trypan blue solution was then removed, and the roots were rinsed four times with 2 mL of distilled water. The samples were observed using a Motic BA210 brightfield microscope (MoticEurope, SLU Barcelona, Spain) and imaged with an integrated Moticam 580INT documentation station (Motic Spain).

4.7. Trait Measurement

We quantified the number of viable seedlings and those producing true leaves at 13 days after germination (5 days after transfer to hydroponic conditions). Twenty days after germination (12 days after transfer to hydroponic conditions), all seedlings from E1 and E2 experiments were carefully placed on a large square Petri dish (245 × 245 × 20 mm) with 150 mL of tap water. Pictures of these seedlings were taken using a photographic bench with a Sony Cyber-shot DSC-H3 8.1 Megapixel digital camera (Sony Corporation, Tokyo, Japan) at a resolution of 3264 × 2448 pixels and saved in JPEG format. Several parameters of the root system of these plants were quantified from these pictures as described elsewhere [89], including primary root length, lateral root number, and the presence of adventitious roots in the hypocotyl.

4.8. Statistical Analyses

Statistical analyses were conducted with StatGraphics Centurion XV software v.16.1.03 (StatPoint Technologies, McKinney, TX, USA). Descriptors such as mean, standard error (SE), maximum, minimum, and correlation values were estimated. Data outliers were identified based on aberrant standard deviation values and were excluded from downstream analyses. To compare the data for a given variable, we conducted multiple testing analyses using either the ANOVA F-test or the Fisher's least significant difference method. We defined significant differences as a 1% level of significance (p-value < 0.01), unless otherwise specified. To establish correlations between different parameters, we conducted multiple correlation tests for the selected parameters.

5. Conclusions

The present study explores the molecular and phenotypic changes in tomato plants in response to salt stress adaptation through the AMM biological stimulant compound. The in vitro hydroponic system enabled observation of the treatment effects on plant viability, growth rates, and root system changes. The AMM compound enhanced tomato plant growth under both control and salt stress conditions. A detailed analysis of the root system revealed a decrease in primary root length and an increase in lateral root length under salt stressed conditions. Notably, the AMM treated plants exhibited no significant variation, although seedling viability was enhanced in comparison to those under salt stress conditions. Furthermore, AMM treatment significantly enhanced AR development, demonstrating a comparable response to salt stress conditions.

The transcriptomic studies conducted in this research facilitated the identification of over 3000 differentially expressed genes. Our findings revealed that several WRKY genes were induced by AMM treatment, including a putative ortholog encoding WRKY53, which was particularly noteworthy for its high level of expression. Furthermore, several other genes, including potential orthologs of *IAA29* and *SlJAZ2*, exhibited increased expression in AMM-treated samples but were repressed by salt treatment. These observations suggest the presence of a cross-regulatory module between JA and auxin in tomato, involving these genes, which could be adversely affected by salinity. Moreover, treatment with the AMM compound led to an increase in the expression of genes such as *GOX3* or *DIR1*, which are associated with the plant's defense response to adverse conditions through the ROS detoxification pathway or the salicylic acid-mediated immune response, respectively. The activation of these genes suggests a priming effect of the AMM biostimulant in tomato that deserves further studies.

Supplementary Materials: The following supporting information can be downloaded at: https://www.mdpi.com/article/10.3390/plants14081198/s1, Figure S1: Experimental design used in this work; Figure S2: Statistics of RNA-Seq analysis; Figure S3: Expression analysis of RNA-Seq data (cont.); Figure S4: Deregulation of genes which are associated with photosynthesis and drought responses; Figure S5: Deregulation of genes which are associated with plant hormone transduction pathways; Figure S6: Validation of RNA-Seq results through RT-qPCR; Figure S7: Expression analysis of ncRNAs; Table S1: Protein coding genes (a) and ncRNAs (b) expressed in our RNA-Seq; Table S2: DEGs found between the different treatments in our RNA-seq; Table S3: Enrichment analysis of DEGs found in our RNA-seq; Table S4: Annotation and expression values of DEGs related to defense responses (a), photosynthesis (b), and phytohormone signaling (c); Table S5: Annotation and expression values of DEGs encoding putative TF genes.

Author Contributions: Conceptualization, J.M.P.-P.; methodology, M.S.J. and E.L.; formal analysis, M.S.J. and E.L.; investigation, M.S.J. and E.L.; resources, E.A.Z.-G. and A.A.-M.; data curation, J.M.P.-P.; writing—original draft preparation, M.S.J. and E.L.; writing—review and editing, J.M.P.-P.; visualization, M.S.J. and E.L.; supervision, J.M.P.-P.; project administration, J.M.P.-P.; funding acquisition, J.M.P.-P. All authors have read and agreed to the published version of the manuscript.

Funding: This research was funded by MCIN/AEI/10.13039/501100011033 (TED2021-132256B-C22), the Conselleria d'Innovació, Universitats, Ciència i Societat Digital of the Generalitat Valenciana (AGROALNEXT/2022/036), by the "ERDF A way of making Europe" and by the European Union NextGenerationEU/PRTR (TED2021-132256B-C22 and AGROALNEXT/2022/036).

Data Availability Statement: All data generated or analyzed during this study are provided in this published article and its supplementary data files or it will be provided at a reasonable request.

Acknowledgments: We would like to thank María Gomáriz Clemente (SATDI-UMH) for her technical support and Santiago García Martínez (UMH, Escuela Politécnica Superior de Orihuela, Spain) for kindly providing the plant material.

Conflicts of Interest: Authors Ernesto Alejandro Zavala-González and Almudena Aranda-Martínez were employed by the company Atlántica Agrícola S.A. The remaining authors declare that the research was conducted in the absence of any commercial or financial relationships that could be construed as a potential conflict of interest.

Abbreviations

The following abbreviations are used in this manuscript:

AMF	arbuscular mycorrhizal fungi
AMM	Atlanticell Micomix
AR	adventitious root
CPM	counts per million
DEG	differential expressed gene
GO	gene ontology
JA	jasmonic acid
LR	lateral root
NCED	9-cis-epoxycarotenoid dioxygenase
ncRNA	non-coding RNA
PGPR	plant growth-promoting rhizobacteria
PR	primary root
ROS	reactive oxygen species
TF	transcription factor

References

1. Kumar, L.; Chhogyel, N.; Gopalakrishnan, T.; Hasan, M.K.; Jayasinghe, S.L.; Kariyawasam, C.S.; Kogo, B.K.; Ratnayake, S. Climate Change and Future of Agri-Food Production. *Future Foods Glob. Trends Oppor. Sustain. Chall.* **2022**, 49–79. [CrossRef]
2. Benitez-Alfonso, Y.; Soanes, B.K.; Zimba, S.; Sinanaj, B.; German, L.; Sharma, V.; Bohra, A.; Kolesnikova, A.; Dunn, J.A.; Martin, A.C.; et al. Enhancing Climate Change Resilience in Agricultural Crops. *Curr. Biol.* **2023**, *33*, R1246–R1261. [CrossRef] [PubMed]
3. Wanders, N.; Wada, Y. Human and Climate Impacts on the 21st Century Hydrological Drought. *J. Hydrol.* **2015**, *526*, 208–220. [CrossRef]
4. Essa, Y.H.; Hirschi, M.; Thiery, W.; El-Kenawy, A.M.; Yang, C. Drought Characteristics in Mediterranean under Future Climate Change. *NPJ Clim. Atmos. Sci.* **2023**, *6*, 133. [CrossRef]
5. Litalien, A.; Zeeb, B. Curing the Earth: A Review of Anthropogenic Soil Salinization and Plant-Based Strategies for Sustainable Mitigation. *Sci. Total Environ.* **2020**, *698*, 134235. [CrossRef]
6. Sahab, S.; Suhani, I.; Srivastava, V.; Chauhan, P.S.; Singh, R.P.; Prasad, V. Potential Risk Assessment of Soil Salinity to Agroecosystem Sustainability: Current Status and Management Strategies. *Sci. Total Environ.* **2021**, *764*, 144164. [CrossRef]
7. Penuelas, J.; Coello, F.; Sardans, J. A Better Use of Fertilizers Is Needed for Global Food Security and Environmental Sustainability. *Agric. Food Secur.* **2023**, *12*, 5. [CrossRef]
8. González Guzmán, M.; Cellini, F.; Fotopoulos, V.; Balestrini, R.; Arbona, V. New Approaches to Improve Crop Tolerance to Biotic and Abiotic Stresses. *Physiol. Plant.* **2022**, *174*, e13547. [CrossRef]
9. du Jardin, P. Plant Biostimulants: Definition, Concept, Main Categories and Regulation. *Sci. Hortic.* **2015**, *196*, 3–14. [CrossRef]
10. Rouphael, Y.; Colla, G. Editorial: Biostimulants in Agriculture. *Front. Plant Sci.* **2020**, *11*, 511937. [CrossRef]
11. Rouphael, Y.; Spíchal, L.; Panzarová, K.; Casa, R.; Colla, G. High-Throughput Plant Phenotyping for Developing Novel Biostimulants: From Lab to Field or from Field to Lab? *Front. Plant Sci.* **2018**, *9*, 408288. [CrossRef] [PubMed]
12. Ricci, M.; Tilbury, L.; Daridon, B.; Sukalac, K. General Principles to Justify Plant Biostimulant Claims. *Front. Plant Sci.* **2019**, *10*, 444124. [CrossRef] [PubMed]
13. Paul, K.; Sorrentino, M.; Lucini, L.; Rouphael, Y.; Cardarelli, M.; Bonini, P.; Miras Moreno, M.B.; Reynaud, H.; Canaguier, R.; Trtílek, M.; et al. A Combined Phenotypic and Metabolomic Approach for Elucidating the Biostimulant Action of a Plant-Derived Protein Hydrolysate on Tomato Grown under Limited Water Availability. *Front. Plant Sci.* **2019**, *10*, 450777. [CrossRef] [PubMed]
14. Della Lucia, M.C.; Baghdadi, A.; Mangione, F.; Borella, M.; Zegada-Lizarazu, W.; Ravi, S.; Deb, S.; Broccanello, C.; Concheri, G.; Monti, A.; et al. Transcriptional and Physiological Analyses to Assess the Effects of a Novel Biostimulant in Tomato. *Front. Plant Sci.* **2022**, *12*, 781993. [CrossRef]
15. Baghdadi, A.; Della Lucia, M.C.; Borella, M.; Bertoldo, G.; Ravi, S.; Zegada-Lizarazu, W.; Chiodi, C.; Pagani, E.; Hermans, C.; Stevanato, P.; et al. A Dual-Omics Approach for Profiling Plant Responses to Biostimulant Applications under Controlled and Field Conditions. *Front. Plant Sci.* **2022**, *13*, 983772. [CrossRef]
16. Sorrentino, M.; Panzarová, K.; Spyroglou, I.; Spíchal, L.; Buffagni, V.; Ganugi, P.; Rouphael, Y.; Colla, G.; Lucini, L.; De Diego, N. Integration of Phenomics and Metabolomics Datasets Reveals Different Mode of Action of Biostimulants Based on Protein Hydrolysates in *Lactuca sativa* L. and *Solanum lycopersicum* L. Under Salinity. *Front. Plant Sci.* **2022**, *12*, 808711. [CrossRef]

17. Lee, E.H.; Eo, J.K.; Ka, K.H.; Eom, A.H. Diversity of Arbuscular Mycorrhizal Fungi and Their Roles in Ecosystems. *Mycobiology* **2013**, *41*, 121–125. [CrossRef]
18. van der Heijden, M.G.A.; Martin, F.M.; Selosse, M.A.; Sanders, I.R. Mycorrhizal Ecology and Evolution: The Past, the Present, and the Future. *New Phytol.* **2015**, *205*, 1406–1423. [CrossRef]
19. Bitterlich, M.; Rouphael, Y.; Graefe, J.; Franken, P. Arbuscular Mycorrhizas: A Promising Component of Plant Production Systems Provided Favorable Conditions for Their Growth. *Front. Plant Sci.* **2018**, *9*, 366362. [CrossRef]
20. FAOSTAT. Available online: https://www.fao.org/faostat/es/#home (accessed on 16 January 2024).
21. Hart, M.; Ehret, D.L.; Krumbein, A.; Leung, C.; Murch, S.; Turi, C.; Franken, P. Inoculation with Arbuscular Mycorrhizal Fungi Improves the Nutritional Value of Tomatoes. *Mycorrhiza* **2015**, *25*, 359–376. [CrossRef]
22. Schubert, R.; Werner, S.; Cirka, H.; Rödel, P.; Moya, Y.T.; Mock, H.P.; Hutter, I.; Kunze, G.; Hause, B. Effects of Arbuscular Mycorrhization on Fruit Quality in Industrialized Tomato Production. *Int. J. Mol. Sci.* **2020**, *21*, 7029. [CrossRef] [PubMed]
23. Volpe, V.; Chitarra, W.; Cascone, P.; Volpe, M.G.; Bartolini, P.; Moneti, G.; Pieraccini, G.; Di Serio, C.; Maserti, B.; Guerrieri, E.; et al. The Association with Two Different Arbuscular Mycorrhizal Fungi Differently Affects Water Stress Tolerance in Tomato. *Front. Plant Sci.* **2018**, *9*, 412387. [CrossRef] [PubMed]
24. Khoshkhatti, N.; Eini, O.; Koolivand, D.; Pogiatzis, A.; Klironomos, J.N.; Pakpour, S. Differential Response of Mycorrhizal Plants to Tomato Bushy Stunt Virus and Tomato Mosaic Virus Infection. *Microorganisms* **2020**, *8*, 2038. [CrossRef] [PubMed]
25. Song, Y.Y.; Ye, M.; Li, C.Y.; Wang, R.L.; Wei, X.C.; Luo, S.M.; Zeng, R. Sen Priming of Anti-Herbivore Defense in Tomato by Arbuscular Mycorrhizal Fungus and Involvement of the Jasmonate Pathway. *J. Chem. Ecol.* **2013**, *39*, 1036–1044. [CrossRef]
26. Rivero, J.; Lidoy, J.; Llopis-Giménez, Á.; Herrero, S.; Flors, V.; Pozo, M.J. Mycorrhizal Symbiosis Primes the Accumulation of Antiherbivore Compounds and Enhances Herbivore Mortality in Tomato. *J. Exp. Bot.* **2021**, *72*, 5038–5050. [CrossRef]
27. Song, Y.; Chen, D.; Lu, K.; Sun, Z.; Zeng, R. Enhanced Tomato Disease Resistance Primed by Arbuscular Mycorrhizal Fungus. *Front. Plant Sci.* **2015**, *6*, 163569. [CrossRef]
28. Chitarra, W.; Pagliarani, C.; Maserti, B.; Lumini, E.; Siciliano, I.; Cascone, P.; Schubert, A.; Gambino, G.; Balestrini, R.; Guerrieri, E. Insights on the Impact of Arbuscular Mycorrhizal Symbiosis on Tomato Tolerance to Water Stress. *Plant Physiol.* **2016**, *171*, 1009–1023. [CrossRef]
29. Rivero, J.; Álvarez, D.; Flors, V.; Azcón-Aguilar, C.; Pozo, M.J. Root Metabolic Plasticity Underlies Functional Diversity in Mycorrhiza-Enhanced Stress Tolerance in Tomato. *New Phytol.* **2018**, *220*, 1322–1336. [CrossRef]
30. Balestrini, R.; Rosso, L.C.; Veronico, P.; Melillo, M.T.; De Luca, F.; Fanelli, E.; Colagiero, M.; Di Fossalunga, A.S.; Ciancio, A.; Pentimone, I. Transcriptomic Responses to Water Deficit and Nematode Infection in Mycorrhizal Tomato Roots. *Front. Microbiol.* **2019**, *10*, 1807. [CrossRef]
31. Duc, N.H.; Szentpéteri, V.; Mayer, Z.; Posta, K. Distinct Impact of Arbuscular Mycorrhizal Isolates on Tomato Plant Tolerance to Drought Combined with Chronic and Acute Heat Stress. *Plant Physiol. Biochem.* **2023**, *201*, 107892. [CrossRef]
32. Leventis, G.; Tsikina, M.; Feka, M.; Ladikou, E.V.; Papadakis, I.E.; Chatzipavlidis, I.; Papadopoulou, K.; Ehaliotis, C. Arbuscular Mycorrhizal Fungi Enhance Growth of Tomato under Normal and Drought Conditions, via Different Water Regulation Mechanisms. *Rhizosphere* **2021**, *19*, 100394. [CrossRef]
33. Negrão, S.; Schmöckel, S.M.; Tester, M. Evaluating Physiological Responses of Plants to Salinity Stress. *Ann. Bot.* **2017**, *119*, 1–11. [CrossRef] [PubMed]
34. Roșca, M.; Mihalache, G.; Stoleru, V. Tomato Responses to Salinity Stress: From Morphological Traits to Genetic Changes. *Front. Plant Sci.* **2023**, *14*, 1118383. [CrossRef] [PubMed]
35. Ghorbani, A.; Omran, V.O.G.; Razavi, S.M.; Pirdashti, H.; Ranjbar, M. Piriformospora Indica Confers Salinity Tolerance on Tomato (*Lycopersicon esculentum* Mill.) through Amelioration of Nutrient Accumulation, K+/Na+ Homeostasis and Water Status. *Plant Cell Rep.* **2019**, *38*, 1151–1163. [CrossRef]
36. Ghorbani, A.; Razavi, S.M.; Ghasemi Omran, V.O.; Pirdashti, H. Piriformospora Indica Inoculation Alleviates the Adverse Effect of NaCl Stress on Growth, Gas Exchange and Chlorophyll Fluorescence in Tomato (*Solanum lycopersicum* L.). *Plant Biol.* **2018**, *20*, 729–736. [CrossRef]
37. Heidarianpour, M.B.; Aliasgharzad, N.; Olsson, P.A. Positive Effects of Co-Inoculation with Rhizophagus Irregularis and Serendipita Indica on Tomato Growth under Saline Conditions, and Their Individual Colonization Estimated by Signature Lipids. *Mycorrhiza* **2020**, *30*, 455–466. [CrossRef]
38. Jung, S.C.; Martinez-Medina, A.; Lopez-Raez, J.A.; Pozo, M.J. Mycorrhiza-Induced Resistance and Priming of Plant Defenses. *J. Chem. Ecol.* **2012**, *38*, 651–664. [CrossRef]
39. Ouziad, F.; Hildebrandt, U.; Schmelzer, E.; Bothe, H. Differential Gene Expressions in Arbuscular Mycorrhizal-Colonized Tomato Grown under Heavy Metal Stress. *J. Plant Physiol.* **2005**, *162*, 634–649. [CrossRef]
40. Kashyap, S.P.; Kumari, N.; Mishra, P.; Prasad Moharana, D.; Aamir, M.; Singh, B.; Prasanna, H.C. Transcriptional Regulation-Mediating ROS Homeostasis and Physio-Biochemical Changes in Wild Tomato (*Solanum chilense*) and Cultivated Tomato (*Solanum lycopersicum*) under High Salinity. *Saudi J. Biol. Sci.* **2020**, *27*, 1999–2009. [CrossRef]

41. Figueira-Galán, D.; Heupel, S.; Duelli, G.; Tomasi Morgano, M.; Stapf, D.; Requena, N. Exploring the Synergistic Effects of Biochar and Arbuscular Mycorrhizal Fungi on Phosphorus Acquisition in Tomato Plants by Using Gene Expression Analyses. *Sci. Total Environ.* **2023**, *884*, 163506. [CrossRef]
42. Turan, M.; Ekinci, M.; Argin, S.; Brinza, M.; Yildirim, E. Drought Stress Amelioration in Tomato (*Solanum lycopersicum* L.) Seedlings by Biostimulant as Regenerative Agent. *Front. Plant Sci.* **2023**, *14*, 1211210. [CrossRef] [PubMed]
43. Patani, A.; Prajapati, D.; Ali, D.; Kalasariya, H.; Yadav, V.K.; Tank, J.; Bagatharia, S.; Joshi, M.; Patel, A. Evaluation of the Growth-Inducing Efficacy of Various Bacillus Species on the Salt-Stressed Tomato (*Lycopersicon esculentum* Mill.). *Front. Plant Sci.* **2023**, *14*, 1168155. [CrossRef] [PubMed]
44. Altimira, F.; Godoy, S.; Arias-Aravena, M.; Vargas, N.; González, E.; Dardón, E.; Montenegro, E.; Viteri, I.; Tapia, E. Reduced Fertilization Supplemented with Bacillus Safensis RGM 2450 and Bacillus Siamensis RGM 2529 Promotes Tomato Production in a Sustainable Way. *Front. Plant Sci.* **2024**, *15*, 1451887. [CrossRef] [PubMed]
45. Dong, H.; Wang, Y.; Di, Y.; Qiu, Y.; Ji, Z.; Zhou, T.; Shen, S.; Du, N.; Zhang, T.; Dong, X.; et al. Plant Growth-Promoting Rhizobacteria Pseudomonas Aeruginosa HG28-5 Improves Salt Tolerance by Regulating Na+/K+ Homeostasis and ABA Signaling Pathway in Tomato. *Microbiol. Res.* **2024**, *283*, 127707. [CrossRef]
46. Giovannini, L.; Pagliarani, C.; Cañizares, E.; Sillo, F.; Chitarra, W.; De Rose, S.; Zampieri, E.; Ioannou, A.; Spanos, A.; Vita, F.; et al. Mycorrhization and Chemical Seed Priming Boost Tomato Stress Tolerance by Changing Primary and Defence Metabolic Pathways. *J. Exp. Bot.* **2024**, erae457. [CrossRef]
47. Mekkaoui, F.; Ait-El-Mokhtar, M.; Zaari Jabri, N.; Amghar, I.; Essadssi, S.; Hmyene, A. The Use of Compost and Arbuscular Mycorrhizal Fungi and Their Combination to Improve Tomato Tolerance to Salt Stress. *Plants* **2024**, *13*, 2225. [CrossRef]
48. Soussani, F.E.; Boutasknit, A.; Ben-Laouane, R.; Benkirane, R.; Baslam, M.; Meddich, A. Arbuscular Mycorrhizal Fungi and Compost-Based Biostimulants Enhance Fitness, Physiological Responses, Yield, and Quality Traits of Drought-Stressed Tomato Plants. *Plants* **2023**, *12*, 1856. [CrossRef]
49. Arcidiacono, M.; Pellegrino, E.; Nuti, M.; Ercoli, L. Field Inoculation by Arbuscular Mycorrhizal Fungi with Contrasting Life-History Strategies Differently Affects Tomato Nutrient Uptake and Residue Decomposition Dynamics. *Plant Soil* **2023**, *500*, 105–127. [CrossRef]
50. Chandrasekaran, M.; Boopathi, T.; Manivannan, P. Comprehensive Assessment of Ameliorative Effects of AMF in Alleviating Abiotic Stress in Tomato Plants. *J. Fungi* **2021**, *7*, 303. [CrossRef]
51. Li, Y.; Laterrière, M.; Lay, C.Y.; Klabi, R.; Masse, J.; St-Arnaud, M.; Yergeau, É.; Lupwayi, N.Z.; Gan, Y.; Hamel, C. Effects of Arbuscular Mycorrhizal Fungi Inoculation and Crop Sequence on Root-Associated Microbiome, Crop Productivity and Nutrient Uptake in Wheat-Based and Flax-Based Cropping Systems. *Appl. Soil Ecol.* **2021**, *168*, 104136. [CrossRef]
52. Zhu, B.; Gao, T.; Zhang, D.; Ding, K.; Li, C.; Ma, F. Functions of Arbuscular Mycorrhizal Fungi in Horticultural Crops. *Sci. Hortic.* **2022**, *303*, 111219. [CrossRef]
53. Feller, C.; Bleiholder, H.; Buhr, L.; Hack, H.; Hess, M.; Klose, R.; Meier, U.; Stauss, R.; Van den Boom, T.; Weber, E. Phänologische Entwicklungsstadien von Gemüsepflanzen: II. Fruchtgemüse Und Hülsenfrüchte. *Nachrichtenblatt Dtsch. Pflanzenschutzdienstes* **1995**, *47*, 217–232.
54. Shui, D.; Sun, J.; Xiong, Z.; Zhang, S.; Shi, J. Comparative Identification of WRKY Transcription Factors and Transcriptional Response to Ralstonia Solanacearum in Tomato. *Gene* **2024**, *912*, 148384. [CrossRef] [PubMed]
55. Carella, P.; Kempthorne, C.J.; Wilson, D.C.; Isaacs, M.; Cameron, R.K. Exploring the Role of DIR1, DIR1-like and Other Lipid Transfer Proteins during Systemic Immunity in Arabidopsis. *Physiol. Mol. Plant Pathol.* **2017**, *97*, 49–57. [CrossRef]
56. Rojas, C.M.; Senthil-Kumar, M.; Wang, K.; Ryu, C.M.; Kaundal, A.; Mysore, K.S. Glycolate Oxidase Modulates Reactive Oxygen Species–Mediated Signal Transduction during Nonhost Resistance in Nicotiana Benthamiana and Arabidopsis. *Plant Cell* **2012**, *24*, 336–352. [CrossRef]
57. Jiang, Y.; Liang, G.; Yang, S.; Yu, D. Arabidopsis WRKY57 Functions as a Node of Convergence for Jasmonic Acid– and Auxin-Mediated Signaling in Jasmonic Acid–Induced Leaf Senescence. *Plant Cell* **2014**, *26*, 230–245. [CrossRef]
58. Gonzalez-Guzman, M.; Rodriguez, L.; Lorenzo-Orts, L.; Pons, C.; Sarrion-Perdigones, A.; Fernandez, M.A.; Peirats-Llobet, M.; Forment, J.; Moreno-Alvero, M.; Cutler, S.R.; et al. Tomato PYR/PYL/RCAR Abscisic Acid Receptors Show High Expression in Root, Differential Sensitivity to the Abscisic Acid Agonist Quinabactin, and the Capability to Enhance Plant Drought Resistance. *J. Exp. Bot.* **2014**, *65*, 4451–4464. [CrossRef]
59. Yu, X.; Chen, G.; Tang, B.; Zhang, J.; Zhou, S.; Hu, Z. The Jasmonate ZIM-Domain Protein Gene SlJAZ2 Regulates Plant Morphology and Accelerates Flower Initiation in *Solanum lycopersicum* Plants. *Plant Sci.* **2018**, *267*, 65–73. [CrossRef]
60. Dere, S. Mitigating the Adverse Effects of Salt Stress on Pepper Plants Through Arbuscular Mycorrhizal Fungi (AMF) and Beneficial Bacterial (PGPR) Inoculation. *Horticulturae* **2024**, *10*, 1150. [CrossRef]
61. Kumawat, K.C.; Sharma, B.; Nagpal, S.; Kumar, A.; Tiwari, S.; Nair, R.M. Plant Growth-Promoting Rhizobacteria: Salt Stress Alleviators to Improve Crop Productivity for Sustainable Agriculture Development. *Front. Plant Sci.* **2023**, *13*, 1101862. [CrossRef]

62. Fasusi, O.A.; Babalola, O.O.; Adejumo, T.O. Harnessing of Plant Growth-Promoting Rhizobacteria and Arbuscular Mycorrhizal Fungi in Agroecosystem Sustainability. *CABI Agric. Biosci.* **2023**, *4*, 26. [CrossRef]
63. Reyes-Impellizzeri, S.; Moreno, A.A. The Endoplasmic Reticulum Role in the Plant Response to Abiotic Stress. *Front. Plant Sci.* **2021**, *12*, 755447. [CrossRef] [PubMed]
64. Van Oosten, M.J.; Di Stasio, E.; Cirillo, V.; Silletti, S.; Ventorino, V.; Pepe, O.; Raimondi, G.; Maggio, A. Root inoculation with *Azotobacter chroococcum* 76A enhances tomato plants adaptation to salt stress under low N conditions. *BMC Plant Biol.* **2018**, *18*, 205. [CrossRef] [PubMed]
65. Larriba, E.; Nicolás-Albujer, M.; Sánchez-García, A.B.; Pérez-Pérez, J.M. Identification of Transcriptional Networks Involved in De Novo Organ Formation in Tomato Hypocotyl Explants. *Int. J. Mol. Sci.* **2022**, *23*, 16112. [CrossRef]
66. Atta, K.; Mondal, S.; Gorai, S.; Singh, A.P.; Kumari, A.; Ghosh, T.; Roy, A.; Hembram, S.; Gaikwad, D.J.; Mondal, S.; et al. Impacts of Salinity Stress on Crop Plants: Improving Salt Tolerance through Genetic and Molecular Dissection. *Front. Plant Sci.* **2023**, *14*, 1241736. [CrossRef]
67. Rivera, P.; Moya, C.; O'Brien, J.A. Low Salt Treatment Results in Plant Growth Enhancement in Tomato Seedlings. *Plants* **2022**, *11*, 807. [CrossRef]
68. Liang, S.M.; Li, Q.S.; Liu, M.Y.; Hashem, A.; Al-Arjani, A.B.F.; Alenazi, M.M.; Abd_Allah, E.F.; Muthuramalingam, P.; Wu, Q.S. Mycorrhizal Effects on Growth and Expressions of Stress-Responsive Genes (Aquaporins and SOSs) of Tomato under Salt Stress. *J. Fungi* **2022**, *8*, 1305. [CrossRef]
69. Chen, G.; Yang, A.; Wang, J.; Ke, L.; Chen, S.; Li, W. Effects of the Synergistic Treatments of Arbuscular Mycorrhizal Fungi and Trehalose on Adaptability to Salt Stress in Tomato Seedlings. *Microbiol. Spectr.* **2024**, *12*, e03404-23. [CrossRef]
70. Zahra, N.; Al Hinai, M.S.; Hafeez, M.B.; Rehman, A.; Wahid, A.; Siddique, K.H.M.; Farooq, M. Regulation of Photosynthesis under Salt Stress and Associated Tolerance Mechanisms. *Plant Physiol. Biochem.* **2022**, *178*, 55–69. [CrossRef]
71. Zhang, Y.; Kaiser, E.; Zhang, Y.; Yang, Q.; Li, T. Short-Term Salt Stress Strongly Affects Dynamic Photosynthesis, but Not Steady-State Photosynthesis, in Tomato (*Solanum lycopersicum*). *Environ. Exp. Bot.* **2018**, *149*, 109–119. [CrossRef]
72. Lu, C.; Li, L.; Liu, X.; Chen, M.; Wan, S.; Li, G. Salt Stress Inhibits Photosynthesis and Destroys Chloroplast Structure by Downregulating Chloroplast Development–Related Genes in Robinia Pseudoacacia Seedlings. *Plants* **2023**, *12*, 1283. [CrossRef] [PubMed]
73. Holsteens, K.; De Jaegere, I.; Wynants, A.; Prinsen, E.L.J.; Van de Poel, B. Mild and Severe Salt Stress Responses Are Age-Dependently Regulated by Abscisic Acid in Tomato. *Front. Plant Sci.* **2022**, *13*, 982622. [CrossRef] [PubMed]
74. Vogel, J.T.; Walter, M.H.; Giavalisco, P.; Lytovchenko, A.; Kohlen, W.; Charnikhova, T.; Simkin, A.J.; Goulet, C.; Strack, D.; Bouwmeester, H.J.; et al. SlCCD7 Controls Strigolactone Biosynthesis, Shoot Branching and Mycorrhiza-Induced Apocarotenoid Formation in Tomato. *Plant J.* **2010**, *61*, 300–311. [CrossRef] [PubMed]
75. Kohlen, W.; Charnikhova, T.; Lammers, M.; Pollina, T.; Tóth, P.; Haider, I.; Pozo, M.J.; Maagd, R.A.; Ruyter-Spira, C.; Bouwmeester, H.J.; et al. The Tomato *CAROTENOID CLEAVAGE DIOXYGENASE 8* (*SlCCD 8*) Regulates Rhizosphere Signaling, Plant Architecture and Affects Reproductive Development through Strigolactone Biosynthesis. *New Phytol.* **2012**, *196*, 535–547. [CrossRef]
76. Hu, Y.; Jiang, Y.; Han, X.; Wang, H.; Pan, J.; Yu, D. Jasmonate Regulates Leaf Senescence and Tolerance to Cold Stress: Crosstalk with Other Phytohormones. *J. Exp. Bot.* **2017**, *68*, 1361–1369. [CrossRef]
77. Jiao, C.; Li, K.; Zuo, Y.; Gong, J.; Guo, Z.; Shen, Y. CALMODULIN1 and WRKY53 Function in Plant Defense by Negatively Regulating the Jasmonic Acid Biosynthesis Pathway in Arabidopsis. *Int. J. Mol. Sci.* **2022**, *23*, 7718. [CrossRef]
78. Chen, Y.; Chen, Y.; Shi, C.; Huang, Z.; Zhang, Y.; Li, S.; Li, Y.; Ye, J.; Yu, C.; Li, Z.; et al. SOAPnuke: A MapReduce acceleration-supported software for integrated quality control and preprocessing of high-throughput sequencing data. *GigaScience* **2018**, *7*, gix120. [CrossRef]
79. Larriba, E.; Sánchez-García, A.B.; Martínez-Andújar, C.; Albacete, A.; Pérez-Pérez, J.M. Tissue-Specific Metabolic Reprogramming during Wound-Induced Organ Formation in Tomato Hypocotyl Explants. *Int. J. Mol. Sci.* **2021**, *22*, 10112. [CrossRef]
80. Hosmani, P.S.; Flores-Gonzalez, M.; van de Geest, H.; Maumus, F.; Bakker, L.V.; Schijlen, E.; van Haarst, J.; Cordewener, J.; Sanchez-Perez, G.; Peters, S.; et al. An Improved de Novo Assembly and Annotation of the Tomato Reference Genome Using Single-Molecule Sequencing, Hi-C Proximity Ligation and Optical Maps. *bioRxiv* **2019**, 767764. [CrossRef]
81. Dobin, A.; Davis, C.A.; Schlesinger, F.; Drenkow, J.; Zaleski, C.; Jha, S.; Batut, P.; Chaisson, M.; Gingeras, T.R. STAR: Ultrafast Universal RNA-Seq Aligner. *Bioinformatics* **2013**, *29*, 15–21. [CrossRef]
82. Liao, Y.; Smyth, G.K.; Shi, W. The R Package Rsubread Is Easier, Faster, Cheaper and Better for Alignment and Quantification of RNA Sequencing Reads. *Nucleic Acids Res.* **2019**, *47*, e47. [CrossRef] [PubMed]
83. Sol Genomics Network. Available online: https://solgenomics.net/ (accessed on 16 January 2024).
84. Ge, S.X.; Son, E.W.; Yao, R. IDEP: An Integrated Web Application for Differential Expression and Pathway Analysis of RNA-Seq Data. *BMC Bioinform.* **2018**, *19*, 534. [CrossRef] [PubMed]

85. Ge, S.X.; Jung, D.; Jung, D.; Yao, R. ShinyGO: A Graphical Gene-Set Enrichment Tool for Animals and Plants. *Bioinformatics* **2020**, *36*, 2628–2629. [CrossRef] [PubMed]
86. Lechner, M.; Findeiß, S.; Steiner, L.; Marz, M.; Stadler, P.F.; Prohaska, S.J. Proteinortho: Detection of (Co-)Orthologs in Large-Scale Analysis. *BMC Bioinform.* **2011**, *12*, 124. [CrossRef]
87. Morpheus. Available online: https://software.broadinstitute.org/morpheus/ (accessed on 16 January 2024).
88. Tseng, K.C.; Wu, N.Y.; Chow, C.N.; Zheng, H.Q.; Chou, C.Y.; Yang, C.W.; Wang, M.J.; Chang, S.B.; Chang, W.C. JustRNA: A Database of Plant Long Noncoding RNA Expression Profiles and Functional Network. *J. Exp. Bot.* **2023**, *74*, 4949–4958. [CrossRef]
89. Justamante, M.S.; Ibáñez, S.; Peidró, A.; Pérez-Pérez, J.M. A Genome-Wide Association Study Identifies New Loci Involved in Wound-Induced Lateral Root Formation in Arabidopsis Thaliana. *Front. Plant Sci.* **2019**, *10*, 311. [CrossRef]

plants

MDPI

Review

Potential of the Use of Biostimulants in Lettuce Production

Talys Moratti Lemos de Oliveira [1], Janyne Soares Braga Pires [2], Vinicius de Souza Oliveira [3], Ana Júlia Câmara Jeveaux Machado [2], Adriano Alves Fernandes [4], Lúcio de Oliveira Arantes [3] and Sara Dousseau-Arantes [3,*]

[1] Federal Institute of Espírito Santo (IFES), Barra de São Francisco Campus, Rodovia ES 320, Km. 118, Zona Rural, Barra de São Francisco 29800-000, ES, Brazil; talys.oliveira@ifes.edu.br

[2] Federal University of Espírito Santo (UFES), Center for Natural Human Sciences (CCHN), Goiabeiras Campus, Vitória/ES. Av. Fernando Ferrari, 514, Vitória 29075-910, ES, Brazil; janynesbraga2@gmail.com (J.S.B.P.); anajucamara@gmail.com (A.J.C.J.M.)

[3] Capixaba Institute of Research, Technical Assistance and Rural Extension, Rodovia BR-101 Norte, Km. 151, Bebedouro, Linhares 29915-140, ES, Brazil; souzaoliveiravini@gmail.com (V.d.S.O.); lucio.arantes@incaper.es.gov.br (L.d.O.A.)

[4] Federal University of Espírito Santo (UFES), Centro Universitário Norte do Espírito Santo (CEUNES), São Mateus Campus, Rodovia BR 101-Norte, Km. 60, Litorâneo, São Mateus 29932-540, ES, Brazil; adriano.fernandes@ufes.br

* Correspondence: saradousseau@gmail.com

Abstract: Lettuce (*Lactuca sativa* L.) is one of the main leafy vegetables in the world, being present in several countries. Due to its composition, which includes a substance with antioxidant action and beneficial effects on health, it is consumed constantly. However, due to ongoing climate change that has had global effects, the crop has been suffering a reduction in productivity and quality. Thus, technologies aiming to mitigate the effects of climate extremes have been developed. In lettuce production, biostimulants make it possible to improve the growth and sustainable development of plants. This is due to their ability to activate physiological and biochemical processes in plants, resulting in an increase in the production of bioactive compounds such as vitamins, amino acids, and antioxidants. In addition, biostimulants contribute to improving the nutritional quality of lettuces, making them more resistant and adapted to different environmental conditions, resulting in a more sustainable development for the crop. This review aims to compile and discuss the available scientific evidence on the use of biostimulants in lettuce cultivation, addressing their mechanisms of action, the types of substances used, the results obtained in different cultivation systems, and their potential to promote more efficient and adaptable agriculture in the face of environmental changes.

Keywords: *Lactuca sativa* L.; humic substances; sustainability

1. Introduction

Lettuce (*Lactuca sativa* L.) is a leafy vegetable belonging to the Asteraceae family; it stands out as one of the crops with the greatest economic and nutritional importance worldwide [1]. Generally, lettuce varieties are grown for leaf production, so that the life cycle is interrupted before flower formation, preventing seed maturation and ensuring the quality of the leaves for consumption [2].

Lettuce is widely known and consumed across many regions of the world [3]. However, the growing of lettuce faces challenges related to climate change, such as heat waves, high temperatures, and nutrient solution salinity, which impact productivity and product quality [3–7]. In view of this scenario, the use of a substance with biostimulant effects appears as a promising avenue to mitigate the effects of abiotic stress on lettuce production [5–7].

Biostimulants are available in various forms, including plant extracts, seaweed extract, protein hydrolysates, vitamins, amino acids, non-microbial compounds, humic acids, fulvic acids, and their derivatives [7–16]. These substances have already shown a beneficial effect on several plant species, boosting growth and development, improving the absorption and use of nutrients, minimizing the effects of biotic and abiotic stress, improving hormone synthesis, and minimizing the effects of free radicals [7–11,16].

Considering these aspects, conducting a review on the use of biostimulants in lettuce cultivation becomes essential to consolidate current knowledge, identify research gaps, and guide future studies aiming to improve crop productivity and sustainability. Thus, the present review aims to deepen the knowledge on lettuce cultivation by addressing the influence of biostimulants on agronomy, aiming to contribute to the development of more sustainable and efficient agricultural vegetables.

2. Biostimulants in Lettuce Cultivation

The lettuce crop has been facing constant challenges, especially in relation to climate extremes [3]. Climate change has caused significant impacts on lettuce production, raising costs, reducing productivity, and limiting supply in the consumer market [3]. Extreme weather events, such as torrential rains and prolonged droughts, have forced farmers to seek alternative solutions, such as adopting protected crops [7]. In this context, the use of technologies that allow the mitigation of the effects of climate on crops through practical measures and allow more sustainable crops becomes necessary to ensure food security where the weather conditions are unfavorable [6–10].

Several compounds are tested in order to generate biostimulant effects in plants and optimize their growth and development [7,10]. These compounds act by stimulating plant metabolism, increasing enzyme activity and protein synthesis. They trigger physiological defense responses, such as increased antioxidant production and the activation of protection systems against environmental stresses [7–10]. Furthermore, they help in hormonal regulation, favoring cell division and tissue stretching, which contributes to a more balanced and efficient development. In addition, biostimulants can play a crucial role in nutrient uptake and translocation, optimizing the utilization of available resources and maximizing the productive potential of crops [7,9–11].

In light of these considerations, the integration of biostimulant technologies into lettuce cultivation emerges not only as an innovative agronomic tool, but also as a strategic response to the increasing vulnerability of crops under climatic pressure. By reinforcing physiological resilience and enhancing nutrient efficiency, biostimulants contribute to stabilizing yields and maintaining crop quality, even under suboptimal environmental conditions (Figure 1). As climate variability continues to intensify, adopting such adaptive measures becomes essential—not merely to improve production outcomes but to sustain agricultural systems in a future marked by uncertainty and a growing demand for resilient, high-quality food.

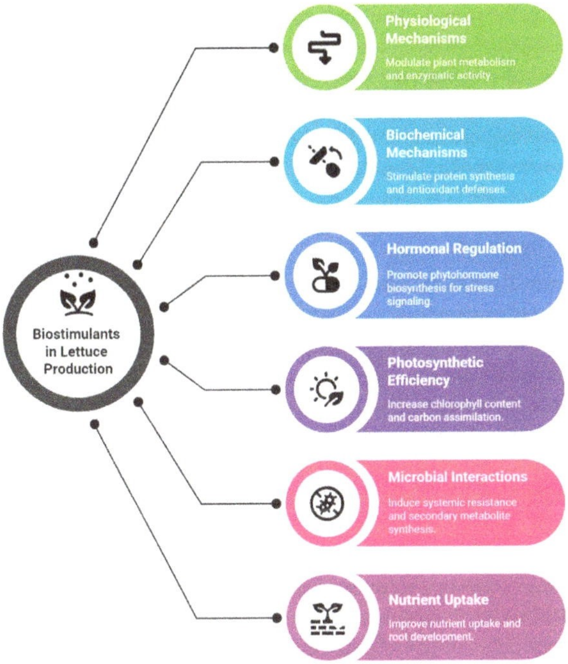

Figure 1. Biostimulants in lettuce production.

2.1. Humic Substances

Humic substances have in their composition organic material resulting from the decomposition of plant and animal remains that are used in the formulation of fertilizers for various crops [10,12,13,15]. They use is related to positive effects both on the characteristics of the substrate and on the plant's metabolism. Because they are made up of humic acids, fulvic acids, and humins, humic substances stimulate the production of hormones important for plants, such as auxin, cytokinins, and gibberellins, which can influence metabolic processes favorably, favoring plant growth and development [5,11,12,14,17–20]. Humic substances act as amphiphilic redox compounds with outstanding chelating properties, contributing to soil health by increasing nutrient bioavailability, stabilizing soil aggregates, and enhancing plant growth by promoting gains in chlorophyll levels, the synthesis of nucleic acids, and improvement in water and nutrient absorption [10,13,14,17].

The application of humic acids, plant growth-promoting bacteria, and their combination positively influenced several agronomic traits of lettuce cv. Vanda [13]. The isolated application of bacteria increased plant height by 18.85%, while the combined treatment yielded the best results for the length of the largest leaf, rosette circumference, and shoot fresh weight, with an increase of up to 33.5% compared to the control [13]. The application of a combined biostimulant containing plant-growth-promoting bacteria (PGPB) and *Chlorella vulgaris* significantly improved the yield and quality of lettuce [14]. When applied every 14 days, this treatment increased the fresh weight of romaine lettuce by 18.9% in the spring crop and leaf lettuce by 22.7% in the summer crop [14]. These biostimulants are low-cost, sustainable technologies with high potential to improve lettuce productivity under field conditions [20].

Taken together, the evidence reinforces the functional versatility of humic substances and microbial biostimulants in enhancing lettuce cultivation [12,14,15]. Their synergistic action—whether through improved nutrient dynamics, hormonal stimulation, or structural

weed whose application in agriculture is successfully reported. It has calcium carbonate, humic and fulvic substances, and more than 20 compounds in its composition [29,30]. The study demonstrated its biostimulant effectiveness in agricultural crops [30].

Chitosan has demonstrated promising effects as a plant biostimulant in lettuce, acting at multiple physiological and biochemical levels [31–35]. The foliar application of chitosan at varying concentrations (0, 75, 150, and 300 ppm) has been shown to significantly improve yield, enhance chlorophyll content, and improve leaf quality [31]. Notably, a 300 ppm concentration resulted in a marked reduction in nitrate accumulation, one of the primary metabolic contaminants in leafy vegetables, as reported in recent studies [31].

Beyond its isolated use, chitosan has also been employed in association with iodine-based compounds such as potassium iodate (KIO_3), forming complexes like chitosan-KIO_3 [32]. This combination not only promotes increased biomass accumulation but also enables iodine biofortification without negatively impacting crop performance. Experimental findings indicate that such complexes maintain lettuce productivity while elevating the iodine concentrations in plant tissues, thus contributing to human nutritional security [32].

During the early stages of development, the application of chitosan hydrolysate for seed treatment has proven to be effective in promoting germination and stimulating root growth [33,34]. Recent data show increases in fresh biomass, chlorophyll, and carotenoid contents, along with enhanced root branching, critical traits for successful crop establishment [32–34]. These effects highlight the role of chitosan as a modulator of primary metabolism and a promoter of photosynthetic efficiency [31–35].

Under saline stress conditions, chitosan continues to function effectively as a biostimulant [34]. Its foliar application supports the maintenance of leaf area and biomass in both shoot and root systems while enhancing physiological stability through the activation of antioxidant enzymes and regulation of ion balance and osmotic adjustment [34]. These findings suggest that chitosan plays a role in the adaptive mechanisms associated with abiotic stress, potentially through hormonal signaling and the induction of secondary metabolic pathways.

Moreover, the combination of plant-growth-promoting bacteria and microalgae can act positively under heat stress, enhancing plant productivity and antioxidant capacity even under high-temperature conditions [14,20,24]. Such combinations have been effective in improving shoot biomass and total carotenoid content, highlighting their potential for sustainable lettuce production during periods of climatic stress [20].

More recently, the use of iodine-associated nanochitosan complexes (NPSCS-i) has shown superior responses compared to conventional formulations [35]. Under controlled conditions, the application of 5 mg L^{-1} of KIO_3 in a nanostructured formulation resulted in a 30.8% increase in lettuce yield, underscoring the potential of combining nanotechnology with natural polymers to sustainably intensify agricultural production [35].

In this context, the application of chitosan has emerges as a promising strategy to optimize the physiological performance of lettuce and enhance its productive efficiency, particularly in agricultural systems aiming for greater sustainability. The wide range of observed effects—from the improved germination to increased biomass and enhanced nutritional quality of leaves—demonstrates that chitosan does not act in isolation [31–35], but rather interacts with multiple physiological and metabolic pathways, enabling the plant to adapt to various environmental conditions.

Its effectiveness under saline stress, for instance, highlights a regulatory role that extends beyond basic nutrition and involving antioxidant defense mechanisms and osmotic balance [34]. Additionally, the formation of iodine-based complexes, especially in nanostructured formulations, represents a significant advancement in both biofortification and the maintenance of crop productivity. This indicates that chitosan, beyond serving as a valuable tool in agronomic management, may also contribute to strategies focused on improving the nutritional and functional quality of food.

The consistency of results across different modes of application—whether via seed treatment, foliar spraying, or in combination with micronutrients—further reinforces its potential as a versatile and efficient biostimulant. Therefore, its use in lettuce cultivation not only supports immediate agronomic gains but also aligns with ecologically based agriculture, reducing the dependence on synthetic inputs and promoting the adoption of clean, integrated technologies.

2.4. Inorganic Compounds

Among the main inorganic compounds that play a beneficial role in plant development are Al, Co, Na, Se, and Si [36,37]. These elements help promote growth but are not essential for plants [37]. The positive effects on plants may be related to silica deposition in the cell wall under stressful conditions such as pathogen attack or high salinity [36,37]. Studies prove that silicon plays a role in mitigating biotic and abiotic stress in plants grown in hydroponic systems [37]. In studies carried out with *Lithothamnium* sp. algal extract, it is reported that a possible cause of the improvement in the quality of lettuce production is related to carbonate-siliciclastic sea sand being incorporated during the extraction of the algae, which contains silicon [29,30].

These findings reinforce the relevance of certain inorganic elements—particularly silicon—in supporting plant development, even though they are not classified as essential nutrients [36,37]. In lettuce cultivation, especially under hydroponic conditions, silicon has shown consistent benefits by enhancing nutrient uptake, strengthening structural defenses, and promoting tolerance to stress factors. The presence of silicon in natural sources such as *Lithothamnium* sp. may further explain improvements in crop quality when these inputs are used [29,30]. Overall, incorporating silicon-based compounds into lettuce production systems appears to be a promising strategy to boost both growth and physiological resilience.

2.5. Fungi

In the case of biostimulants that include fungi in their composition, *Trichoderma*-based products are the most used in Brazil and Latin America. The practical application of *Trichoderma* lies in the fungus's ability to manage disease, compete with pathogenic agents for energy sources, and produce antibiotics that inhibit the proliferation of the pathogen [38]. Another important function of *Trichoderma* is to stimulate the production of plant hormones, such as jasmonic acid, salicylic acid, and ethylene, which induce plant resistance [19,20,39–41].

Successful cases using *Trichoderma* as a biostimulant were reported with a 23.7% increase in the production of commercial pepper fruits (*Capsicum annuum* L.) [19]. It was also possible to observe positive effects on vegetative and reproductive growth; the synthesis of hormones such as gibberellins, auxin, and cytokinin; and the induction of the formation of secondary compounds such as carotenoids, saponin, and phenolic compounds [19]. In carrot crops, the application of *Trichoderma* to the seeds promoted greater plant growth, in addition to presenting similar effectiveness to treatment with agrochemicals in terms of reducing the incidence of pathogenic fungi such as *Alternaria dauci*, *Alternaria radicina*, *Rhizoctonia solani*, and *Sclerotinia sclerotiorum* [19]. For lettuce crops, the addition of *Trichoderma* provided an improvement in the nutritional status of the leaf, with a greater accumulation of potassium and magnesium, in addition to a higher yield of fresh and dry biomass. It was also found that the application of *Trichoderma* provided greater antioxidant activity, increased total ascorbic acid and total phenols, and lowered the nitrate content [19,40].

In a study with lettuce plants, a biostimulan derived from the solid-state fermentation of green residues and wood chips with *Trichoderma harzianum* was evaluated, with and without L-tryptophan supplementation [40]. The product improved lettuce growth parameters such as fresh weight, plant height, leaf area, and antioxidant compounds, especially under proper irrigation [40].

Altogether, the promising results observed with *Trichoderma*-based formulations, as well as their combinations with other microbial agents, reinforce their value as multifunctional biostimulants capable of enhancing both productivity and crop quality [19,20,38–41]. Their ability to induce resistance mechanisms, promote hormonal balance, and stimulate the accumulation of secondary metabolites places these fungi-based products among the most versatile tools available in sustainable horticulture. In lettuce cultivation specifically, their effects go beyond simple yield increases, contributing meaningfully to the nutritional and functional quality of the harvested product. As agricultural systems face increasing environmental and economic pressures, expanding the understanding and use of biostimulants such as *Trichoderma* has become a necessary step toward resilient and efficient production models.

3. Perspectives

Continued advancements in the use of biostimulants for lettuce cultivation opens up promising avenues for more sustainable and resilient agricultural systems. Although considerable progress has been made in identifying compounds that improve growth, quality, and stress tolerance in lettuce, the complexity of interactions between bioactive molecules, environmental factors, and plant physiology still demands more in-depth investigation. The current body of research—summarized in Table 1—demonstrates the diversity of the available biostimulant classes, from humic substances and protein hydrolysates to seaweed extracts, microbial consortia, and enzyme-based formulations, each offering distinct physiological benefits to the crop.

Table 1. Comparative analysis of biostimulant applications in lettuce based on published scientific literature.

Class	Product	Authors	Effects on Lettuce
Humic and Fulvic Acids	BLACKJAK® Humic and Fulvic acids	[12] [18,20,28]	Increased biomass, antioxidant. Nutrient absorption, stress resistance.
Protein Hydrolysates	Trainer®, Vegamin® Viva®, Radifarm®, Megafol®	[16,17]	Leaf area, number of leaves, shoot fresh weight (SFW), and chlorophyll content.
	PHs	[16,39]	Antioxidant activity and secondary metabolites. Shoot fresh weight, dry weight, leaf area, and yield.
Algal Extracts	*Ascophyllum nodosum* *Sargassum* spp. *Lithothamnium* sp.	[26–28] [27] [30]	Increase in quality, shelf life, protective enzymes, and biomass under salt stress. Accumulated dry biomass. Root development.
Bacteria + Microalgae	PGPR + *Chlorella vulgaris*	[13,14]	Shoot biomass and carotenoid content under heat stress. Lettuce yield.
Enzymes	B-glucosidase + xylanase	[23]	Increase in fresh and dry biomass and bioactive compounds.
Plant Extracts	Moringa Leaf Extract (MLE)	[26]	Increase in head size and number of leaves.
Fungi-based	*Trichoderma* spp.	[19,38,40,41]	Increase in biomass, antioxidants, ascorbic acid, morphological and biomass parameters.

Source: Author's own elaboration based on published studies, 2025.

Although the beneficial effects of biostimulants in mitigating abiotic stress and enhancing plant growth are well recognized, their effective adoption in lettuce cultivation requires a careful and context-specific approach. The physiological mechanisms activated by biostimulants vary widely depending on the type of compound used—such as humic substances, seaweed extracts, amino acids, protein hydrolysates, or growth-promoting microorganisms—as well as factors like application rate, timing, and environmental conditions. This complexity often limits the consistency of agronomic outcomes, especially when transitioning from controlled experimental conditions to the variability of open-field environments. Moving forward, integrating biostimulants into commercial production systems will require a more precise understanding of dose–response relationships, application timing, and synergistic effects between compounds. Furthermore, as climate change continues to intensify abiotic stresses such as salinity, heat, and water scarcity, the role of biostimulants in mitigating these effects is becoming increasingly relevant.

In this context, despite aligning with sustainable agricultural practices and the goal of reducing chemical input dependence, the technical and economic feasibility of biostimulants in lettuce production depends on further research. Large-scale field validation is necessary to confirm the promising effects observed in preliminary studies. A critical perspective suggests that the widespread integration of biostimulants will require investment in technological innovation to improve formulations, optimize delivery systems, and deepen understanding of the interactions between bioactive compounds, abiotic stressors, and plant physiology. Beyond yield enhancement, biostimulants also hold the potential to improve nutritional quality, increase stress tolerance, and promote adaptive responses, positioning them as valuable tools for modern horticulture—though still in need of consistent and reproducible scientific validation across diverse agricultural systems.

Investment in research and innovation is therefore essential, not only to validate emerging products under diverse environmental and cultivation conditions but also to refine formulations and delivery technologies. The demand for more functional, nutrient-rich, and environmentally sustainable vegetables also adds pressure for solutions that improve both productivity and food quality. In this sense, the use of biostimulants aligns with global trends toward clean agriculture, a reduced reliance on synthetic inputs, and an enhanced nutritional efficiency.

4. Conclusions

In lettuce cultivation, biostimulants have stood out as a promising strategy, promoting more sustainable growth, higher productivity, and greater plant resistance to abiotic stresses. The use of these substances improves nutrient absorption and optimizes physiological processes such as photosynthesis and the synthesis of bioactive compounds. However, there are still gaps to be filled, such as the lack of specific regulation for the use of biostimulants and the need for more in-depth studies on their effects. Therefore, it is essential to invest in research that investigates the efficiency of different types of biostimulants, their mechanisms of action, and the economic and environmental impacts of their application in lettuce. This will enable the development of more accessible and regulated technologies, ensuring that the benefits of biostimulants are known and exploited.

Author Contributions: Original draft preparation: T.M.L.d.O.; Literature review: T.M.L.d.O., A.J.C.J.M., J.S.B.P. and V.d.S.O.; Editing: T.M.L.d.O., A.J.C.J.M. and S.D.-A.; Figures preparation: A.J.C.J.M.; Supervision: S.D.-A. and A.A.F.; Conceptualization: S.D.-A. and L.d.O.A. All authors have read and agreed to the published version of the manuscript.

Funding: The APC for this research was partially funded by the Federal Institute of Espírito Santo.

Data Availability Statement: The original contributions presented in the study are included in the article; further inquiries can be directed to the corresponding authors.

Acknowledgments: The authors would like to thank the Instituto Federal do Espírito Santo (Ifes) for the financial support granted for the development of this study. We also acknowledge institutions such as Instituto Capixaba de Pesquisa and Assistência Técnica e Extensão Rural and the individuals who provided technical, administrative, or logistical assistance throughout the preparation of this manuscript.

Conflicts of Interest: The authors declare no conflict of interest.

References

1. Mitsigiorgi, K.; Ntroumpogianni, G.C.; Katsifas, E.A.; Hatzinikolaou, D.G.; Chassapis, K.; Skampa, E.; Stefi, A.L.; Christodoulakis, N.S. Lettuce (*Lactuca sativa* L.) Cultures and the Bioactivity of Their Root Microflora Are Affected by Amended Soil. *Plants* **2024**, *13*, 13. [CrossRef] [PubMed]
2. Xu, L.; Geelen, D. Developing biostimulants from agro-food and industrial by-products. *Front. Plant Sci.* **2018**, *9*, 1567. [CrossRef]
3. He, S.; Huo, G.; Ge, G.; Du, L.; Long, D.; Luo, L.; Long, P.; Wei, S. Identification and Preliminary Functional Analysis of Responsive Genes to Short-term Heat Stress in Lettuce. *J. Am. Soc. Hortic. Sci.* **2024**, *149*, 327–336. [CrossRef]
4. Szparaga, A.; Kuboń, M.; Kocira, S.; Czerwińska, E.; Pawłowska, A.; Hara, P.; Kobus, Z.; Kwaśniewski, D. Towards Sustainable Agriculture—Agronomic and Economic Effects of Biostimulant Use in Common Bean Cultivation. *Sustainability* **2019**, *11*, 4575. [CrossRef]
5. Velasco-Clares, D.; Navarro-León, E.; Izquierdo-Ramos, M.J.; Blasco, B.; Ruiz, J.M. Enhancing Drought Tolerance in Lettuce: The Efficacy of the Seaweed-Derived Biostimulant Cytolan® Stress Applied at Different Growth Stages. *Horticulturae* **2025**, *11*, 157. [CrossRef]
6. Zuzunaga-Rosas, J.; Calone, R.; Mircea, D.M.; Shakya, R.; Ibáñez-Asensio, S.; Boscaiu, M.; Fita, A.; Moreno-Ramón, H.; Vicente, O. Mitigation of salt stress in lettuce by a biostimulant that protects the root absorption zone and improves biochemical responses. *Front. Plant Sci.* **2024**, *15*, 1341714. [CrossRef]
7. Singh, M.; Subahan, G.M.; Sharma, S.; Singh, G.; Sharma, N.; Sharma, U.; Kumar, V. Enhancing Horticultural Sustainability in the Face of Climate Change: Harnessing Biostimulants for Environmental Stress Alleviation in Crops. *Stresses* **2025**, *5*, 23. [CrossRef]
8. Khan, S.; Yu, H.; Li, Q.; Gao, Y.; Sallam, B.N.; Wang, H.; Liu, P.; Jiang, W. Exogenous application of amino acids improves lettuce growth and yield by increasing photosynthetic assimilation and nutrient availability. *Agronomy* **2019**, *9*, 266. [CrossRef]
9. Abdelkader, M.; Oliveira, L.; Oliveira, L.; Oliveira, O.; Oliveira, M.; Puchkov, M.; Oliveira, E.; Amantayev, B.; Kipshakbaeva, A.; Arinov, B. Biostimulants-Based Amino Acids Augment Physio-Biochemical Responses and Promote Salinity Tolerance of Lettuce Plants (*Lactuca sativa* L.). *Horticulturae* **2023**, *9*, 807. [CrossRef]
10. Keskin, B.; Akhoundnejad, Y.; Dasgan, H.Y.; Gruda, N.S. Fulvic Acid, Amino Acids, and Vermicompost Enhanced Yield and Improved Nutrient Profile of Soilless Iceberg Lettuce. *Plants* **2025**, *14*, 609. [CrossRef]
11. Shahrajabian, M.H.; Chaski, C.; Polyzos, N.; Petropoulos, S.A. Biostimulants Application: A Low Input Cropping Management Tool for Sustainable Farming of Vegetables. *Biomolecules* **2021**, *11*, 698. [CrossRef]
12. Atero-Calvo, S.; Magro, F.; Masetti, G.; Navarro-León, E.; Rios, J.J.; Ruiz, J.M. Assaying the Use of a Leonardite-Suspension Concentrate-Based Product as a Potential Biostimulant to Enhance Growth, NPK Use Efficiency, and Antioxidant Capacity in *Lactuca sativa* L. *Agronomy* **2024**, *14*, 64. [CrossRef]
13. Meirelles, A.F.M.; Baldotto, M.A.; Baldotto, L.E.B. Productivity of lettuce (*Lactuca sativa* L.) in response to the application of humic acids and diazotrophic bacteria, under field conditions. *Rev. Ceres* **2017**, *64*, 5. [CrossRef]
14. Kopta, T.; Sękara, A.; Kalisz, A.; Poniedziałek, M.; Hura, T. Effect of bacterial-algal biostimulant on the yield and internal quality of lettuce (*Lactuca sativa* L.) produced for spring and summer crop. *Not. Bot. Horti Agrobot. Cluj-Napoca* **2018**, *46*, 2. [CrossRef]
15. Wang, Y.; Lu, Y.; Wang, L.; Song, G.; Ni, L.; Xu, M.; Nie, C.; Li, B.; Bai, Y. Analysis of the molecular composition of humic substance-sand their effects on physiological metabolism in maize based on untar-geted metabolomics. *Front. Plant Sci.* **2023**, *14*, 1122621. [CrossRef]
16. Sorrentino, M.; De Luca, G.; Rouphael, Y.; Colla, G.; Francini, A.; Marabottini, R.; De Palma, M.; Paradisi, D.; Lucini, L.; Canale, A.; et al. Integration of phenomics and metabolomics datasets reveals different mode of action of biostimulants based on protein hydrolysates in *Lactuca sativa* L. and *Solanum lycopersicum* L. under salinity. *Front. Plant Sci.* **2022**, *12*, 1150964. [CrossRef]
17. Zahra, A.M.; Sinaga, A.N.K.; Nugroho, B.D.A.; Masithoh, R.E. Effect of Plant Biostimulants on Red and Green Romaine Lettuce (*Lactuca sativa*) Growth in Indoor Farming. *IOP Conf. Ser. Earth Environ. Sci.* **2024**, *1297*, 012008. [CrossRef]

18. Atero-Calvo, S.; Magro, F.; Masetti, G.; Navarro-León, E.; Albacete, A.; Ruiz, J.M. The effects of humic substances application on thephytohormone profile in *Lactuca sativa* L. *Ann. Appl. Biol.* **2024**, *42*, 115–124. [CrossRef]
19. Bonini, P.; Rouphael, Y.; Miras-Moreno, B.; Lee, B.; Cardarelli, M.; Erice, G.; Cirino, V.; Lucini, L.; Colla, G. A Microbial-Based Biostimulant Enhances Sweet Pepper Performance by Metabolic Reprogramming of Phytohormone Profile and Secondary Metabolism. *Front. Plant Sci.* **2020**, *11*, 567388. [CrossRef]
20. Clément, J.; Delisle-Houde, M.; Nguyen, T.T.A.; Dorais, M.; Tweddell, R.J. Effect of Biostimulants on Leafy Vegetables (Baby Leaf Lettuce and Batavia Lettuce) Exposed to Abiotic or Biotic Stress under Two Different Growing Systems. *Agronomy* **2023**, *13*, 879. [CrossRef]
21. Rouphael, Y.; Carillo, P.; Garcia-Perez, P.; Cardarelli, M.; Senizza, B.; Miras-Moreno, B.; Colla, G.; Lucini, L. Plant biostimulants from seaweeds or vegetal proteins enhance the salinity tolerance in greenhouse lettuce by modulating plant metabolism in a distinctive manner. *Sci. Hortic.* **2022**, *305*, 111368. [CrossRef]
22. Navarro-León, E.; Lópes-Moreno, F.J.; Borda, E.; Marín, C.; Blasco, N.S.B.; Ruiz, J.M. Effect of L-amino acid-based biostimulants on nitrogen use efficiency (NUE) in lettuce plants. *J. Sci. Food Agric.* **2022**, *102*, 15. [CrossRef]
23. Tamburino, R.; Docimo, T.; Sannino, L.; Gualtieri, L.; Palomba, F.; Valletta, A.; Ruocco, M.; Scotti, N. Enzyme-Based Biostimulants Influence Physiological and Biochemical Responses of *Lactuca sativa* L. *Biomolecules* **2023**, *13*, 1765. [CrossRef]
24. Yassen, A.A.; Hájos, M.T. The potential role of moringa leaf extract as bio-stimulant to improve some quality parameters of different lettuce (*Lactuca sativa* L.) genotypes. *Sarhad J. Agric.* **2021**, *37*, 4. [CrossRef]
25. Alou, D.; Kalleli, F.; Abid, G.; Karmous, C.; Manaa, M.; M' Hamdi, M.M. Marine Algal extract as a biostimulant to improve tolerance to salinity in lettuce plants. *Emir. J. Food Agric.* **2023**, *35*, 1–18. [CrossRef]
26. Chrysargyris, A.; Xylia, P.; Anastasiou, M.; Pantelides, I.; Tzortzakis, N. Effects of *Ascophyllum nodosum* seaweed extracts on the growth, physiology, and storage of minimally cut salads under potassium deficiency. *J. Sci. Food Agric.* **2018**, *98*, 5861–5872. [CrossRef] [PubMed]
27. Ramdin, T.M.; Ali, O.; Ramsubhag, A.; Jayaraman, J. Sargassum spp. and Ascophyllum nodosum seaweed extract formulations improve plant growth and alter the bacterial dynamics in hydroponically cultivated lettuce. *J. Appl. Phycol.* **2024**, *36*, 3769–3782. [CrossRef]
28. Sandepogu, M.; Shukla, P.S.; Asiedu, S.; Yurgel, S.; Prithiviraj, B. Combination of *Ascophyllum nodosum* Extract and Humic Acid Improve Early Growth and Reduces Post-Harvest Loss of Lettuce and Spinach. *Agriculture* **2019**, *9*, 240. [CrossRef]
29. Oliveira, L.F.C.; Pereira, B.A.; Silva, T.M.L.E.G. Adsorption of metal ions in red seaweed *Lithothamnium calcareum* in the treatment of industrial effluents. *Eng. Sanitária E Ambient.* **2021**, *26*, 2.
30. Ramos, E.P.; Ferreira, T.R.; Aguiar, D.B.; Alves, F.L.; Dousseau-Arantes, S. *Lithothamnion* sp. as biostimulant in plant cultivation. *Pesq. Agropec. Trop.* **2023**, *53*, e76273. [CrossRef]
31. Ovali, G.; Özdamar Ünlü, H.O. Effects of Chitosan Application on the Yield and Quality of Lettuce. *BioResources* **2024**, *19*, 985–997. [CrossRef]
32. Dávila Rangel, I.E.; Trejo Téllez, L.I.; Ortega Ortiz, H.; Juárez Maldonado, A.; González Morales, S.; Companioni González, B.; Cabrera De la Fuente, M.; Benavides Mendoza, A. Comparison of Iodide, Iodate, and Iodine-Chitosan Complexes for the Biofortification of Lettuce. *Appl. Sci.* **2020**, *10*, 2378. [CrossRef]
33. Lyalina, T.; Shagdarova, B.; Zhuikova, Y.; Il'ina, A.; Lunkov, A.; Varlamov, V. Effect of Seed Priming with Chitosan Hydrolysate on Lettuce (*Lactuca sativa*) Growth Parameters. *Molecules* **2023**, *28*, 1915. [CrossRef]
34. Zhang, G.; Wang, Y.; Wu, K.; Zhang, Q.; Feng, Y.; Miao, Y.; Yan, Z. Exogenous Application of Chitosan Alleviate Salinity Stress in Lettuce (*Lactuca sativa* L.). *Horticulturae* **2021**, *7*, 342. [CrossRef]
35. Treviño-Ruiz, K.S.; Ortega-Ortiz, H.; Benavides-Mendoza, A.; González-Morales, S. Effect of nanochitosan-iodine application in lettuce on biofortification, growth and yield. *Ecosist. Recur. Agropec.* **2024**, *11*, 3615. [CrossRef]
36. Carneiro, P.B.M.; Silva, A.C.; Souza, R.L.; Fernandes, M.T.; Oliveira, J.F.; Santos, P.H. Structure, growth and production of $CaCO_3$ in a shallow bed of rhodoliths of a highly energetic siliciclastic-carbonate coast in the equatorial southwest of the Atlantic Ocean. *Mar. Environ. Res.* **2021**, *166*, 105280. [CrossRef]
37. Villa e Vila, V.; Oliveira, P.A.A.; Gomes, T.M.; Oliveira, A.F.; Oliveira, V.G.; Wenneck, G.S.; Franco, L.B. Deficit Irrigation with Silicon Application as Strategy to Increase Yield, Photosynthesis and Water Productivity in Lettuce Crops. *Plants* **2024**, *13*, 1029. [CrossRef]
38. Rouphael, Y.; Carillo, P.; Colla, G.; Fiorentino, N.; Sabatino, L.; El-Nakhel, C.; Giordano, M.; Carvalho, A.; Cirillo, V.; Shabani, E.; et al. Evaluation of Combined Applications of *Trichoderma virens* and a Biopolymer-Based Biostimulant in Agronomic, Physiological and Qualitative Properties of Lettuce under Variable N Regimes. *Agronomy* **2020**, *10*, 196. [CrossRef]
39. Cristofano, F.; El-Nakhel, C.; Pannico, A.; Giordano, M.; Colla, G.; Rouphael, Y. Foliar and Root Applications of Vegetal-Derived Protein Hydrolysates Differentially Enhance the Yield and Qualitative Attributes of Two Lettuce Cultivars Grown in Floating System. *Agronomy* **2021**, *11*, 1194. [CrossRef]

40. Porras, R.C.S.; Ghoreishi, G.; Sánchez, A.; Barrena, R.; Font, X.; Ballardo, C.; Artola, A. Solid-state fermentation of green waste for the production of biostimulants to enhance lettuce (*Lactuca sativa* L.) cultivation under water stress: Closing the organic waste cycle. *Chemosphere* **2025**, *370*, 143919. [CrossRef]
41. Gutiérrez-Chávez, A.; Robles-Hernández, L.; Guerrero, B.I.; González-Franco, A.C.; Medina-Pérez, G.; Acevedo-Barrera, A.A.; Hernández-Huerta, J. Potential of *Trichoderma asperellum* as a Growth Promoter in Hydroponic Lettuce Cultivated in a Floating-Root System. *Plants* **2025**, *14*, 382. [CrossRef]

plants

MDPI

Article

An Efficient Agrobacterium-Mediated Transient Transformation System Using In Vitro Embryo-Derived Seedlings for Gene Function Elucidation in *Paeonia ostii*

Yuhui Zhai †, Xinrong Xie †, Liping Zhang, Xuefei Wang, Zixuan Zhang, Lixin Niu * and Yanlong Zhang *

College of Landscape Architecture and Arts, Northwest A&F University, Yangling 712100, China; zhaiyuhui@nwafu.edu.cn (Y.Z.); xiexinrong@nwafu.edu.cn (X.X.); zlp0501@nwafu.edu.cn (L.Z.); wxuefei@nwafu.edu.cn (X.W.); 2024057025@nwafu.edu.cn (Z.Z.)

* Correspondence: niulixin@nwafu.edu.cn (L.N.); zhangyanlong@nwafu.edu.cn (Y.Z.)

† These authors contributed equally to this work.

Abstract: *Paeonia ostii* is an economically significant species serving as an ornamental, medicinal herb, and woody oilseed crop. Gene function elucidation and molecular breeding are hindered by the lack of efficient, stable transformation methods due to tissue culture challenges. To enable year-round functional studies without material constraints, we established a novel transient transformation system mediated by Agrobacterium using in vitro embryo-derived seedlings (TTAES) in *P. ostii*. By optimizing embryo germination media, we achieved consistent seedling production. Orthogonal experiments with a GUS reporter identified optimal conditions: OD_{600} = 1.0, 200 μM of acetosyringone, six negative-pressure treatments, and 2 h infection. Under this optimized system, maximum transformation efficiency was achieved at 35 days after germination. With this system, we demonstrated its application in investigating transcription factor-mediated regulation of target gene promoters using GUS as a reporter gene. To achieve non-destructive identification of transiently transformed plants, we employed GFP as a reporter gene. Using transient expression of VIGS (knockdown) and 35S constructs (overexpression), we characterized gene functions, thereby confirming the system's effectiveness for functional analysis. This system facilitates the acquisition of plant experimental materials and significantly improves research efficiency for year-round gene function elucidation in *P. ostii*.

Keywords: *Paeonia ostii*; in vitro embryo-derived seedlings; Agrobacterium; transient transformation system

1. Introduction

Tree peony, a crop of significant economic importance that has been cultivated in China for over 2000 years [1], is highly prized for both its ornamental value and medicinal properties [2,3]. The recent advancements witnessed in the cut flower markets and seed oil production of peonies have further fueled growth within this sector [4,5]. The fleshy root bark of *Paeonia ostii* is a key source of Moutan Cortex in herbal medicine [6]. Meanwhile, *P. ostii* is recognized as a promising woody oilseed crop due to its high seed oil content and rich unsaturated fatty acid profile [7,8]. Following the completion of the chromosome-level genome assembly of *P. ostii* [9], research on the molecular mechanisms underlying genetic improvement has significantly expanded [10–12]. However, peony tissue culture remains plagued by persistent challenges, including severe explant browning, vitrification, contamination, difficulties in differentiation, limited regeneration efficiency, and suboptimal rooting [13,14]. Consequently, the lack of stable genetic transformation systems has made transient transformation crucial for elucidating gene functions and promoting molecular breeding in peony [15,16].

Transient transformation is a rapid and convenient approach to gene function analysis. Currently, the three predominant methodologies for transient genetic transformation in plants are gene gun, protoplast transformation, and Agrobacterium-mediated infiltration [17,18]. Among these techniques, Agrobacterium-mediated infiltration has been proven particularly efficacious due to its simplicity, speed, and efficiency in facilitating transient, high-level expression of target genes, thus making it the favored method for plant gene characterization [19]. This technique involves the introduction of suspension-cultured Agrobacterium cells into intact plant organs. Notably, the transformation efficiency of this technique is influenced by several parameters, including bacterial concentration (OD_{600}), infiltration duration, and acetosyringone (AS) concentration [20]. Agrobacterium-mediated transient overexpression is widely used for gene function verification in peony [21]. Additionally, tobacco rattle virus (TRV)-mediated virus-induced gene silencing (VIGS) represents a commonly employed method for gene silencing analysis [10]. Agrobacterium-mediated delivery facilitates the introduction of TRV vectors into diverse plant tissues (e.g., buds, leaves, roots, and seeds) to achieve transient gene silencing [22–25]. However, a critical limitation is that fresh tissue materials required for these experiments are often seasonally restricted, such as being limited to specific growth stages. Consequently, elucidating gene function in peony takes significantly longer compared with model plants exhibiting year-round growth.

Besides their large quantity, *P. ostii* seeds are easy to preserve after harvest and show high germination rates [26,27]. However, under natural conditions, germination and rooting require approximately three months to complete, and the quality of open-field planted seedlings is unstable [28]. Embryo culture technology has been demonstrated to facilitate accelerated seedling development from *P. ostii* embryos, thereby reducing the breeding cycle and enhancing germination rates [29]. Nevertheless, this technique continues to encounter difficulties in terms of achieving optimal rooting efficiency. Given the short experimental duration of the transient transformation system, it remains to be determined whether transient transformation assays can be exclusively performed using in vitro embryo-derived seedlings without prior rooting. Recently, an Agrobacterium-mediated transient genetic transformation system has been established in vitro embryo-derived seedlings of *P. lactiflora* [30]. Nonetheless, the system investigated the transient transformation of *P. lactiflora* seedlings cultivated in vitro, neglecting to examine the growth conditions of in vitro embryo-derived seedlings. In addition, the transient transformation system developed for *P. lactiflora* relies solely on GUS labeling, which inflicts irreversible damage on plant materials following detection, thereby limiting its utility for in vivo studies [31]. The development of an efficient genetic transformation system for *P. ostii* (tree peony) presents distinct and significant challenges compared to *P. lactiflora* (herbaceous peony), primarily stemming from its woody perennial nature. Woody plant tissues, including seeds and embryos from *P. ostii*, are often more difficult to surface-sterilize effectively due to crevices and protective structures, leading to higher contamination risks in tissue culture compared to *P. lactiflora* [32,33]. Like many woody plants, *P. ostii* produces abundant secondary metabolites, such as paeonol, which exhibits antibacterial activity against diverse bacteria [34]. This increases the complexity of the Agrobacterium-mediated transient transformation system, a complexity that is generally less pronounced in herbaceous plants [35].

The present application of Agrobacterium-mediated transient genetic transformation in peonies is predominantly focused on gene transient overexpression and silencing. There has been little investigation into subcellular protein localization, protein–protein interactions, or transcription factor-target promoter relationships. It is imperative to acknowledge that, as a woody crop, *P. ostii* is distinct from the herbaceous crop *P. lactiflora*, and thus exhibits unique optimal germination and transient transformation requirements.

In this study, we established a novel transient transformation system mediated by Agrobacterium using in vitro embryo-derived seedlings in *P. ostii*. This system was designated as the transient transformation system mediated by Agrobacterium using in vitro embryo-derived seedlings (TTAES). By systematically screening embryo germination media, we achieved a consistent annual supply of experimental materials, overcoming seasonal limitations. Orthogonal experiments were conducted using the GUS gene as a reporter, with the objective of optimizing key parameters. The following factors were identified as the key elements for optimizing the Agrobacterium infection conditions: the bacterial concentration (OD_{600}), AS concentration, negative pressure treatment frequency, and infection duration. The infection efficiency was then assessed at distinct developmental stages of axenic embryogenic seedlings under optimized conditions. Finally, the system was validated during the verification of the transcription factor-activated target gene promoter, and its applications in gene silencing and overexpression were discussed with GFP as a marker.

2. Materials and Methods

2.1. Plant Materials and Growth Conditions

P. ostii were cultivated in the Peony Resource Nursery of Northwest A&F University. Uniform-genotype seeds were produced by controlled cross-pollination between selected mother and father plants. Mature seeds were collected 120 days after artificial pollination and air-dried for use as experimental materials. A refined protocol for *P. ostii* seed embryo germination was applied as follows [14,30,36,37]: Large and plump seeds were selected and soaked in water for 48 h. Subsequently, seeds that sank to the container bottom were treated with a 1 g/L GA_3 solution for 24 h. After removing the seed coat, seeds were rinsed repeatedly under tap water for 1 h, followed by half an hour in detergent water. Thereafter, they were rinsed again under tap water for 15 min. Seeds were then transferred to the benchtop, where they were disinfected with 75% ethanol for 30 s, washed twice with sterile water, treated with 2% sodium hypochlorite for 6 min, washed twice with sterile water, and subjected to a second 2% sodium hypochlorite treatment for 6 min, followed by six rinses with sterile water. Finally, seed embryos were excised using a scalpel and dissection needle and inoculated onto embryo germination media (Figure 1).

The embryo germination media used included Murashige and Skoog (MS) medium and Woody Plant Medium (WPM) [17,30]. WPM is characterized by lower overall nutrient levels—specifically reduced nitrogen and potassium—that may better suit woody plants' growth requirements, and by relatively higher calcium content that may enhance cell wall stability and stress tolerance [38].MS medium: MS supplemented with 30 g/L sucrose, 0.5 mg/L 6-BA, and 1.0 mg/L GA_3 (pH 5.6); WPM: WPM supplemented with 30 g/L sucrose, 0.5 mg/L 6-BA, and 1.0 mg/L GA_3 (pH 5.6). Inoculated embryos were then cultured in an artificial growth chamber under a 16 h light (2000 Lx)/8 h dark photoperiod at 24 ± 2 °C. Embryo germination rates in both media were recorded every five days starting from 5 days after inoculation. The germination rate was calculated as (number of germinated seedlings/number of inoculated embryos) × 100%, with data collected at five-day intervals. Each medium was inoculated with 30 embryos, with three biological replicates per medium.

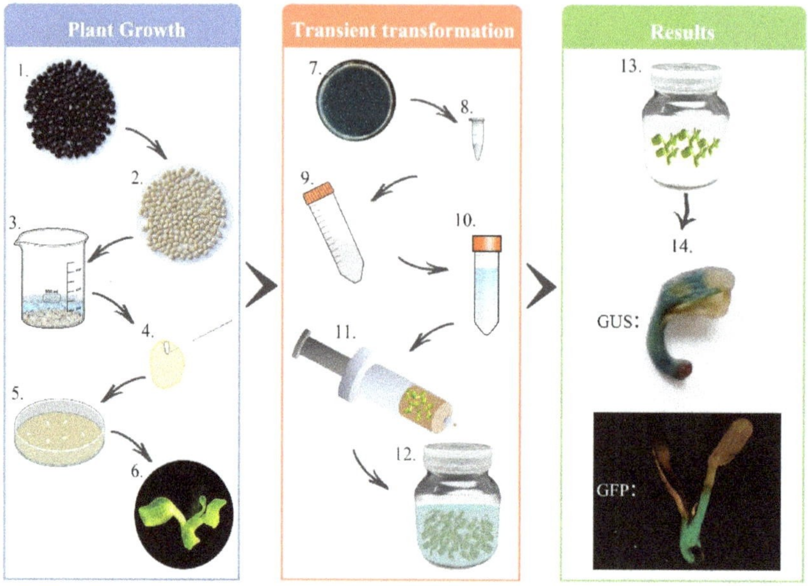

Figure 1. Overview of TTAES protocol in *P. ostii*. The protocol is divided into three sections: Plant Growth, Transient Transformation, and Results. Plant growth: (1) select large, plump seeds, then water selection and GA_3 treatment; (2) remove the seed coat; (3) sterilize the seeds in a laminar flow cabinet; (4) isolate the seed embryos using a dissection needle; (5) inoculate the seed embryos onto embryo germination medium; (6) culture in a tissue culture chamber. Transient transformation: (7) transform the target gene vector into Agrobacterium GV3101; (8) inoculate a single colony into 200 μL of YEP liquid medium; (9) inoculate the bacterial culture into fresh YEP medium at a 1:100 dilution for a second round of cultivation; (10) collect the bacterial cells and resuspend them; (11) subject the embryogenic seedlings to vacuum infiltration using a syringe; (12) co-cultivate the vacuum-infiltrated seedlings with the resuspended bacterial solution at 28 °C, 100 rpm, in the dark. Results: (13) after air-drying the seedlings on sterile filter paper, inoculate them onto co-culture medium; (14) perform GUS staining or GFP fluorescence observation according to the tag carried by the vector.

2.2. RNA Extraction and Gene Expression Profiling

In this experiment, total RNA extraction was conducted in accordance with the instructions provided in RNAprep Pure Plant Kit (TIANGEN, Beijing, China). Using the PrimScript^{TM}RT reagent Kit with gDNA Eraser (Takara, Kusatsu, Shiga, Japan), synthesize first-strand cDNA. 18S-26S ITS RNA was used as the internal standard to assess mRNA abundances following the $2^{-\Delta\Delta Ct}$ method. RT-qPCR was performed using SYBR® Premix Ex TaqTM II (Takara, Japan) on a StepOne Real Time PCR system. Primers used for vector construction and RT-qPCR are listed in Table 1.

2.3. Construction of Recombinant Plasmids

The cDNA from *P. ostii* seeds was used as the template, and primers 35S:PoABI5-F/R, PoABI5-pTRV2-F/R, and PoABI5-GFP-F/R were used for PCR amplification. The full-length CDS of PoABI5 was cloned into the pCAMBIA1300, pTRV2, and pCAMBIA2300-eGFP vectors using one-step cloning. Notably, the fragment amplified with PoABI5-GFP-F/R primers lacked a stop codon. Subsequently, genomic DNA was extracted from *P. ostii* using the CTAB method and used as a template to amplify the target fragment with ProPoFAD3-GUS-F/R primers. The resulting PCR product was then cloned into the pBI121 vector. All PCR primers used in this study are listed in Table 1.

Table 1. Primers used in this study.

Primer Name	Primer Sequence (5′-3′)	Annotation
ProPoFAD3-GUS-F	ACCATGATTACGCCAAGCTTTGTTCGTCTCGTCGTCCTCTTC	Vector Construction
ProPoFAD3-GUS-R	GACTGACCACCCGGGGATCCTTGGCCTTCGGTTCACAGAT	
35S:PoABI5-F	ACGGGGGACGAGCTCGGTACCATGGTTGTTCCTGAGTCCGAAA	
35S:PoABI5-R	GCTCTGCAGGTCGACTCTAGATCAAAAAGGGGAACTCAAAGTCC	
PoABI5-pTRV2-F	GAAGGCCTCCATGGGGATCCTAAACCAGCGCGGTTATCCA	
PoABI5-pTRV2-R	GGACATGCCCGGGCCTCGAGGGATTGTCTTCCCAATGATGCA	
PoABI5-GFP-F	ACGGGGGACGAGCTCGGTACCATGGTTGTTCCTGAGTCCGAAA	
PoABI5-GFP-R	CTTGCTCACCATGGTGTCGACAAAAGGGGAACTCAAAGTCCTCC	
PoABI5-RT-F	TGGAAGCAGAGCTGAACCAG	RT-PCR /qRT-PCR
PoABI5-RT-R	CGAACACCTCAAAAAGGGGA	
PoFAD3-RT-F	GGCAGCCACATTTCCGTCTT	
PoFAD3-RT-R	AGTAGAGTGGCCAAACAGCC	
TRV1-RT-F	CAGTCTATACACAGAAACAGA	
TRV1-RT-R	GACGTGTGTACTCAAGGGTT	
TRV2-RT-F	GGCTAACAGTGCTCTTGGTG	
TRV2-RT-R	GTATCGGACCTCCACTCGC	
18S-26S ITS RNA-F	ACCGTTGATTCGCACAATTGGTCATCG	
18S-26S ITS RNA-R	TACTGCGGGTCGGCAATCGGACG	

2.4. *Agrobacterium Transformation, Cultivation, and Preparation of Infiltration Suspension*

Given the versatility of GV3101 and its demonstrated efficacy in the transient transformation of *P. ostii* tissues in prior studies, this work employed the GV3101 strain for Agrobacterium-mediated transformation [10,22,23]. The Agrobacterium strain GV3101 competent cells (Weidi Biotech, Shanghai, China), stored at −80 °C, were thawed on ice. Subsequently, 1 μL of target plasmids was added, and the mixture was incubated on ice for 5 min, quick-frozen in liquid nitrogen for 5 min, and then incubated at 37 °C for 5 min. After a further 5 min incubation on ice, 700 μL of fresh Yeast Extract Peptone (YEP) liquid medium was added. The culture was incubated with shaking at 28 °C and 200 rpm for 3 h. Subsequently, 200 μL of the bacterial culture was plated on a YEP agar plate containing rifampicin (25 mg/L), gentamicin (50 mg/L), and kanamycin (50 mg/L). The plate was then incubated at 28 °C in the dark for 48 h.

A single colony was picked and inoculated into 200 μL YEP liquid medium with 50 μg/mL kanamycin, 50 μg/mL gentamicin, and 25 μg/mL rifampicin. It was incubated at 28 °C with shaking for 5 h. Then, the bacterial culture was diluted and regrown at 28 °C, 200 rpm for 10–12 h until the target OD was reached. Cells were collected by centrifugation at 6000 rpm for 10 min at 4 °C, then resuspended in AS solutions (10 mM $MgCl_2$ + 10 mM MES + AS) at varying concentrations (Table 2).

Table 2. Orthogonal experimental design for Agrobacterium-mediated transient transformation.

Test Number	OD_{600}	AS Concentration (μmol·L^{-1})	Number of Negative Pressure	Infection Time (h)
1	0.8	150	2	2
2	0.8	200	4	4
3	0.8	250	6	8
4	1.0	150	4	8
5	1.0	200	6	2
6	1.0	250	2	4
7	1.2	150	6	4
8	1.2	200	2	8
9	1.2	250	4	2

2.5. Infiltration of the P. ostii Seedlings

The 50 mL sterile syringe was taken, and the injection port was closed, after which 20 mL of Agrobacterium suspension and an appropriate number of embryo seedlings were added in turn. The syringe was then extracted for negative pressure treatment. The duration of each treatment was set at five seconds, and negative pressure treatment was administered in accordance with the target number (Table 2). Subsequently, the seedlings and Agrobacterium suspension were transferred into a tissue culture bottle and shaken at 100 rpm at 28 °C for the target duration. Following infection, the seedlings were excised, air-dried on sterile filter paper, and inoculated onto WPM medium that had been supplemented with 30 g/L sucrose, 0.5 mg/L 6-BA, 1.0 mg/L GA_3, and 200 µmol/L AS (pH = 5.6). The seedlings were cultivated in the dark at 24 ± 2 °C for a 24 h period, then co-cultured under a 16 h light (2000 Lx)/8 h dark photoperiod for 2 days. Detailed information is provided in Table 2.

2.6. Orthogonal Experimental Design for Agrobacterium-Mediated Transient Genetic Transformation

In Agrobacterium-mediated transient genetic transformation, multiple factors can impact transformation efficiency. To identify the optimal infection conditions, this study conducted orthogonal experiments. Four factors were selected, each having three levels. Using the L9(3^4) orthogonal array as a template, the fifth factor was sequentially assigned to the last column at levels 1, 2, and 3, generating nine experimental treatments. Each treatment consisted of 30 embryogenic seedlings with three biological replicates (Table 2).

2.7. GUS Histochemical Analysis and Enzyme Activity Assay

The embryogenic seedlings that had undergone co-cultivation were rinsed three times with sterile water and then placed in GUS staining solution for staining at 37 °C in the dark for 18 h. After staining, the seedlings were transferred to 75% ethanol for decolorization for 5 h, followed by a further decolorization in 95% ethanol for 10 h. The ethanol could be changed 2–3 times until all pigments were removed. The seedlings were photographed, and the transient expression rate of GUS was recorded and calculated according to the following criteria:

The GUS transient expression was categorized into three levels (GUS-1, GUS-2, and GUS-3) based on the staining intensity and area in the embryogenic seedlings, to compare and evaluate the effectiveness of different infection condition combinations. The ratio of stained area to total seedling area was quantified using ImageJ 1.53a. GUS-1: Light staining or obvious GUS spots, but with an area less than 50% of the total seedling area. GUS-2: Distinct staining with an area ≥ 50% of the total seedling area. GUS-3: Very distinct deep blue staining with an area > 50% of the total seedling area.

The transient expression rates of GUS were calculated as follows:

GUS transient expression rate (%) = Number of seedlings showing GUS staining/Total number of stained seedlings × 100; GUS-1/2/3 grade rate (%) = Number of seedlings showing GUS-1/2/3 level staining/Total number of stained seedlings × 100

GUS activity was quantified fluorometrically. Tissue samples were flash-frozen in liquid nitrogen, homogenized in extraction buffer, and centrifuged at 12,000× *g* for 15 min to collect the supernatant. These were utilized for GUS assays, with MUG serving as the substrate. Following the incubation period, the reactions were halted, and the GUS activity was measured at 365/455 nm using a fluorometer. The specific activity was expressed as nmol of 4 MU/mg protein·min^{-1} [39].

2.8. GFP Fluorescence Observation

After co-cultivation, embryogenic seedlings were examined for GFP fluorescence using a portable UV analyzer under total darkness, and images were captured. Fluorescent tissues were embedded in agarose gel and sectioned with a vibratome (VT1000S, Leica, Berlin, Germany) to generate temporary slides. Subsequently, GFP fluorescence was visualized under a confocal laser scanning microscope (CLSM, TCS SP8, Leica, Berlin, Germany) at an excitation wavelength of 488 nm and an emission wavelength range of 500–560 nm.

2.9. Virus-Induced Gene Silencing of PoABI5

The pTRV2:GFP and pTRV2:PoABI5 constructs were transformed into Agrobacterium tumefaciens strain GV3101. Single colonies were cultured overnight at 28 °C in YEP medium supplemented with rifampicin (25 mg/L), gentamicin (50 mg/L), and kanamycin (50 mg/L) until reaching an OD_{600} of 1.0. Bacterial cells were harvested by centrifugation at 6000 r/min for 10 min and resuspended in an equal volume of infiltration buffer (10 mM $MgCl_2$, 10 mM MES, 200 µM AS, pH = 5.6). Two separate mixtures were prepared: Equal volumes of pTRV1 and pTRV2:GFP suspensions; Equal volumes of pTRV1 and pTRV2:PoABI5 suspensions. Each mixture was incubated at 28 °C in the dark for 3–4 h. Embryogenic seedlings were subjected to vacuum infiltration using a 50 mL syringe with the needle removed. A volume of 20 mL of the Agrobacterium mixture and seedlings were placed in the syringe, and a vacuum of −0.8 MPa was applied six times for 5 s each. The seedlings and suspension were then transferred to a tissue culture flask and co-cultivated at 28 °C with gentle shaking (100 rpm) for 2 h. Following incubation, seedlings were blotted dry on sterile filter paper and transferred to WPM basal medium supplemented with 30 g/L sucrose, 0.5 mg/L 6-BA, 1.0 mg/L GA_3, and 200 µM AS (pH 5.6). Samples were first incubated in the dark at 24 ± 2 °C for 1 day, followed by a 16 h light (2×10^3 lx)/8 h dark photoperiod for 2 days before analysis.

2.10. FA Extraction and Analysis

The present study methodically extracts the fatty acids from the seedlings of *P. ostii*, according to the method described by Ma et al. [40]. In summary, 100 mg of seedlings were placed into 4 mL of a methanol solution containing 2.5% (*v*/*v*) sulfuric acid, and the mixture was heated at 80 °C for 2 h. Following the transmethylation reaction, 2 mL of 0.9% NaCl and 1 mL of hexane were added for gas chromatography (GC) analysis. The analysis of FAs by GC was conducted in accordance with the established protocol [25].

3. Results

3.1. Optimal Germination Medium Screening for P. ostii Embryos

Mature seeds were harvested, air-dried, and used as experimental materials in this study. A refined embryo germination protocol for *P. ostii* seeds was employed to obtain in vitro embryo-derived seedlings for use in the transient transformation system (Figure 1). To further optimize the system, it is necessary to determine the optimal medium for initiating embryo germination in *P. ostii*. Based on previous studies, the germination rates in MS and WPM were evaluated [17,30]. Non-germinated embryos exhibited brownish-yellow spotting and lacked significant cotyledon expansion (Figure 2a), whereas successfully germinated ones appeared creamy white with expanded cotyledons (Figure 2b). Germination rates were regularly recorded (Figure 2c), showing a substantial difference as early as 5 days after inoculation (10.44% in MS vs. 18.7% in WPM), with the most conducive germination period occurring at 5–10 days, when 24.33% (MS) and 62.59% (WPM) of seeds germinated. Thereafter, germination levels stabilized at 38.88% (MS) and 78.88% (WPM, 2.03-fold higher) by day 20, with no significant differences observed between 20 and 25 days. Consequently, WPM was selected as the optimal medium for initiating embryo germination in *P. ostii* for subsequent experiments.

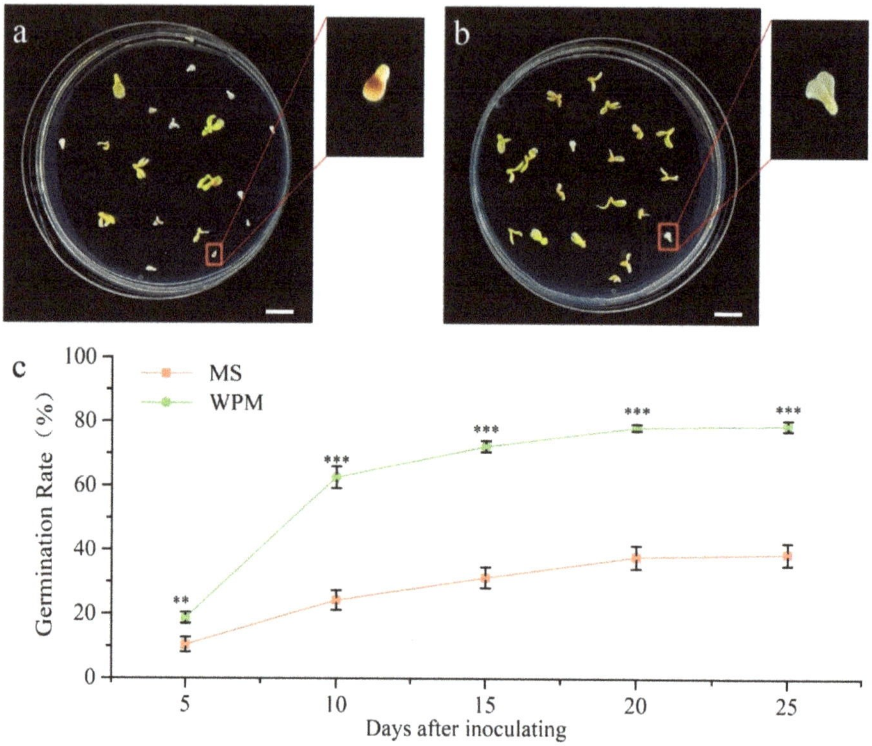

Figure 2. Screening of optimal germination medium for *P. ostii* embryos. (**a**). Germination status of *P. ostii* embryos on MS 15 days after inoculation, Bars = 1 cm; (**b**). Germination status of *P. ostii* embryos on WPM 15 days post-inoculation, Bars = 1 cm; (**c**). Time-course of embryo germination rates on MS and WPM media. A *t-test* was performed between all biological replicates and the control sample. Samples with significant differences are marked with asterisks (** $p < 0.01$, *** $p < 0.001$).

3.2. Orthogonal Optimization of Agrobacterium-Mediated Transient Transformation

In this study, a homologous protocol has been developed for the in vitro embryo-derived seedlings of *P. ostii*, drawing on the established transient transformation system of *P. lactiflora* (Figure 3a). To optimize Agrobacterium-mediated transient transformation efficiency, a L9(3^4) orthogonal array was used to investigate four three-level factors—Agrobacterium suspension OD_{600}, AS concentration, vacuum infiltration cycles, and infection duration, yielding nine treatment combinations (Table 2). The transformation efficiency of each group was assessed using GUS staining, with the results of the partial staining for the nine experimental groups presented in Figure 2b. The findings of the analysis demonstrated that Treatment 5 exhibited the highest GUS transient expression rate at 89.74%, with Treatment 1 following closely behind at 81.59% (Figure 3b). A further hierarchical comparison of GUS staining intensity demonstrated that for the strongest staining category (GUS-3), Treatments 2, 5, and 8 exhibited optimal performance, with GUS-3 proportions of 15.85%, 18.26%, and 16.67%, respectively (Figure 3e). Within the GUS-2 category, Treatments 1 and 5 exhibited the highest proportions, at 42.1% and 50%, respectively (Figure 3d). In the category of weakest staining (GUS-1), treatments 6, 7, and 9 exhibited higher proportions of 37.43%, 41.58%, and 39.96%, respectively (Figure 3c).

Collectively, Treatment 5 displayed the highest GUS transient expression rate and the highest proportions in both GUS-2 and GUS-3 categories. The following protocol has been determined to be the most effective Agrobacterium-mediated transient transformation: the OD_{600} of the culture was measured to be 1.0, AS concentration in the resuspension buffer was 200 μM, the negative pressure treatment was performed six times, and the infection duration was 2 h.

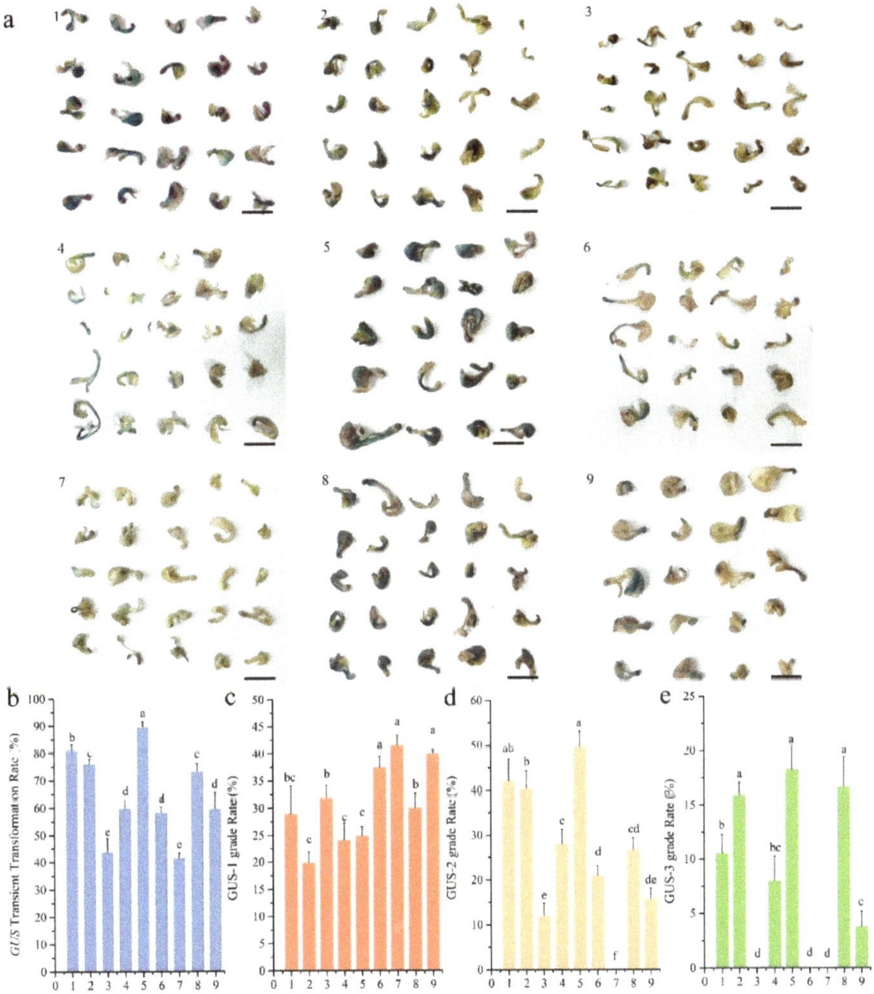

Figure 3. Orthogonal optimization of Agrobacterium-mediated transient transformation. (**a**). GUS staining phenotypes of seedlings, 1–9 represent nine experimental treatment groups designed based on the orthogonal experiment, Bars = 1 cm; (**b**). Transient transformation rate of GUS; (**c**). GUS-1 transient expression rate of seedlings; (**d**). GUS-2 transient expression rate of seedlings; (**e**). GUS-3 transient expression rate of seedlings. Error bar = standard deviation. Different letters above each bar indicate significant differences according to one-way ANOVA with Duncan's multiple comparisons test ($p < 0.05$).

3.3. Impact of Growth Status of In Vitro Embryo-Derived Seedlings in P. ostii on Agrobacterium Infection Efficiency

Given the close correlation between infection efficiency and plant growth status, the effect of in vitro embryo-derived seedling growth status on Agrobacterium infection efficiency was further investigated. Germinated seedlings were monitored from the inoculation day, designated as day after germination (DAG), with growth phenotypes documented via periodic observations and photography at 5-day intervals (Figure 4a). Before DAG25, seedlings showed a creamy white hue and sluggish vegetative growth, while chlorophyll accumulation initiated greening with accelerated growth at DAG25-30; callus formation first appeared at the base by DAG35, followed by true leaf emergence at DAG40, after which seedlings entered a senescence phase (growth arrest and yellowing) post-DAG40. Based on these observations, DAG30-40 seedlings were identified as optimal for Agrobacterium-mediated transient transformation.

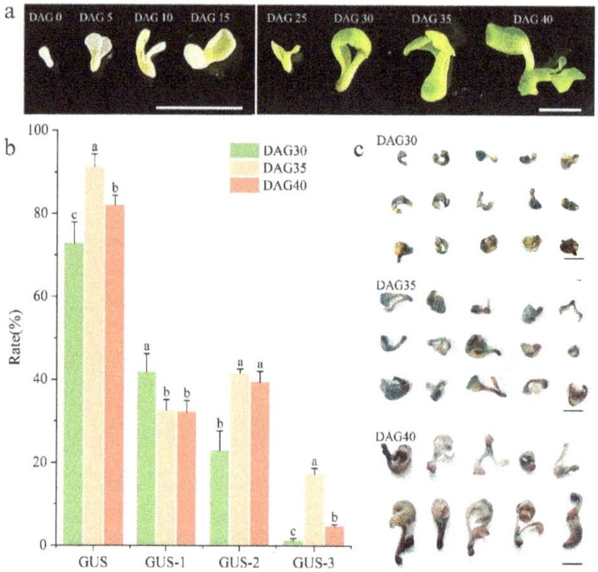

Figure 4. Effects of in vitro embryo-derived seedling Growth Status on Agrobacterium Infection Efficiency. (**a**). Morphological stages of in vitro embryo-derived seedling, Bars = 1 cm; (**b**). GUS transient transformation rate efficiency across seedling growth states, Error bar = standard deviation. Different letters above each bar indicate significant differences according to one-way ANOVA with Duncan's multiple comparisons test ($p < 0.05$); (**c**). GUS Staining phenotypes of in vitro embryo-derived seedlings, Bars = 1 cm.

Previously, DAG30 seedlings were selected as experimental material. To assess stage-specific infection efficiency variations, DAG30, 35, and 40 seedlings were infected using pre-identified optimal Agrobacterium parameters. GUS staining phenotypes were analyzed (Figure 4c), revealing that DAG35 seedlings exhibited the highest transient expression rate (91.23%), with 41.57% showing moderate (GUS-2) and 17.39% strong (GUS-3) staining, significantly higher than DAG30 and DAG40 counterparts (Figure 4b). Comparative analysis further showed DAG40 outperformed DAG30 in GUS staining efficiency. Collectively, these results identified DAG35 as the optimal developmental stage for *P. ostii* in vitro embryo derived seedlings, and TTAES was established following the optimization of Agrobacterium infection parameters and determination of the optimal stage.

3.4. Elucidation of TTAES for Assessing Transcription Factor-Mediated Target Gene Promoter Activation

To validate these results, TTAES was employed to assess the activation of target gene promoters mediated by transcription factors. The extant evidence demonstrates that PoABI5 has the capacity to upregulate *PoFAD3*, thereby contributing to the regulation of ALA synthesis in *P. ostii* seeds [22]. In this study, *PoABI5* and the *PoFAD3* promoter (ProPoFAD3) were successfully cloned into the pCAMBIA1300 and pBI121 vectors, respectively. TTAES was performed to assess GUS reporter expression and enzymatic activity across a range of experimental conditions. It has been demonstrated that the positive control group (35S:GUS + 35S:PoABI5) exhibits optimal GUS staining outcomes (Figure 5a) and the most elevated levels of GUS enzyme activity (Figure 5b), with a value of 93.55 nmol·mg^{-1}·min^{-1}. In comparison with the negative control group (ProPoFAD3:GUS + 35S), the experimental group (ProPoFAD3:GUS + 35S:PoABI5) demonstrated a more pronounced level of GUS staining (Figure 5a), with GUS enzyme activity reaching 67.13 nmol·mg^{-1}·min^{-1}, which was 1.69 times higher than that of the negative control group. This finding suggested that PoABI5 significantly activates the transcriptional activity of *PoFAD3* (Figure 5b). This finding was consistent with previous results [22], indicating that TTAES can be applied to verify the activation of target gene promoters by transcription factors.

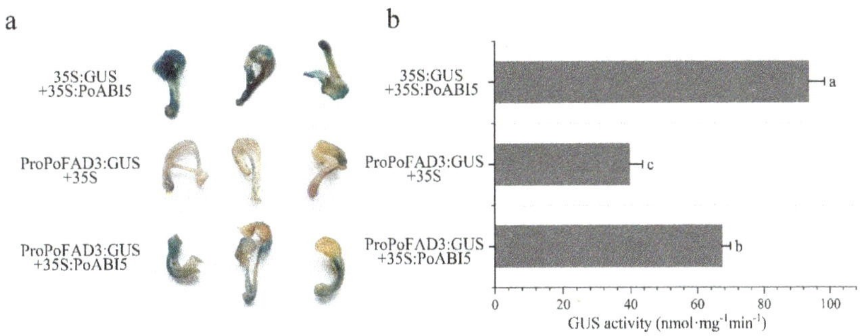

Figure 5. Transcriptional activation of the PoFAD3 promoter by PoABI5 in *P. ostii*. (**a**): GUS staining elucidation of *PoFAD3* promoter activation by PoABI5; (**b**): GUS enzyme activity assay for *PoFAD3* promoter transcription induced by PoABI5, Error bar = standard deviation. Different letters above each bar indicate significant differences according to one-way ANOVA with Duncan's multiple comparisons test ($p < 0.05$).

3.5. Elucidation of TTAES as a GFP-Marked System for Gene Function Analysis

Considering the irreversible damage of GUS staining to experimental materials, GFP tagging is commonly used for transformation detection in gene function elucidation [31,41]. TTAES of *P. ostii* was evaluated through two experiments using GFP as a tag and employing TRV-mediated VIGS and transient overexpression methods. The VIGS technique was employed to knock down *PoABI5* expression. Following the conclusion of the treatment, a period of one week was allowed to elapse before the plants were subjected to visualization under a portable UV lamp. The untreated wild-type (WT) sample displayed no fluorescence, whereas both pTRV2:GFP and pTRV2:PoABI5 tissues exhibited green fluorescence, indicating successful GFP protein expression and confirming effective agroinfiltration (Figure 6b). The efficiency of the VIGS system was validated by conducting semi-quantitative analysis of the fluorescent tissues. The presence of both *TRV1* and *TRV2* transcripts in the pTRV2:GFP and pTRV2:PoABI5 samples confirmed successful viral vector delivery (Figure 6a). qRT-PCR analysis revealed that *PoABI5* expression in the pTRV2: PoABI5 sample was reduced to 0.19-fold relative to the pTRV2: GFP sample, thereby demonstrating significant gene silencing (Figure 6c). Concomitantly, *PoFAD3* expression in pTRV2:PoABI5 was downregulated to 0.42-fold compared to the control (Figure 6c). Fatty acid content of WT, pTRV2:GFP, and pTRV2:PoABI5 tissues demonstrated a significant decrease in C18:3 (α-linolenic acid) and an increase in C18:2 (linoleic acid) in the silenced line (Figure 6d). These findings demonstrated the efficacy of TTAES in facilitating gene function elucidation of *P. ostii* through the implementation of GFP-tagged VIGS assays. This system provided a robust platform for gene function elucidation in *P. ostii*.

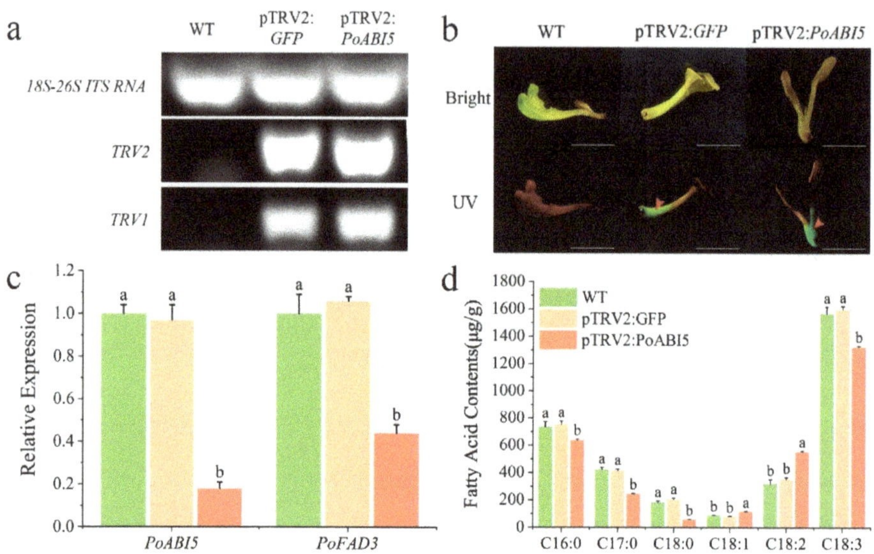

Figure 6. Elucidation of *PoABI5* silencing via VIGS in vitro seedlings of *P. ostii*. (**a**). Semi-quantitative RT-PCR analysis of *PoABI5* VIGS silencing; (**b**). UV imaging of seedlings after *PoABI5* VIGS silencing, Bars = 1 cm; (**c**). qRT-PCR quantification of *PoABI5* silencing efficiency; (**d**). Fatty acid composition changes in seedlings. Error bar = standard deviation. Different letters above each bar indicate significant differences according to one-way ANOVA with Duncan's multiple comparisons test ($p < 0.05$).

PoABI5 was cloned into the pCAMBIA2300-GFP vector for TTAES. Within one week after transformation, the plants were visualized under a portable UV lamp, and the resulting green fluorescence confirmed the successful transient expression of both 35S:GFP and 35S:PoABI5 (Figure 7a). Subsequent analysis using laser confocal microscopy further validated that GFP signals were detected exclusively in the 35S:GFP and 35S:PoABI5 samples, a finding that supported the efficacy of the transient expression process (Figure 7b). qRT-PCR analysis revealed that PoABI5 expression in the 35S:PoABI5 group was significantly upregulated compared to the 35S:GFP control, with *PoFAD3* expression in the overexpressing line reaching 2.43-fold the level of the control group, as presented (Figure 7c). The study revealed that the expression of *PoABI5*, which was observed as being significantly increased, was accompanied by a transient increase in the concentration of C18:3, whilst concurrently there was a decrease in the level of C18:2 (Figure 7d). Altogether, these results demonstrated that TTAES of *P. ostii* is effective for elucidating GFP-tagged gene function, thereby offering a robust platform for molecular research in this species.

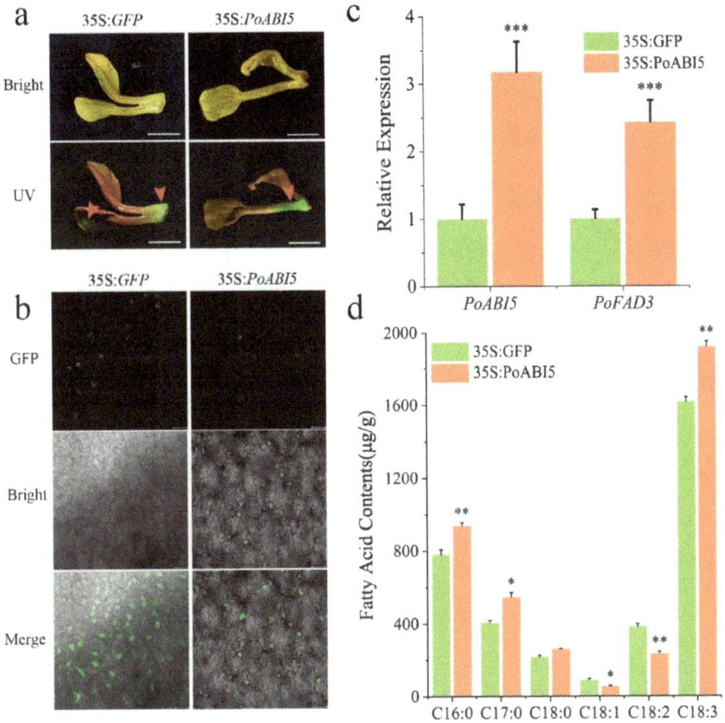

Figure 7. Elucidation of PoABI5 overexpression in seedlings of *P. ostii*. (**a**). UV fluorescence imaging of *PoABI5* overexpressing seedlings, Bars = 1 cm; (**b**). Confocal microscopy analysis of *PoABI5* overexpressing seedlings; (**c**). qRT-PCR quantification of *PoABI5* overexpression efficiency; (**d**). Fatty acid composition changes in *PoABI5* overexpressing seedlings. Samples with significant differences are marked with asterisks (* $p < 0.05$, ** $p < 0.01$, *** $p < 0.001$).

4. Discussion

Seed germination in *P. ostii*, strictly regulated by morphophysiological dormancy, requires harvesting and sowing at dormancy-specific optimal timing to avoid a significant reduction in germination rates [42]. Previous studies investigating germination requirements using mature embryos as explants employed modified MS medium to examine the effects of different hormone concentrations on embryo germination and identified 0.5 mg/L BA + 1.0 mg/L GA_3 as the optimal treatment [43]. Consequently, this hormone combination was adopted in our study. However, previous research utilized exclusively MS as the basal medium for embryonic germination. To evaluate the impact of different basal media, a comparison was made between MS and WPM. Studies on embryo maturation have shown that WPM promotes higher adventitious bud induction than MS during subsequent shoot development [16,37,44]. Our results demonstrate that, under identical hormone conditions, WPM resulted in a significantly higher embryo germination rate (Figure 2c). This phenomenon may be attributed to the composition of WPM, which is characterized by lower overall nutrient levels, specifically reduced nitrogen and potassium concentrations. These lower nutrient levels may better suit the growth requirements of woody plants. In addition, the relatively higher calcium content of WPM may enhance cell wall stability and stress tolerance [38]. In contrast, MS medium, distinguished by its high nitrate concentration and extensive use in plant culture, exhibits elevated salt ion levels. This inherent property has been demonstrated to exacerbate explant browning [13,16], a phenomenon that is consistent with the observed browning and yellowing of non-germinated embryos on MS medium in the present study (Figure 2a).

Agrobacterium-mediated transformation efficiency is influenced by a number of factors, including bacterial density (OD_{600}) [17,45], AS concentration [18,46], the number of vacuum infiltration cycles [45,47], infection duration [17,48], and the co-culture period [49]. In the present study, a combination of parameters that have been demonstrated to be optimal was established. The experimental conditions comprised an Agrobacterium density at an OD_{600} of 1.0, AS supplementation at a rate of 200 μmol/L in the bacterial suspension, six negative pressure treatments, and an infection time of 2 h (Figure 3). It is evident that there is variability in the optimal Agrobacterium density (OD_{600}) across peony species and explant types. For instance, studies report OD_{600} = 1.0 for cut flowers of Itoh peony 'Bartzella' [4], OD_{600} = 0.6 for flower buds of tree peony (*P. suffruticosa*) 'Yulouchun' [24], and OD_{600} = 1.5 for tuberous roots of *P. lactiflora* 'Da Fugui' [50]. The findings of the present study identified OD_{600} = 1.0 as optimal. Suboptimal bacterial density may compromise transformation efficiency due to insufficient bacterial-explant contact, while excessive concentrations can induce explant browning or necrosis, thereby reducing transformation frequency [20]. In a similar manner, the duration of infection and co-culture has been demonstrated to exert an influence on efficiency by modulating the period of interaction between Agrobacterium and the explant. While the duration of contact is known to have a positive effect on transformation rates, prolonged exposure has been shown to cause explant damage, resulting in diminished transient expression efficiency [20,30]. As a phenolic virulence inducer, AS activates the vir gene region of Agrobacterium, thereby improving T-DNA transfer efficiency. However, explants exhibit substantial variation in their sensitivity to AS [51]. Our optimal AS concentration (200 μmol/L) aligns with numerous peony Agrobacterium-mediated transient transformation protocols [10,22,24,50,52].

Previous studies have shown that the developmental stage of plant tissues significantly influences the outcomes of transformation processes. Transient leaf transformation studies demonstrate that there is a decline in transformation efficiency that is correlated with an increase in leaf age [53,54]. The maturation process is characterized by an increase in vein density, resulting in the partitioning of the mesophyll apoplast. Consequently, upon traversing the epidermal barrier, the bacterial suspension encounters restricted diffusion within the leaf, thereby diminishing its efficiency. In accordance with this phenomenon, the present study observed a decline in efficiency in later developmental stages of sterile tissue-cultured embryogenic seedlings, characterized by growth retardation, cotyledon senescence, and emergence of true leaves, mirroring the pattern seen in mature leaves. In the study of the instantaneous transformation of the strawberry, it was ascertained that infection at the fruit receptacle was difficult in the early stage of fruit firmness. The optimum infection period was identified as occurring before the rapid decrease in fruit hardness and the onset of red coloration [55]. This limitation can be attributed to reduced intercellular space and denser cell packing during the initial developmental stages, which physically impedes suspension diffusion and compromises transformation [53]. The lower transformation efficiency observed at DAG30 compared to DAG35 in the present system may be attributable to this developmental constraint.

Compared to existing Agrobacterium-mediated transient transformation systems for *P. ostii*, the TTAES approach offers distinct advantages. First, as a high-yielding woody oil crop, *P. ostii* produces substantially more seeds than other species and varieties in the same genus—mature 7-year-old plants yield up to 10 fruits and 200–500 seeds per plant, enabling the procurement of abundant material of the same genotype for functional validation [1,26,56]. Second, unlike experiments requiring buds, leaves, or floral tissues with strict seasonal/temperature constraints, TTAES utilizes storable seeds that permit year-round seedling cultivation within 12 months post-harvest [10,22,25,57]. Finally, TTAES employs GFP as a reporter for non-destructive transgenic identification while laying a foundation for further exploring the application of this system in subcellular localization, BiFC, or dual-label imaging.

To further validate TTAES, we applied it to analyze the regulatory relationship between a target transcription factor and its cognate gene. ALA, an essential polyunsaturated fatty acid for humans, is a focus of ongoing research on its biosynthetic pathway. It has been demonstrated that ABA regulates lipid synthesis-related gene transcription by activating ABI5 [5,22,58]. Specifically, PoABI5 is responsible for activating the transcription of *PoFAD3*, thus promoting ALA synthesis [22]. While the transcriptional activation of the *PoFAD3* promoter by PoABI5 was initially validated in a tobacco heterologous system, we further confirmed this interaction in *P. ostii* using TTAES. It was determined that PoABI5 activated the *GUS* driven by the *PoFAD3* promoter, resulting in a 1.69-fold increase in GUS enzyme activity relative to the negative control (Figure 5b). These findings are consistent with the conclusions drawn from the tobacco system [22]. This finding enables the verification of the relationship between the transcription factors of *P. ostii* and the target gene promoters in the native system. Consequently, this ensures the accuracy and reliability of the experimental results.

In gene function studies, phenotypic changes induced by gene silencing or overexpression are often assessed using reporter tags. Although GUS tags offer high sensitivity and stability, GUS staining or activity assays inevitably damage experimental materials, compromising subsequent phenotypic analysis [41]. In contrast, GFP enables non-destructive in vivo observation, making it preferable for elucidating transient silencing and overexpression [31,41]. In this study, we integrated GFP into TTAES and performed VIGS-mediated silencing (pTRV2 vector) and transient overexpression of *PoABI5*. Expression levels of its target gene (*PoFAD3*) and ALA content were quantified by qRT-PCR and GC-MS, respectively. This approach provides a more rapid and non-destructive alternative to traditional GUS-based methods for gene function elucidation in *P. ostii*.

5. Conclusions

This study established an efficient transient transformation system, TTAES in *P. ostii*, which overcomes seasonal constraints in gene function elucidation by optimizing embryo germination protocols to enable year-round in vitro seedling propagation. Orthogonal experimental design identified optimal parameters: OD_{600} = 1.0, 200 μM AS, six negative pressure treatments, and 2 h infection duration, which achieved peak transformation efficiency in vitro embryo-derived seedlings of *P. ostii* 35 days after germination. Using GUS and GFP reporter systems, the platform confirmed transcription factor–promoter interactions and enabled VIGS and overexpression analyses, demonstrating its utility for functional genomics. This system eliminates seasonal restrictions, accelerates molecular breeding workflows, and provides a robust framework for elucidating gene functions in *P. ostii*, advancing its development as a dual-purpose medicinal and oilseed crop. However, as a transient transformation system, TTAES exhibits phenotypic variability due to transformation efficiency fluctuations. Moreover, because phenotypic assessment commences as early as 3 days post-transformation, it is unsuitable for functional validation requiring prolonged transgene expression to manifest phenotypes. For CRISPR-mediated knockout or stable transgenic line generation, optimization of seedling rooting protocols and post-transformation screening regimes remains necessary.

Author Contributions: Y.Z. (Yanlong Zhang) and L.N. conceived and designed the experiments. Y.Z. (Yuhui Zhai), X.X., L.Z. and Z.Z. did the experiments. Y.Z. (Yuhui Zhai), X.X. and X.W. analyzed the data. Y.Z. (Yuhui Zhai) and X.X. wrote the manuscript and prepared the figures. All authors have read and agreed to the published version of the manuscript.

Funding: This research was supported by the Key Technologies Research and Development Program of China (Project Numbers 2023YFD2300905) and the 2024 Shaanxi Provincial Fiscal Agricultural Special Fund Project.

Data Availability Statement: All data generated or analyzed during this study are included in this published article.

Conflicts of Interest: The authors declare no conflicts of interest.

References

1. Sun, X.; Niu, L.; Zhang, M.; Zhang, H.; Liu, H.; Zhao, M.; Zhang, X.; Zhang, Q.; Zhang, Y. Application of Carbon-Based Nutrient Fertilizer Improved Soil Fertility and Seed Yield of *Paeonia ostii* 'Feng Dan'. *Ind. Crops Prod.* **2024**, *212*, 118348. [CrossRef]
2. Wang, X.; Liang, H.; Guo, D.; Guo, L.; Duan, X.; Jia, Q.; Hou, X. Integrated Analysis of Transcriptomic and Proteomic Data from Tree Peony (*P. ostii*) Seeds Reveals Key Developmental Stages and Candidate Genes Related to Oil Biosynthesis and Fatty Acid Metabolism. *Hortic. Res.* **2019**, *6*, 111. [CrossRef] [PubMed]
3. Zhang, X.-X.; Zuo, J.-Q.; Wang, Y.-T.; Duan, H.-Y.; Zhou, M.-H.; Li, H.-J.; Hu, Y.-H.; Yuan, J.-H. PoDPBT, a BAHD Acyltransferase, Catalyses the Benzoylation in Paeoniflorin Biosynthesis in *Paeonia ostii*. *Plant Biotechnol. J.* **2023**, *21*, 14–16. [CrossRef] [PubMed]
4. Shi, T.; Shang, S.; Fan, L.; Zhou, S.; Gao, S.; Shaaban, M.; Fan, B.; Shi, G.; Wang, Z. The Role of *IpSAG12* in Regulating Florescence and Senescence of Cut Itoh Peony Flowers. *Postharvest Biol. Technol.* **2025**, *222*, 113352. [CrossRef]
5. Li, Y.; Wang, X.; Wu, S.; Mei, X.; Song, W.; Shu, Q. Metabolic Pathway for Evaluation of Polyunsaturated Fatty Acids Content Variation and Regulation of Its Biosynthesis in *Paeonia ostii* through Abscisic Acid and ABSCISIC ACID-INSENSITIVE 5 (PoABI5). *Ind. Crops Prod.* **2025**, *227*, 120758. [CrossRef]
6. Sun, J.; Guo, H.; Liu, M.; Chen, M.; Zhu, M.; Liu, D.; Tao, J. Histology and Transcriptomic Profiling Reveal the Dynamics of Seed Coat and Endosperm Formation in Tree Peony (*Paeonia ostii*). *Hortic. Res.* **2022**, *9*, uhac106. [CrossRef]
7. Liu, R.; Ci, H.; Ren, X.; Gao, J.; Wang, S.; Zhang, X. Optimization of Callus Induction from Immature Embryo and Establishment Regeneration System of *Paeonia ostii* 'Fengdan'. *Acta Hortic. Sin.* **2022**, *49*, 166. [CrossRef]
8. Duan, S.; Xin, R.; Guan, S.; Li, X.; Fei, R.; Cheng, W.; Pan, Q.; Sun, X. Optimization of Callus Induction and Proliferation of *Paeonia lactiflora* Pall. and *Agrobacterium*-Mediated Genetic Transformation. *Front. Plant Sci.* **2022**, *13*, 996690. [CrossRef]
9. Liu, P.; Zhang, L.; Wang, X.; Gao, J.; Yi, J.; Deng, R. Characterization of *Paeonia ostii* Seed and Oil Sourced from Different Cultivation Areas in China. *Ind. Crops Prod.* **2019**, *133*, 63–71. [CrossRef]
10. Luan, Y.; Chen, Z.; Fang, Z.; Meng, J.; Tao, J.; Zhao, D. PoWRKY69-PoVQ11 Module Positively Regulates Drought Tolerance by Accumulating Fructose in *Paeonia ostii*. *Plant J.* **2024**, *119*, 1782–1799. [CrossRef] [PubMed]
11. Li, L.; Wang, Z.; Li, Y.; Wang, D.; Xiu, Y.; Wang, H. Characterization of Genes Encoding ω-6 Desaturase PoFAD2 and PoFAD6, and ω-3 Desaturase PoFAD3 for ALA Accumulation in Developing Seeds of Oil Crop *Paeonia ostii* Var. *Lishizhenii*. *Plant Sci.* **2021**, *312*, 111029. [CrossRef]
12. Yuan, J.; Jiang, S.; Jian, J.; Liu, M.; Yue, Z.; Xu, J.; Li, J.; Xu, C.; Lin, L.; Jing, Y.; et al. Genomic Basis of the Giga-Chromosomes and Giga-Genome of Tree Peony *Paeonia ostii*. *Nat. Commun.* **2022**, *13*, 7328. [CrossRef]
13. Gao, J.; Xue, J.; Xue, Y.; Liu, R.; Ren, X.; Wang, S.; Zhang, X. Transcriptome Sequencing and Identification of Key Callus Browning-Related Genes from Petiole Callus of Tree Peony (*Paeonia suffruticosa* Cv. Kao) Cultured on Media with Three Browning Inhibitors. *Plant Physiol. Biochem.* **2020**, *149*, 36–49. [CrossRef]
14. Liu, R.; Xue, Y.; Ci, H.; Gao, J.; Wang, S.; Zhang, X. Establishment of Highly Efficient Plant Regeneration of *Paeonia ostii* 'Fengdan' through Optimization of Callus, Adventitious Shoot, and Rooting Induction. *Hortic. Plant J.* **2022**, *8*, 777–786. [CrossRef]
15. Demei, N.; Fang, L.; Linqiang, G.; Huailong, Z.; Naibin, L.; Lu, Z.; Yanchao, Y.; Chunying, L.; Shupeng, G.; Yuxi, Z. MADS-Domain Transcription Factor AGAMOUS LIKE-9 Participates in the Gibberellin Pathway to Promote Bud Dormancy Release of Tree Peony. *Hortic. Res.* **2025**, *12*, uhaf043. [CrossRef] [PubMed]
16. Xin, R.; Song, Y.; Zhang, X.; Kang, X.; Zhou, X.; Song, W.; Guan, S.; Sun, X. Optimization Strategy for Establishing an Efficient Mature Embryo Reproduction System of *Paeonia lactiflora* Pall. *Sci. Hortic.* **2025**, *340*, 113952. [CrossRef]
17. Liu, S.; Ma, J.; Liu, H.; Guo, Y.; Li, W.; Niu, S. An Efficient System for *Agrobacterium*-Mediated Transient Transformation in *Pinus tabuliformis*. *Plant Methods* **2020**, *16*, 52. [CrossRef] [PubMed]
18. Zheng, L.; Yang, J.; Chen, Y.; Ding, L.; Wei, J.; Wang, H. An Improved and Efficient Method of *Agrobacterium* Syringe Infiltration for Transient Transformation and Its Application in the Elucidation of Gene Function in Poplar. *BMC Plant Biol.* **2021**, *21*, 54. [CrossRef]

19. Liu, K.; Yang, Q.; Yang, T.; Wu, Y.; Wang, G.; Yang, F.; Wang, R.; Lin, X.; Li, G. Development of Agrobacterium-Mediated Transient Expression System in Caragana Intermedia and Characterization of CiDREB1C in Stress Response. *BMC Plant Biol.* **2019**, *19*, 237. [CrossRef] [PubMed]
20. Xian, B.; Xi, Z.; Ren, C.; Yan, J.; Chen, J.; Pei, J. The Establishment of Transient Expression Systems and Their Application for Gene Function Analysis of Flavonoid Biosynthesis in *Carthamus tinctorius* L. *BMC Plant Biol.* **2023**, *23*, 186. [CrossRef]
21. Zhao, Q.; Li, Y.; Gu, L.; Yang, Y.; He, D.; Luo, J.; Zhang, Y. The PlMYB73-PlMYB70-PlMYB108 Complex Regulates PlTPS1 to Promote Geraniol Biosynthesis in *Paeonia lactiflora*. *Hortic. Res.* **2025**, *12*, uhaf141. [CrossRef]
22. Li, Y.; Wang, X.; Zhang, X.; Liu, Z.; Peng, L.; Hao, Q.; Liu, Z.; Men, S.; Tong, N.; Shu, Q. ABSCISIC ACID-INSENSITIVE 5-ω3 FATTY ACID DESATURASE3 Module Regulates Unsaturated Fatty Acids Biosynthesis in *Paeonia ostii*. *Plant Sci.* **2022**, *317*, 111189. [CrossRef]
23. Xie, L.; Zhang, Q.; Sun, D.; Yang, W.; Hu, J.; Niu, L.; Zhang, Y. Virus-Induced Gene Silencing in the Perennial Woody *Paeonia ostii*. *PeerJ* **2019**, *7*, e7001. [CrossRef]
24. Yuan, Y.; Zeng, L.; Kong, D.; Mao, Y.; Xu, Y.; Wang, M.; Zhao, Y.; Jiang, C.-Z.; Zhang, Y.; Sun, D. Abscisic Acid–Induced Transcription Factor PsMYB306 Negatively Regulates Tree Peony Bud Dormancy Release. *Plant Physiol.* **2024**, *194*, 2449–2471. [CrossRef] [PubMed]
25. Yang, W.; Xin, Z.; Hu, J.; Zhang, Y.; Zhang, Q.; Niu, L. The Tree Peony Nuclear Factor Y Transcription Factor PrNF-YC2 Promotes Seed Oil Accumulation. *Plant J.* **2023**, *115*, 546–562. [CrossRef] [PubMed]
26. Ren, L. A Primary Study on the Selection in Field Forbreeding New Oil-Used *Paeonia ostii* "Fendan" Varieties. Master's Thesis, Northwest A&F University, Yangling, China, 2016.
27. Ren, H. Study on the Germination of Oil Tree Peony Seeds and Container Nursery. Master's Thesis, Northwest A&F University, Yangling, China, 2016.
28. Cheng, F.; Du, X. Effects of Chilling and Gibberellic Acid on the Seed Germination and Seedling Growth in *Paeonia ostii* 'Feng Dan'. *Acta Hortic. Sin.* **2008**, *35*, 553–558.
29. Li, X.U.; Fang-Yun, C.; Yuan, Z. Study on Rapid Seedling-Raising Technology of Tree Peony Embryo Culture. *Bull. Bot. Res.* **2017**, *37*, 690. [CrossRef]
30. Guan, S.; Kang, X.; Ge, J.; Fei, R.; Duan, S.; Sun, X. An Efficient *Agrobacterium*-Mediated Transient Transformation System and Its Application in Gene Function Elucidation in *Paeonia lactiflora* Pall. *Front. Plant Sci.* **2022**, *13*, 999433. [CrossRef]
31. Pinski, A.; Betekhtin, A. Efficient Agrobacterium-Mediated Transformation and Genome Editing of *Fagopyrum tataricum*. *Front. Plant Sci.* **2023**, *14*, 1270150. [CrossRef]
32. Shen, M.; Wang, Q.; Yu, X.; Teixeira da Silva, J.A. Micropropagation of Herbaceous Peony (*Paeonia lactiflora* Pall.). *Sci. Hortic.* **2012**, *148*, 30–38. [CrossRef]
33. Teixeira da Silva, J.A.; Shen, M.; Yu, X. Tissue Culture and Micropropagation of Tree Peony (*Paeonia suffruticosa* Andr.). *J. Crop Sci. Biotechnol.* **2012**, *15*, 159–168. [CrossRef]
34. Zhang, Y.; Ge, M.; Wang, X.; Zeng, Y.; Liao, Y.; Wang, Y.; Zeng, Z. Discrimination of Moutan Cortex from Different Sources and Geographical Origins for Quality Evaluation Using Microsatellite Markers Coupled with Chemical Analysis. *Biochem. Syst. Ecol.* **2020**, *89*, 104005. [CrossRef]
35. Zhai, C.; Meng, X.; Fu, J.; Qin, Z.; Meng, Y. Research Progress of Peony Bark in Modern Pharmacy. *Inf. Tradit. Chin. Med.* **2020**, *37*, 109–114.
36. Chen, M.; Yang, L.; Yang, Z.; Cai, Y.; Yin, L.; Zhang, Y. Callus Induction, Proliferation and Histological Observation of *Paeoniaostii* Cv. 'Phoenix White' Mature Embryo. *Non-Wood For. Res.* **2021**, *39*, 9–16.
37. Liu, Z.; Zhou, H.; Cai, X.-f.; Li, J.; Hu, Y.-h. Paeonia ostii 'Feng Dan' Embryo Culture and Adventitious Buds Induction. In *Advances in Chinese Ornamental Horticulture*; China Forestry Press: Beijing, China, 2011; pp. 371–374.
38. Ioannidis, K.; Koropouli, P. Effects of Different Media and Their Strengths in in Vitro Culture of Three Different *Cistus creticus* L. Populations and Their Genetic Assessment Using Simple Sequence Repeat Molecular Markers. *Horticulturae* **2024**, *10*, 104. [CrossRef]
39. Berumen-Varela, G.; Ochoa-Jiménez, V.-A.; Burgara-Estrella, A.; Trillo-Hernández, E.-A.; Ojeda-Contreras, Á.-J.; Orozco-Avitia, A.; Rivera-Domínguez, M.; Troncoso-Rojas, R.; Báez-Sañudo, R.; Datsenka, T.; et al. Functional Analysis of a Tomato (*Solanum lycopersicum* L.) Rhamnogalacturonan Lyase Promoter. *J. Plant Physiol.* **2018**, *229*, 175–184. [CrossRef]
40. Ma, S.; Du, C.; Ohlrogge, J.; Zhang, M. Accelerating Gene Function Discovery by Rapid Phenotyping of Fatty Acid Composition and Oil Content of Single Transgenic T 1 Arabidopsis and Camelina Seeds. *Plant Direct.* **2020**, *4*, e00253. [CrossRef]
41. Dedow, L.K.; Oren, E.; Braybrook, S.A. Fake News Blues: A GUS Staining Protocol to Reduce False-Negative Data. *Plant Direct.* **2022**, *6*, e367. [CrossRef]
42. Xue, J.; Wang, S.; Zhang, P.; Zhu, F.; Ren, X.; Liu, C.; Zhang, X. On the Role of Physiological Substances, Abscisic Acid and Its Biosynthetic Genes in Seed Maturation and Dormancy of Tree Peony (*Paeonia ostii* 'Feng Dan'). *Sci. Hortic.* **2015**, *182*, 92–101. [CrossRef]

43. Li, X. Study on in Vitro Regeneration via Meristematic Nodule Culture and Embryoculture in Tree Peony. Doctoral Dissertation, Beijing Forestry University, Beijing, China, 2022.
44. An, B. Studies on the Establishment of in Vitro Regeneration System of Peonies Suffruticosa ANDR. Master's Thesis, Northeast Forestry University, Harbin, China, 2005.
45. Singh, P.; Khan, S.; Kumar, S.; ur Rahman, L. Establishment of an Efficient Agrobacterium-Mediated Genetic Transformation System in Pelargonium Graveolens: An Important Aromatic Plant. *Plant Cell Tissue Organ Cult.* **2017**, *129*, 35–44. [CrossRef]
46. Yao, W.; Kong, L.; Lei, D.; Zhao, B.; Tang, H.; Zhou, X.; Lin, Y.; Zhang, Y.; Wang, Y.; He, W.; et al. An Effective Method for Establishing a Regeneration and Genetic Transformation System for Actinidia Arguta. *Front. Plant Sci.* **2023**, *14*, 1204267. [CrossRef]
47. Gu, X.F.; Meng, H.; Qi, G.; Zhang, J.R. Agrobacterium-Mediated Transformation of the Winter Jujube (*Zizyphus Jujuba* Mill.). *Plant Cell Tissue Organ Cult.* **2008**, *94*, 23–32. [CrossRef]
48. Lin, Y.-Y.; Wu, S.-H.; Chen, J.; Yi, X.-G.; Wang, X.-R.; Li, M. Efficient Plant Regeneration and Transient Genetic Transformation System of Prunus Xueluoensis via an Agrobacterium-Mediated Method. *Int. J. Mol. Sci.* **2025**, *26*, 3588. [CrossRef] [PubMed]
49. Faizal, A.; Geelen, D. Agroinfiltration of Intact Leaves as a Method for the Transient and Stable Transformation of Saponin Producing Maesa Lanceolata. *Plant Cell Rep.* **2012**, *31*, 1517–1526. [CrossRef]
50. Wang, X.; Li, J.; Guo, J.; Qiao, Q.; Guo, X.; Ma, Y. The WRKY Transcription Factor PlWRKY65 Enhances the Resistance of Paeonia Lactiflora (*Herbaceous peony*) to Alternaria Tenuissima. *Hort. Res.* **2020**, *7*, 57. [CrossRef]
51. Baros, C.J.; Beerkens, J.; Ludwig, M. Agrobacterium-Mediated Transient Transformation of *Flaveria bidentis* Leaves: A Novel Method to Examine the Evolution of C4 Photosynthesis. *Plant Methods* **2024**, *20*, 193. [CrossRef]
52. Gao, L.; Niu, D.; Chi, T.; Yuan, Y.; Liu, C.; Gai, S.; Zhang, Y. PsRGL1 Negatively Regulates Chilling- and Gibberellin-Induced Dormancy Release by PsF-Box1-Mediated Targeting for Proteolytic Degradation in Tree Peony. *Hortic. Res.* **2023**, *10*, uhad044. [CrossRef] [PubMed]
53. Xie, J.; He, C.; Li, Z.; Li, M.; He, S.; Qian, J.; Tan, B.; Zheng, X.; Cheng, J.; Wang, W.; et al. A Rapid and Efficient Agrobacterium-Mediated Transient Transformation System in Grape Berries. *Protoplasma* **2024**, *261*, 819–830. [CrossRef] [PubMed]
54. Acanda, Y.; Welker, S.; Orbović, V.; Levy, A. A Simple and Efficient Agroinfiltration Method for Transient Gene Expression in Citrus. *Plant Cell Rep.* **2021**, *40*, 1171–1179. [CrossRef]
55. Pi, M.; Gao, Q.; Kang, C. Transient Expression Assay in Strawberry Fruits. *Bio Protoc.* **2019**, *9*, e3249. [CrossRef]
56. Qi, Q.; Li, Y.; Xing, G.; Guo, J.; Guo, X. Fertility Variation among *Paeonia lactiflora* Genotypes and Fatty Acid Composition of Seed Oil. *Ind. Crops Prod.* **2020**, *152*, 112540. [CrossRef]
57. Mao, Y.; Ji, X.; Meng, Q.; Xu, Z.; Yuan, Y.; Li, M.; Niu, L.; Zhang, Y.; Sun, D. Contribution of Anthocyanin and Polyunsaturated Fatty Acid Biosynthesis to Cold Tolerance during Bud Sprouting in Tree Peony. *Ind. Crops Prod.* **2022**, *188*, 115563. [CrossRef]
58. Yeap, W.-C.; Lee, F.-C.; Shabari Shan, D.K.; Musa, H.; Appleton, D.R.; Kulaveerasingam, H. WRI1-1, ABI5, NF-YA3 and NF-YC2 Increase Oil Biosynthesis in Coordination with Hormonal Signaling during Fruit Development in Oil Palm. *Plant J.* **2017**, *91*, 97–113. [CrossRef] [PubMed]

plants

MDPI

Article

Spraying Foliar Fertilizer Affect the Physiological Function of Leaf and Improve the Quality of '*Snick*' Apple

Hong-Fu Xu, Shi-Mei Li, Wei-Feng Ma, Shi-Xiong Lu, Zhi-Yuan Bian, Guo-Ping Liang and Juan Mao *

The College of Horticulture, Gansu Agricultural University, Lanzhou 730070, China; 18893151879@163.com (H.-F.X.); 15214038359@163.com (S.-M.L.); 18409490212@163.com (W.-F.M.); 18893912407@163.com (S.-X.L.); lzubzy@163.com (Z.-Y.B.); lianggp@gsau.edu.cn (G.-P.L.)

* Correspondence: maojuan@gsau.edu.cn

Abstract: Foliar fertilizers are efficient in enhancing nutrient utilization. This experiment aims to improve leaf physiological functions, enhance fruit quality, increase yield, and boost orchard productivity through the screening of foliar fertilizers suitable for apple trees. The 6-year-old apple trees of the 'Snick' were used as experiment material. The results of measurements amino acids, calcium, boron, and potassium indicate that different foliar fertilizers can improve fruit quality and aroma by enhancing leaf physiological functions. In apple fruit, amino acid foliar fertilizer increased the tartaric acid content by 44.26%. Calcium foliar fertilizer resulted in a 32.39% increase in vitamin C, a 19.71% increase in sucrose compared to the control, with a total aroma substance increase of 13.41%. Boron foliar fertilizer elevated flavonoid content in the peel to 3.67 $mg \cdot g^{-1}$, a 70.69% increase over the CK. Potassium foliar fertilizer significantly improved fruit appearance, phenolic substances in the peel, soluble protein content by 25.39%, and glucose content by 55.91%. Therefore, mineral source fulvic acid potassium foliar fertilizer was demonstrated the best overall effect, effectively enhancing fruit quality and flavor. These results provide a theoretical basis and scientific reference for improving apple quality.

Keywords: 'Snick' apple; foliar fertilizer; leaf physiological; fruit quality; aroma components

1. Introduction

Apple (*Malus domestica* Mill.) belongs to the Rosaceae apple genus [1], the world planting area is about 4 to 5 million hectares, the output is about 79 million tons [2]. The Loess Plateau (Latitude ranges from 33°41′ N to 41°16′ N, and longitude spans from 100°52′ E to 114°33′ E) has become a central apple-producing area globally and serves as a key region for the high-quality cultivation of apples in the world [3], it accounts for over 25% of China's apple production [4]. It is a densely cultivated fruit [5], mainly freshly eaten. Its market value is affected by factors such as color, shape, hardness [6]. At the same time, it has become one of the most important fruit crops in the economy and nutritional value due to its excellent taste and rich functional components [7].

Foliar fertilizers, as environmentally friendly fertilizers that are not directly applied to the soil, according to the fertilizer brands primarily contain growth regulators, small-molecule organic acids, and other components [8]. Compared to root fertilization, foliar fertilizers are applied by spraying plant leaves, enabling faster nutrient absorption [9]. Under optimal conditions, they can effectively improve the fruit quality and yield of various horticultural crops [10], but may cause damage to plant leaves. It is recommended to use low-concentration fertilizers [11].

Photosynthesis is the most important chemical reaction in plants [12] and is influenced by external environmental conditions such as light, temperature, humidity, carbon dioxide concentration [13], mineral elements [14], hormones [15], and leaf development [16]. Research shows that increasing the rate of net photosynthesis is crucial to crop yield and quality, so optimizing the photosynthetic mechanism is a key strategy for improving fruit quality with foliar fertilizers [17].

Fruit quality encompasses external attributes such as weight, vertical and horizontal diameters, peel colour, and internal attributes like soluble solids content, vitamin C, soluble sugars, and titratable acidity. Various factors, including nutrient elements such as boron, calcium, potassium, and amino acids, influence this quality. Amino acid foliar fertilisers, high in amino acids and trace elements, can rapidly penetrate plant tissues. Consequently, they promote growth and enhance the content of vitamin C, total phenolic compounds, total flavonoids, total sugars, and soluble solids in fruits [18]. Calcium, an activator of many essential enzymes in plants, enhances enzyme activity and promotes various chemical reactions, thereby regulating plant growth and development. Calcium sprays significantly affected the single fruit weight, fruit flesh firmness, peel color, soluble solids content, soluble sugar, titratable acid, soluble protein, and vitamin C content of the fruit [19]. Plants supplied with sufficient boron can increase the supply of organic matter in various organs, enabling normal crop growth and increasing seed and fruit setting rates. The results showed that spraying boron increased fruit size (diameter, length, weight) and fruit quality (colour and firmness) [20]. Moreover, it promotes the accumulation of soluble solids, protein, vitamin C, total phenol, flavonoids, esters, and other characteristic compounds. As free ions, potassium plays a crucial role in plants by enhancing carbohydrate and nitrogen metabolism [21]. Potassium fertilizer has increased fruit weight, soluble solids, total sugar content, sugar-acid ratio, vitamin C, and phenolic substances. Therefore, it is evident that different nutrient elements affect fruit quality differently [22].

The quality of apple directly impacts their commercial value, and market competitiveness. To maximize economic benefits, improving apple quality such as size, shape, colour, absence of bruising and other is essential. In this study, 'Snick' apples were used as the experiment materials, based on the type and cost of the foliar fertilizer, different concentrations and application timings were used for leaf spraying, followed by analyzing the fruits for external appearance and nutritional quality. Additionally, the aromatic compounds in the fruits were examined. This study aims to select foliar fertilizers suitable for 'Snick' apples that can improve fruit quality, and establish a foundation for further research into the efficacy and mechanisms of foliar fertilizer applications. Moreover, we will also focus on the development and investigation of novel foliar fertilizers, including nano-formulated, specialty, and chelated foliar fertilizers.

2. Results

2.1. *Effect of Different Foliar Fertilizers on Physiological Characteristics of Shoots and Leaves of 'Snick' Apple*

The results showed that as the number of days after full bloom (DAFB) increased, both the length and diameter of new shoots exhibited a gradually increasing trend. By 90 DAFB, under the mineral source potassium fulvic acid (T4), the new shoot length reached 36.06 cm, representing a significant increase of 50.81% compared with the CK treatment; the new shoot diameter was 6.71 mm, showing a significant increase of 11.22% relative to CK. From 60 to 90 DAFB, leaf area displayed an upward trend, with the maximum leaf area observed under the sugar alcohol boron (T3) at 90 DAFB (3263.18 mm^2), which was 5.03% higher than CK, but the leaf area under the T4 was smaller than that of the CK between 60 and 90 DAFB. By 90 DAFB, the leaf area in T4 had decreased by 2.81% compared to CK, suggesting that the T4 may partially inhibit leaf area expansion. This possibly results from a balance between shoot elongation and leaf expansion. This was followed by the effect of fluid calcium (T2), with a 3.83% increase compared to CK. At 60 DAFB, the SPAD value of leaves under the Ye Xiumei (T1) was 53.84%, 9.18% higher than CK. Notably, the T2 achieved the highest SPAD values at both 75 DAFB and 90 DAFB, with increases of 4.03% and 2.46% over CK, respectively. These results indicate that the T2 treatment may enhance the photosynthetic potential of the leaves (Figure 1D).

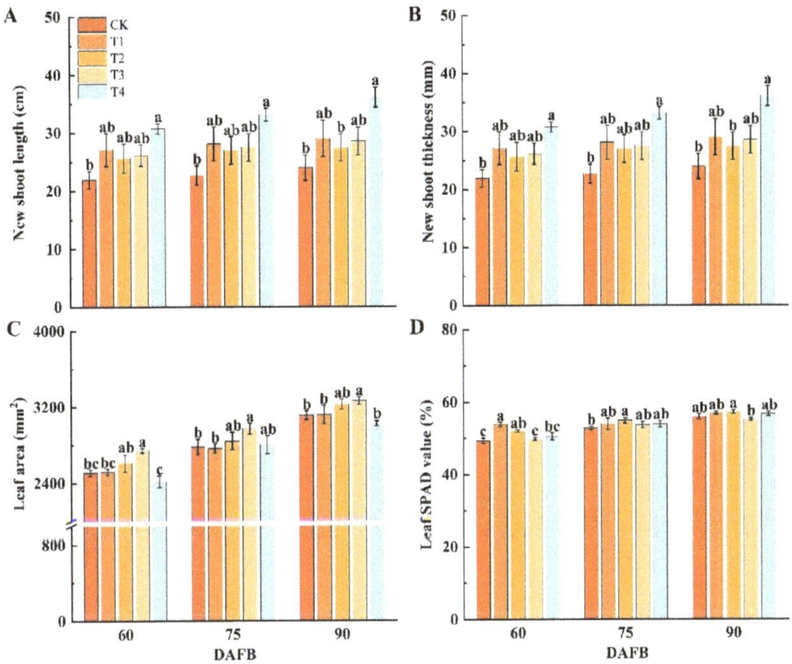

Figure 1. Effect of different foliar fertilizers on physiological characteristics of shoots and leaves of 'Snick' apple. DAFB: Days after full bloom. (**A**): New shoot length. (**B**): New shoot thickness. (**C**): Leaf area. (**D**): Leaf SPAD value. CK: Clear water control. T1: Amino acid foliar fertilizer. T2: Calcium foliar fertilizer. T3: Boron foliar fertilizer. T4: Potassium foliar fertilizer. Error bars represent the mean ± SE from three biological repeats. Different lowercase letters denote significant differences, whereas the same lowercase letters indicate no statistical difference ($p < 0.05$).

2.2. *Effect of Different Foliar Fertilizers on Photosynthesis in 'Snick' Apple Leaves*

The results showed that from 60 to 90 DAFB, the net photosynthetic rate, transpiration rate, stomatal conductance, and intercellular CO_2 concentration of leaves under all treatments exhibited an upward trend (Figure 2). According to the Figure 2A, at 90 DAFB, the net photosynthetic rate of the T2 reached 19.42 $\mu mol \cdot m^{-2} \cdot s^{-1}$, representing a 3.59% increase compared with the CK. The results indicate that the T2 treatment significantly enhanced the net photosynthetic rate. The transpiration rate reached 9.38 $mmol \cdot m^{-2} \cdot s^{-1}$, a 7.97% increase over CK, and the stomatal conductance was 0.29 $mol \cdot m^{-2} \cdot s^{-1}$, an increase of 4.81% relative to CK, calculation of intrinsic water-use efficiency (PN/gs) revealed a significant decrease in intrinsic water-use efficiency as the days after flowering increased. However, the T4 treatment effectively delayed this decline, exhibiting a reduction rate that was 22.41% slower compared to the CK. The intercellular CO_2 concentration was 344.96 $\mu mol \cdot mol^{-1}$, an increase of 7.51% compared with CK.

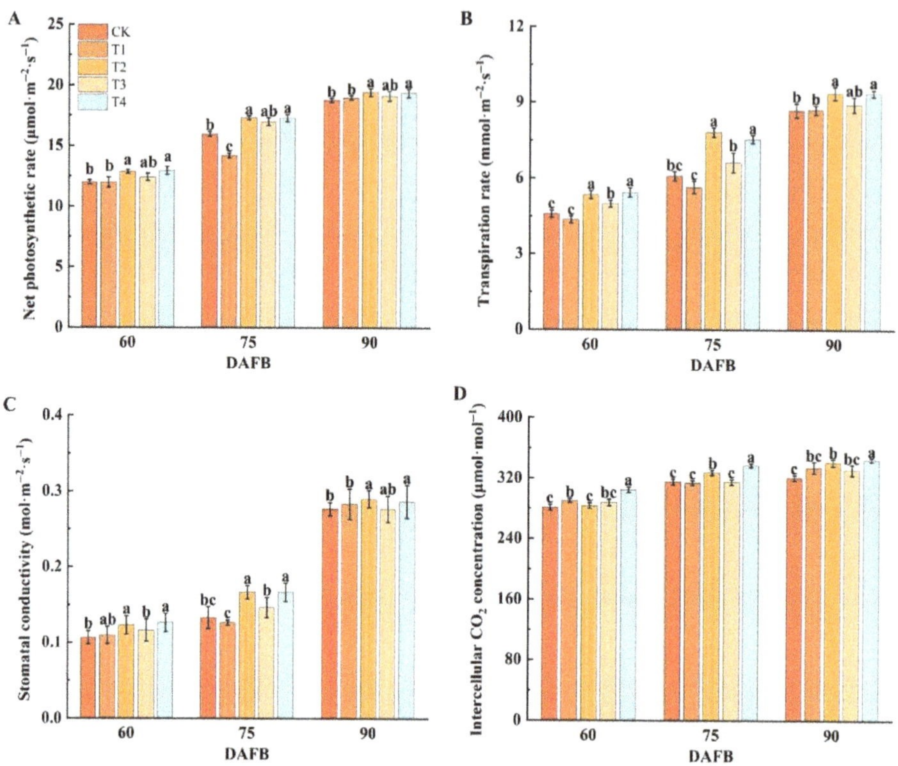

Figure 2. The effect of different foliar fertilizers on photosynthesis in 'Snick' apple leaves. DAFB: Days after full bloom. (**A**): Net photosynthetic rate. (**B**): Transpiration rate. (**C**): Stomatal conductivity. (**D**): Intercellular CO_2 concentration. CK: Clear water control. T1: Amino acid foliar fertilizer. T2: Calcium foliar fertilizer. T3: Boron foliar fertilizer. T4: Potassium foliar fertilizer. Error bars represent the mean ± SE from three biological repeats. Different lowercase letters denote significant differences, whereas the same lowercase letters indicate no statistical difference ($p < 0.05$).

2.3. *Effect of Different Foliar Fertilizers on Fruit Appearance Quality at Harvest Time*

The results showed that the single fruit weight (224.27 g), hardness (10.51 kg·cm^{-2}), longitudinal diameter (73.26 mm), and transverse diameter (79.80 mm) of T4 treatment were significantly higher than those of CK (Figure 3), increasing by 16.69%, 17.83%, 13.89%, and 3.58%, respectively. The fruit shape index of T1 and T4 was significantly higher than that of other treatments. The fruit shape indexes of CK and T3 ranged between 0.8 and 0.9, indicating nearly round fruits, while those of T1, T2, and T4 ranged between 0.9 and 1.0, indicating more elongated fruits. The highest L* value of the fruit peel was observed in T3, followed by T4, the L* value of CK was higher than those of T2 and T1. The highest a* value component in T4 and the lowest in T3. The b* values for T2, T3, and T4 were higher than CK, whereas the b* value for T1 treatment was lower than CK.

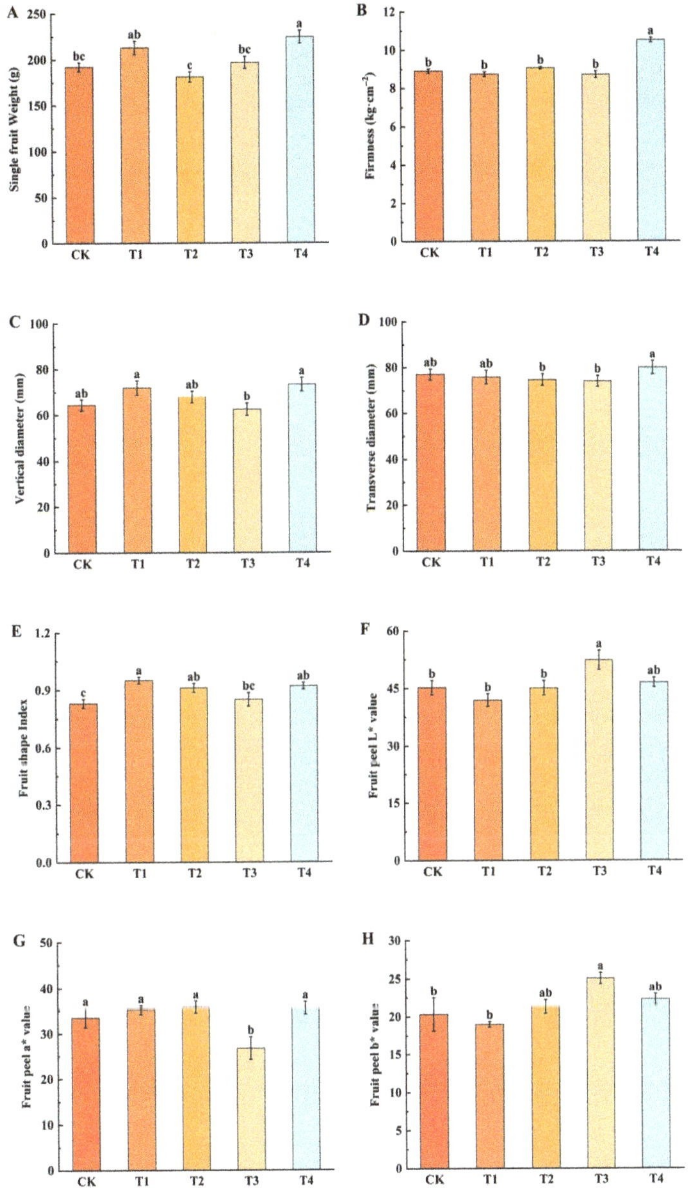

Figure 3. Effect of different foliar fertilizers on fruit appearance quality. (**A**): Single fruit Weight. (**B**): Firmness. (**C**): Vertical diameter. (**D**): Transverse diameter. (**E**): Fruit shape Index. (**F**): Fruit peel L* value. (**G**): Fruit peel a* value. (**H**): Fruit peel b* value. CK: Clear water control. T1: Amino acid foliar fertilizer. T2: Calcium foliar fertilizer. T3: Boron foliar fertilizer. T4: Potassium foliar fertilizer. Error bars represent the mean ± SE from three biological repeats. Different lowercase letters denote significant differences, whereas the same lowercase letters indicate no statistical difference ($p < 0.05$).

2.4. Effect of Different Foliar Fertilizers on Fruit Internal Quality

2.4.1. Effect of Different Foliar Fertilisers on the Soluble Solids, Soluble Sugar, Titratable Acid and Sugar-Acid Ratios of Fruits

The results indicated that applying foliar fertilizer increased the soluble solids content, with the T3 showing the most significant effect at 13.71%, which was 12.01% higher than the CK (Figure 4A). Under the T1, the apple fruits had the highest soluble sugar content and the lowest titratable acid content, suggesting that T1 effectively promotes the accumulation of soluble sugars while reducing titratable acid levels (Figure 4B,C). Further analysis of the sugar-acid ratio revealed that the T1 resulted in a sugar-acid ratio of 54.07, a 21.56% increase compared to the CK (Figure 4D). These results indicated that amino acid foliar fertilizer could promote the accumulation of sugar and the increase of sugar acid ratio in fruit, boron foliar fertilizer could increase the accumulation of soluble solids, and they all reduced the accumulation of titratable acid.

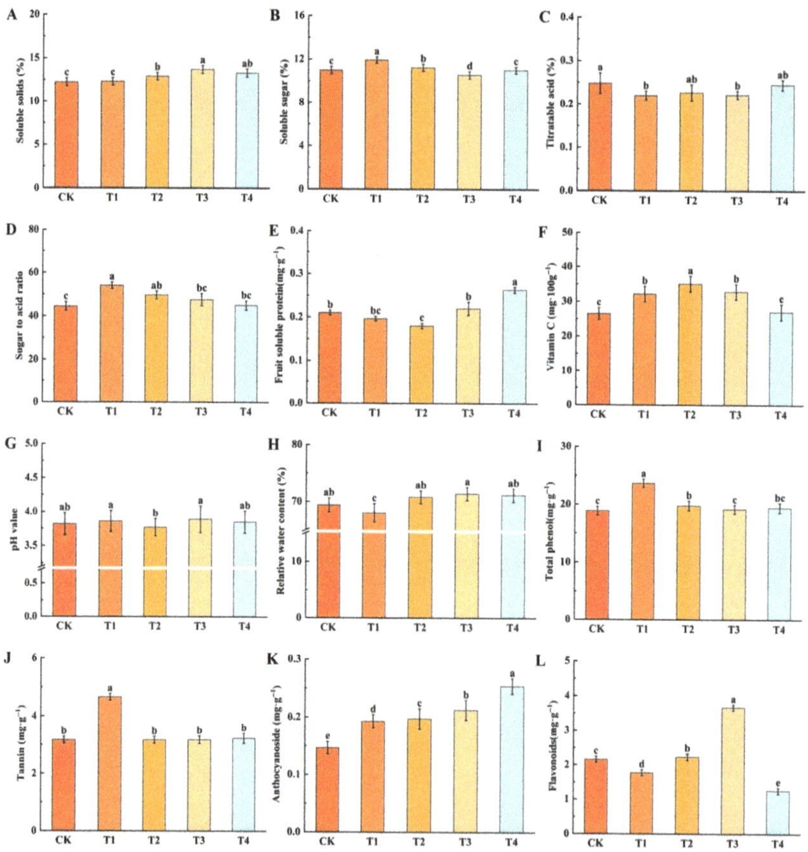

Figure 4. Effect of different foliar fertilizers on fruit internal quality. (**A**): Soluble solids content. (**B**): Soluble sugar content. (**C**): Titratable acid content. (**D**): Ratio of sugar to acid. (**E**): Fruit soluble protein content. (**F**): Vitamin C content. (**G**): pH value. (**H**): Relative water content. (**I**): Total phenolic content. (**J**): Tannin content. (**K**): Anthocyanin content. (**L**): Flavonoid content. CK: Clear water control. T1: Amino acid foliar fertilizer. T2: Calcium foliar fertilizer. T3: Boron foliar fertilizer. T4: Potassium foliar fertilizer. Error bars represent the mean ± SE from three biological repeats. Different lowercase letters denote significant differences, whereas the same lowercase letters indicate no statistical difference ($p < 0.05$).

2.4.2. Effect of Different Foliar Fertilizers on Soluble Protein, Vitamin C Content, pH Value and Relative Water Content of Fruits

Further investigation into the impact of various foliar fertilizers on soluble protein content revealed that T3 and T4 had higher levels than the CK, whereas T1 and T2 led to reduced soluble protein concentrations (Figure 4E). Specifically, the soluble protein content under T4 was 0.26%, marking a significant 25.39% increase relative to CK. Additionally, application of foliar fertilizers enhanced vitamin C content in the fruit, the most effective treatment was T2, followed by T3, T1, and T4, with all of these being more effective than CK, representing increases of 32.39%, 23.91%, 21.42%, and 1.91% over CK, respectively(Figure 4F). Regarding pH values and relative water content, T3 exhibited the highest values at 3.89 and 71.42%, respectively, though these differences were not statistically significant compared to the CK (Figure 4G,H).

2.4.3. Effect of Different Foliar Fertilizers on the Concentrations of Total Phenolic Compounds, Tannins, Anthocyanins, and Flavonoids

The changes of total phenol and tannin content under different foliar fertilizer treatments were analyzed. The T1 resulted in the highest total phenol content (23.66 $mg \cdot g^{-1}$) and tannin content (4.66 $mg \cdot g^{-1}$), which were 25.38% and 47.01% higher than the CK, respectively (Figure 4I,J). In contrast, no significant differences in total phenol and tannin contents were observed under other treatments compared to the CK. Following the application of foliar fertilizers, all treatments exhibited increased anthocyanin content relative to the CK, with the T4 yielding the highest anthocyanin content at 0.25 $mg \cdot g^{-1}$, representing a 73.09% increase over the CK (Figure 4K). Regarding flavonoid content, different fertilizers had varying effects: the T3 resulted in the highest content (3.67 $mg \cdot g^{-1}$), marking a 70.69% increase from the CK, the T4 produced the lowest content (1.26 $mg \cdot g^{-1}$) (Figure 4L).

2.5. *Effect of Different Foliar Fertilizers on the Sugar and Acid Components of Fruits*

Analysis of the sugar components revealed that fructose content was the highest, followed by sucrose and glucose, with sorbitol content being the lowest. The effects of different foliar fertilizers on these sugar components indicated no significant differences in their impact on fructose levels. Among the treatments, T4 showed the highest fructose content at 62.29 $mg \cdot g^{-1}$, a 3.42% increase over the CK. Foliar fertilizers increased glucose, sucrose, and sorbitol contents in the fruit. Specifically, T4 resulted in the highest glucose content at 18.99 $mg \cdot g^{-1}$, representing a 55.91% increase compared to the control. Additionally, T2 yielded the highest sucrose (45.50 $mg \cdot g^{-1}$) and sorbitol (2.40 $mg \cdot g^{-1}$), which were 19.71% and 324.32% higher than the CK, respectively (Figure 5A–D).

Further analysis of the impact of various foliar fertilizers on fruit acid composition revealed that oxalic acid (0.51 $mg \cdot g^{-1}$) levels were highest under T4, an increase of 15.91% compared to the CK. Malic acid is a widely prevalent organic acid, and was consistently identified as one of the primary organic acids across all analyzed samples in this study. Under T4 treatment, its content reached 4.39 $mg \cdot g^{-1}$, representing a 11.99% increase over the CK. Conversely, tartaric acid content peaked under T1 at 1.76 $mg \cdot g^{-1}$, marking a 44.26% elevation over CK. In contrast, quinic acid and citric acid exhibited declining trends relative to CK, with their contents under T1 being the lowest, at 2.34 $mg \cdot g^{-1}$ for quinic acid and 0.03 $mg \cdot g^{-1}$ for citric acid (Figure 5E–I).

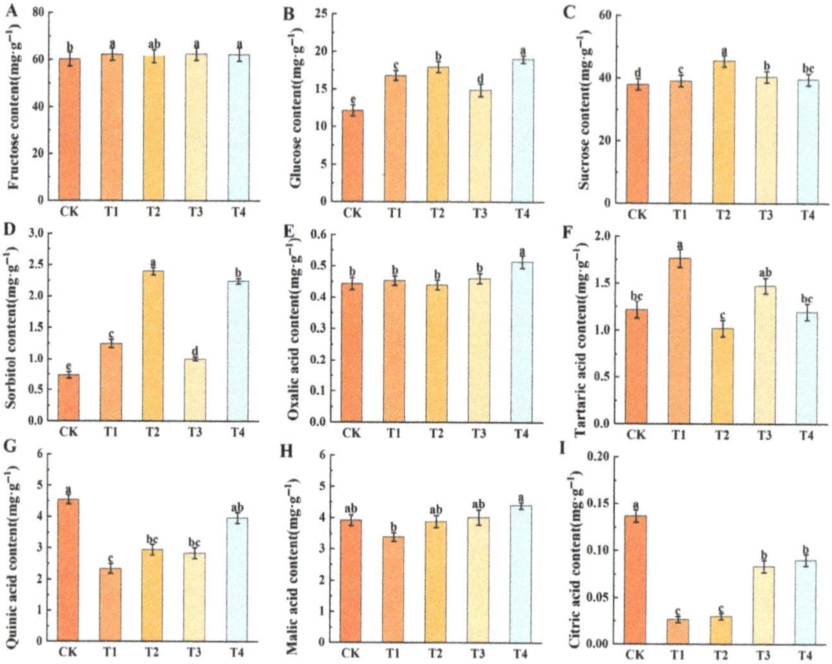

Figure 5. Effect of different foliar fertilizers on the sugar and acid components of fruits. (**A**): Fructose content. (**B**): Glucose content. (**C**): Sucrose content. (**D**): Sorbitol content. (**E**): Oxalic acid content. (**F**): Tartaric acid content. (**G**): Quinic acid content. (**H**): Malic acid content. (**I**): Citric acid content. CK: Clear water control. T1: Amino acid foliar fertilizer. T2: Calcium foliar fertilizer. T3: Boron foliar fertilizer. T4: Potassium foliar fertilizer. Error bars represent the mean ± SE from three biological repeats. Different lowercase letters denote significant differences, whereas the same lowercase letters indicate no statistical difference ($p < 0.05$).

2.6. Effect of Different Foliar Fertilizers on Aroma Composition of Apple Fruits

2.6.1. Effects of Different Foliar Fertilizers on the Contents of Aldehydes, Esters, Alcohols and Other Substances in Aroma Components

Through the analysis of aroma substances in fruits, 25 primary aroma compounds were identified and categorized into six groups: eight aldehydes, six esters, six alcohols, one aromatic hydrocarbon, one ketone, and one acid. These aroma substances were distributed across different treatments: 23.89%, 16.87%, 27.11%, 14.95%, and 17.18% of the total aroma content, respectively. Specifically, the proportions of aldehydes, esters, alcohols, aromatic hydrocarbons, ketones, and acids within each treatment relative to the total aroma content were 87.76%, 2.12%, 9.92%, 0.03%, 0.12%, and 0.05%, respectively (Figure 6).

Analysis of various aromatic compounds in apple fruit revealed that hexanal and 2-hexanal were the most abundant aldehydes. Among these, CK exhibited the highest hexanal content, whereas hexanal levels in other treatment groups were significantly lower, indicating that foliar fertilizer application reduces hexanal production in apple fruit. The 2-hexenal content in T2 was significantly higher than in CK, reaching 101.48 $\mu g \cdot kg^{-1}$, a value 43.03% greater than the control group. Ethyl butyrate, ethyl 2-methylbutyrate, and ethyl hexanoate were the most abundant ester compounds in apple fruit. Ethyl 2-methylbutyrate, the most abundant ester, was not detected in T3. The ethyl butyrate content in T1 was not significantly different from CK, whereas it was lower in other treatment groups. In T2, ethyl hexanoate content was significantly higher than in CK but significantly lower than CK in other treatment groups. n-Hexanol was the most abundant alcohol in apple fruit, though it was not detected in the T1 treatment group. The n-hexanol content in T2 was significantly increased by 42.23% compared to CK, whereas it was significantly lower than CK in other treatment groups. In addition to aldehydes, esters, alcohols, apple fruit also contains olefins, ketones, and acids. Following the application of foliar fertilizer. Dihydromyrcene content increased in T1 and T4, while methyl heptanone content rose significantly in the T2, reaching 0.37 μg·kg^{-1}, a 131.25% increase compared to the CK. Hexanoic acid was present only in the CK and T3, with the T3 showing a lower content than the CK.

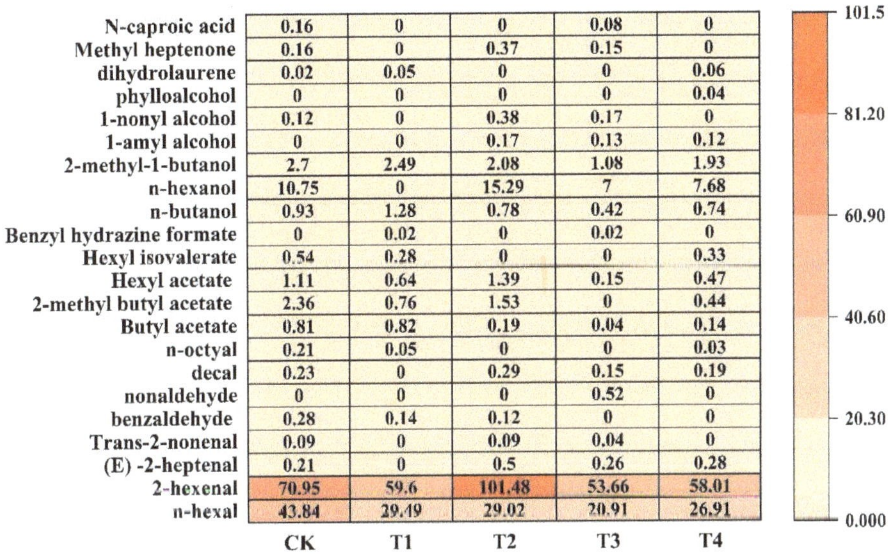

Figure 6. The content of aroma substances in fruits treated with different foliar fertilizers, the abscissa is the treatment, and the ordinate is the name of the aroma substance. The depth of the color represents the level of aroma content. The number in the figure represents the volatile aroma content, and the unit is μg·kg^{-1}. CK: Clear water control. T1: Amino acid foliar fertilizer. T2: Calcium foliar fertilizer. T3: Boron foliar fertilizer. T4: Potassium foliar fertilizer.

2.6.2. Changes of Total Aroma Substance Content and Characteristic Aroma Under Different Nutrient Foliar Fertilizer Treatments

The primary aroma substances in apple fruit were aldehydes, which accounted for the highest proportion, followed by alcohols and then aromatic hydrocarbons (Figure 7A). Among these, T2 significantly increased the total content of apple aroma substances by 13.41% compared to the CK. Specifically, under T2, the contents of aldehydes, alcohols, and ketones increased significantly, while other aroma substances decreased compared to the CK.

The aroma components with a value greater than 1 are considered characteristic of the fruit and play a significant role in its overall aroma profile. n-hexanal and 2-hexenal are key characteristic aroma components within the aldehyde category, exhibiting aroma values exceeding one across all treatments, including the CK (Figure 7B). Aldehydes and alcohols were the predominant volatile aroma substances in apples subjected to various foliar fertilizer treatments. Esters typically imparted sweet aromas, while aldehydes and alcohols contributed to green aromas. Notably, 'Snick' apple fruits displayed the highest aldehyde content, it is speculated that the variety is mainly fresh green aroma.

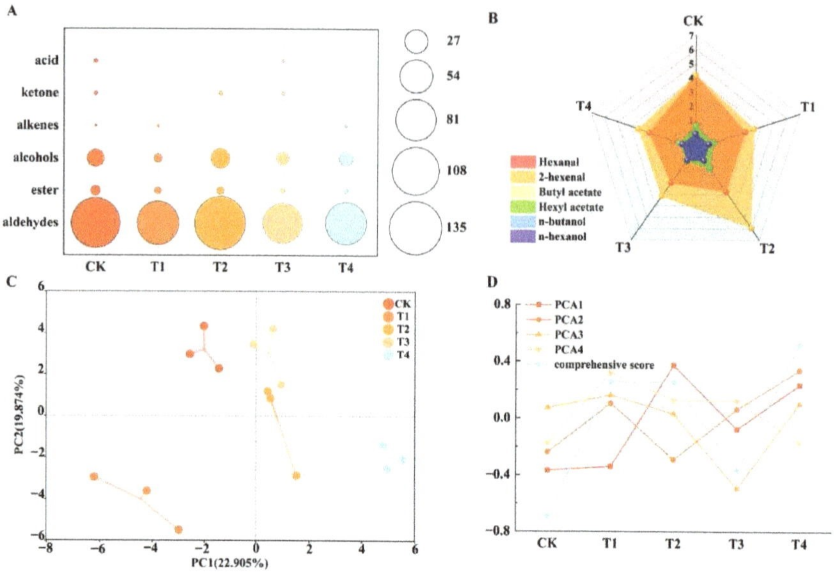

Figure 7. Analysis of aroma content and principal component analysis of determination indexes. (**A**): The total content of aldehydes, esters, alcohols, olefins, ketones and acids aroma substances under different nutrient elements foliar fertilizer treatment, the size of the circle represents the content of aroma substances. (**B**): The content of characteristic aroma under different nutrient element foliar fertilizer treatment. (**C**): Centroid PCA analysis of determination index. (**D**): The scores and comprehensive scores of each treatment on different main analyses. CK: Clear water control. T1: Amino acid foliar fertilizer. T2: Calcium foliar fertilizer. T3: Boron foliar fertilizer. T4: Potassium foliar fertilizer.

2.7. Principal Component Analysis and Correlation Analysis

2.7.1. Principal Component Analysis

To comprehensively assess the impact of various foliar nutrient fertilizers on 'Snick' apples, Principal Component Analysis was employed on the 42 measured indicators (Figure 7C). This analysis yielded four principal components (PC1–PC4), which collectively explained 100% of the variance, thereby validating their use as a condensed representation of the original dataset for evaluating foliar fertilizer effects. Each principal component's eigenvalue served as the weight for calculating the principal component score, facilitating the computation of a comprehensive correlation score. This score enabled ranking of treatments based on their overall enhancement of apple quality and aromatic profiles, with higher scores indicating superior treatment efficacy. Following the application of principal component analysis and subsequent calculations, the effectiveness ranking of various foliar fertilisers on snick apple quality was determined as follows: among all treatments, T4 exhibited the highest improvement effect, followed by T1, then T2 and T3, though all treatments significantly outperformed the CK (Figure 7D).

2.7.2. Analysis of Relationship

Analysis of leaf physiological parameters revealed significant positive correlations: between new shoot length and new shoot thickness (r = 0.92, * $p < 0.05$), between net photosynthetic rate and transpiration rate (r = 0.96, * $p < 0.05$). Further analysis of fruit quality showed highly significant positive correlations (between peel L* value and peel b* value, r = 0.98, ** $p < 0.01$), a highly significant negative correlation (between pH value and aroma compounds, r = −0.97, ** $p < 0.01$), and a highly significant positive correlation (between total phenols and tannins, r = 0.99, ** $p < 0.01$).

Investigations into correlations between leaf and fruit physiological indicators (Figure 8) further revealed: a significant positive correlation between new shoot length and anthocyanins (r = 0.95, * $p < 0.05$); a significant negative correlation between leaf area and transverse diameter (r = 0.96, * $p < 0.05$). These leaf-fruit correlations indicate that foliar fertilizer application influenced leaf physiological functions, which in turn affected fruit quality.

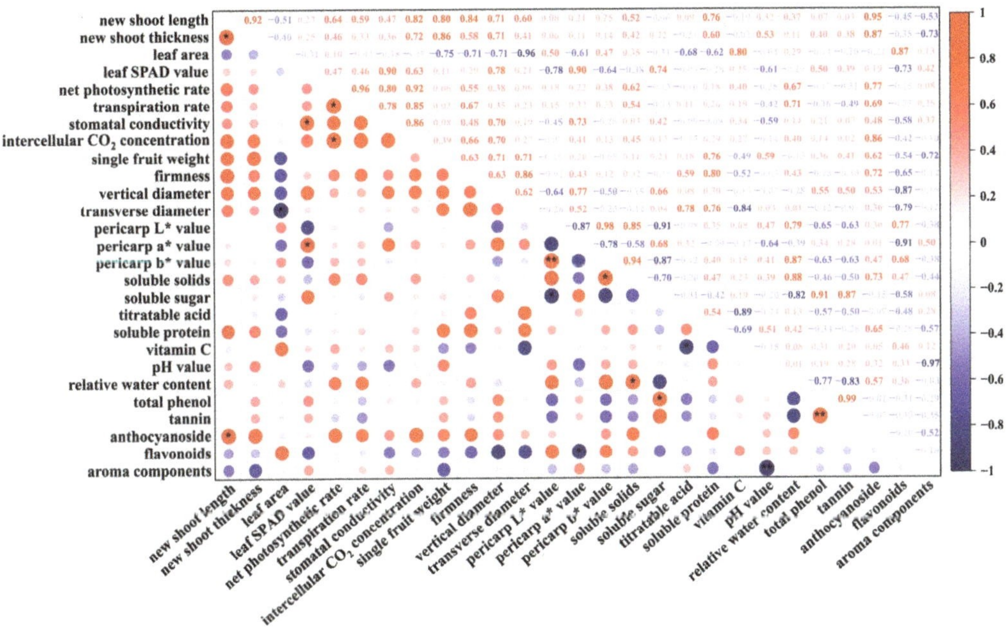

Figure 8. Correlation heat map. The intra-group correlation between main leaves and quality. Red represents a positive correlation, blue represents a negative correlation, and the depth of the color represents the size of the correlation coefficient. The size of the circle represents the strength of the correlation. Pearson correlation coefficient (r): measures the linear correlation between two continuous variables, from negative to positive. (−1 < r < 1) p: To test whether the r value is significant; * $p < 0.05$, significant correlation, ** $p < 0.01$, extremely significant correlation.

3. Discussion

Foliar fertilization primarily stimulates plant growth and boosts resilience against environmental stresses, ultimately improving crop yield and quality [23]. Key determinants of fruit sensory quality include the concentrations of soluble solids, titratable acidity, soluble sugars, and vitamin C [24]. Additionally, flavonoids, phenolic compounds, and soluble proteins are pivotal indicators associated with peel coloration, antioxidant capacity, and physiological metabolic activity [25].

Leaves are one of the nutritional organs of vascular plants. Their morphological structure is highly susceptible to changes in ecological conditions, enabling adaptation to their environment [26]. Their primary functions include carrying out photosynthesis to synthesize organic matter and performing transpiration, with the latter providing the driving force for the root system to absorb water and mineral nutrients from the soil [27]. In previous research, humic acid combined with nano-NPK fertilizer promoted leaf area growth in apple trees [28], and phenylalanine, nano-potassium fertilizer, or potassium sulfate boosted chlorophyll content in apple leaves [29], while foliar application of zinc and boron significantly promoted photosynthesis in apple leaves [30]. In this experiment, the application of foliar fertilizers significantly promoted the growth of new shoots, boron-containing foliar fertilizers notably enhanced leaf area expansion. Meanwhile, calcium-based foliar fertilizers improved the SPAD value of leaves while boosting net photosynthetic rate, transpiration rate, stomatal conductance, and intercellular CO_2 concentration. These results were generally consistent with previous research findings.

Soluble sugars, as one of the leading indicators for evaluating apple fruit quality, is affected by many factors. In this study, the soluble sugar content following amino acid foliar fertilizer treatment increased by 8.53% compared to the control, a result which is largely consistent with previous findings [31]. Additionally, the contents of glucose, sucrose, and sorbitol were also elevated under calcium foliar fertilizer treatment relative to the control [32]. Furthermore, Khan found that compared with the unfertilized control group, the application of amino acid foliar fertilizer at the blooming and fruit-sitting periods had a promoting effect on the content of reducing sugars in Golden tangerine [33], and the total soluble solids, total sugars and total carbohydrates of mango fertilized with amino acid foliar fertilizer were higher than those of the control group by 3.0%, 5.8% and 15.0%. Calcium spraying increased soluble sugar and soluble solids content in mandarin fruits and promoted the accumulation of soluble sugars in the fruits [19]. Cherry variety 'Sunburst' produced maximum sugar content after boron spraying [34]. These conclusions are generally consistent with the results of this experiment, even if these fruits have rather different development and maturation features.

Organic acids are important in determining the flavor of fruits, mainly malic, citric, tartaric, oxalic, and quinic acids, to which vitamin C also belongs. After spraying amino acid foliar fertilizers, titratable acid content and vitamin C content of citrus were increased [31]. It was found that application of calcium fertilizer increased vitamin C content and decreased titratable acid content in tomato fruit [35], and others found that application of calcium fertilizers increased vitamin C content in fruits [36,37]; application of boron increased vitamin C and titratable acid content in strawberries, and sweet oranges [38]; and Aye et al. found that foliar spraying of potassium fertilizer reduced the titratable acid content of citrus fruits [39], but others found that potassium fertilizer spray promoted the accumulation of vitamin C and total acid content in strawberries [40], while improving the acidity level of apple fruits [29]. In this experiment, the contents of vitamin C, oxalic acid, and malic acid increased under potassium fertilizer treatment. In the present research, the content of vitamin C under calcium foliar fertilizer treatment was 32.39% higher than that of the control group. Amino acid treatment increased the contents of vitamin C and oxalic acid in fruits, which was generally consistent with the findings of Ali et al. [41].

The natural antioxidant effect of fruits is closely related to phenolics, which help improve plants organs colour and flavor due to their extreme antioxidant activity. The results of this experiment demonstrated that the contents of total phenol, tannin, and anthocyanin increased in 'Snick' apples across all treatments. Specifically, the total phenol and tannin contents in the amino acid foliar fertilizer treatment were 25.38% and 47.01% higher than those in the CK, respectively; the anthocyanin content in the potassium fertilizer treatment was 73.09% higher than that in CK; and the flavonoid content in the boron fertilizer treatment increased by 70.25%. This finding is consistent with the results reported by Gao Xiang et al. [4], who observed that spraying mixed amino acids at a concentration of 75 $mg \cdot L^{-1}$ promoted the total phenolic content, total flavonoid content, and total antioxidant activity in celery plants. Other researchers have similar conclusions, such as spraying of boron fertilizer increased the antioxidant properties (total phenols) of strawberry fruits and similarly spraying of potash fertilizer promoted an increase in phenolic compounds and antioxidant activity in strawberry fruits [42].

This study investigated the effects of different nutritional foliar fertilizers on the physiological functions of apple leaves, fruit quality, and the content of aromatic compounds. Among them, mineral source fulvic acid potassium has the most significant improvement on fruit quality. This could be attributed to potassium (T4) involvement in over 60 enzyme system activations, photosynthesis, the transport of assimilated products, carbohydrate metabolism, and protein synthesis [43], thereby promoting significant improvements in fruit quality. But because the absorption capacity and sensitivity of apple trees to different nutrients are different, and the mechanism of related nutrients in different physiological metabolic pathways is different, the effects of amino acid, calcium, boron and potassium foliar fertilizers on the quality of apples are also different.

Principal component analysis and correlation analysis indicate that foliar fertilizers act on crops through leaf absorption and influence fruit quality. The core mechanisms involve direct supplementation of key nutrients, regulation of physiological metabolism, and enhancement of stress resistance [44,45]. Ultimately, these processes optimize both fruit internal components (e.g., sugars, vitamins, and proteins) and external traits (e.g., color, size, and firmness) across three interconnected dimensions: "nutrition supply–physiological activity–environmental adaptation [46]." However, the specific metabolic pathways and regulatory mechanisms remain unknown and have not been thoroughly investigated. In subsequent research, we will place greater emphasis on elucidating the mechanisms of action of foliar fertilizers, as well as evaluating the efficacy of novel nano-formulated foliar fertilizers.

4. Materials and Methods

4.1. Experimental Site and Experimental Materials

This study was conducted at the Apple Production Dispatch Centre in Ning County, located in the southern part of Qingyang City, Gansu Province. The region experiences a warm temperate continental monsoon climate, characterized by an altitude of 1074 m, an average annual temperature of 8.7 °C, and an average annual precipitation of 565.9 mm. The climate is semi-arid with rainfall concentrated primarily in July, August, and September. Additionally, the area has an average annual evaporation rate of 1442.6 mm, a frost-free period lasting 168 d, and 2369 h of sunshine annually. The terrain of the experimental site is flat, providing favorable conditions for drainage and irrigation. During the experiment, fertilizer application, irrigation water quantity, and irrigation timing were strictly adhered in accordance with the guidelines in the "Fertilizer Application Plan and Drip Irrigation Quota" developed by the Ning County Apple Production Dispatch Centre.

The test material consisted of 6-year-old 'Snick' apples planted at a spacing of 1×3.5 m, exhibiting a spindle shape and rootstock of M9T337. These trees had identical growth potential and were managed uniformly regarding pruning, fertilization, water management, soil care, and pest and disease control. Composted organic fertilizer (derived from crop residues, cow dung, straw, etc.) was applied as basal fertilizer in autumn at a rate of 45,000 $kg \cdot ha^{-1}$, along with urea (45 kg) and superphosphate (120 kg). An additional topdressing was applied before budburst, consisting of diammonium phosphate (45 $kg \cdot ha^{-1}$) administered via hole application. A second topdressing was applied during the fruit expansion stage, which included potassium chloride (60 $kg \cdot ha^{-1}$) and diammonium phosphate (7.5 $kg \cdot ha^{-1}$), also applied by hole placement.

4.2. Experimental Treatment and Design

The experimental treatment is shown in Table 1:

The experiment commenced in June 2022 during the fruit expansion period. The initial spraying occurred at 55 days after full bloom, and was followed by subsequent applications every 15 days, totaling three sprays. The spraying solution was diluted to the appropriate concentration and applied evenly to the foliage from top to bottom until runoff. Applications were conducted at 10:00 a.m. on sunny, windless days to ensure no rainfall within 48 h. A single-factor, completely randomized block design was employed, with 15 healthy and uniformly grown trees assigned to each treatment. Each treatment had three replicates, each consisting of five trees, for a total of 225 labeled trees.

Table 1. Basic information of experimental treatments.

Treatment	Foliar Fertilizer	Spray Concentration	Active Principle	Producer	Price (1000 g)
T1	Ye Xiumei	$0.2\ g{\cdot}L^{-1}$	amino acid content of $100\ g{\cdot}L^{-1}$, Cu + Fe + Mn + Zn + B $\geq 20\ g{\cdot}L^{-1}$	Qingdao Qianhechun Biotechnology Co., Ltd., Qingdao, China	50 yuan
T2	Fluid calcium	$0.067\ g{\cdot}L^{-1}$	Ca $\geq 180\ g{\cdot}L^{-1}$, sugar alcohol $\geq 50\ g{\cdot}L^{-1}$	Shandong Ruipu Biotechnology Co., Ltd., Zouping, China	30 yuan
T3	Sugar alcohol boron	$0.05\ g{\cdot}L^{-1}$	B $\geq 160\ g{\cdot}L^{-1}$	Shandong Ruipu Biotechnology Co., Ltd., Zouping, China	45 yuan
T4	Mineral source potassium fulvic acid	$0.05\ g{\cdot}L^{-1}$	soluble potassium fulvic acid $\geq$ 50%, and $K_2O \geq 12\%$	Shandong Ruipu Biotechnology Co., Ltd., Zouping, China	75 yuan
CK	water	/	/	/	/

'/' means 'none' or 'not applicable'.

Fruit collection was conducted on 20 August. Fruits of uniform size and without mechanical damage were selected from the four cardinal directions (east, west, south, and north) at the midpoint of the tree crown to minimize sampling errors. For each treatment, 10 fruits were sampled, with three replicates, resulting in a total of 150 fruits. These were immediately stored in a cold storage facility (0 °C) at the Ningxian Apple Production and Distribution Center. Subsequently, the peel and pulp were separated, frozen in liquid nitrogen, transported to the laboratory, and stored in an ultra-low temperature freezer at −80 °C.

4.3. *Growth Index of New Shoots*

Ten new shoot branches of uniform height were selected for each treatment, oriented in four different directions: east, west, south, and north, with three replicates. The length of the branches was kept consistent. The branches were labeled, and a vernier caliper was used to measure the growth length of the new shoots and basal diameter of new shoot. Each measurement was repeated three times per treatment.

4.4. *Indices of Leaf Growth and Photosynthetic Characteristics*

For each treatment, ten leaves were randomly selected with three replicates. The 7th to 9th leaves from the base of the 1-year-old sunny branches were chosen, avoiding the main veins and leaf edges. The SPAD value of the leaves was measured using a hand-held TYS-B portable SPAD chlorophyll content tester (Zhejiang Topyun Agricultural Science and Technology Co., Ltd., Hangzhou, China), with three measurements taken per leaf. Leaf area was measured using a TPYX-A instrument (Zhejiang Tuopu Yunnong Science and

Technology Co., Ltd., Hangzhou, China), selecting ten leaves per treatment and repeating the process three times.

Three fruit trees were selected for each treatment, with measurements were made on ten functional leaves still attached to the branch. Photosynthetic parameters, including net photosynthetic rate (Pn), transpiration rate (E), stomatal conductance (gs), and intercellular CO_2 concentration, were measured using a 3051D portable photosynthetic analyzer (Zhejiang Topyun Agricultural Science and Technology Co., Ltd., Hangzhou, China).

4.5. Determination of Fruit Appearance Quality

The weight of individual fruits was measured using an electronic balance, and the weight of each fruit was recorded. An electronic vernier caliper measured the fruit length and transverse diameters, and the fruit shape index was the ratio of the maximum longitudinal diameter to the transverse diameter. The GY-4-J digital hardness tester measured fruit firmness. The L*, a* and b* values were measured by colorimeter (3nh) (Shenzhen 3NH Technology Co., Ltd., Shenzhen, China). The L* value (0–100) represents the brightness of the fruit. The higher the L* value, the whiter and brighter the fruit surface. The a* value (−80–100) represents the red-green hue, with positive values indicating red and negative values indicating green. The larger the absolute value of a*, the darker the fruit surface color. The b* value (−80–70) represents the yellow-blue hue, with positive values indicating yellow and negative values indicating blue.

4.6. Determination of Soluble Solids, Soluble Sugar, Titratable Acidity, and Sugar-Acid Ratio

Soluble solids contents were measured using a handheld PAL-1 digital refractometer (Atago Co., Ltd., Tokyo, Japan). The determination of total soluble sugars was performed using the modified anthrone colorimetric method [47]. The titratable acid content was determined using the sodium hydroxide titration method [48]. Then the sugar-acid ratio was calculated as soluble sugar content/titratable acid content.

4.7. Determination of Soluble Protein, Vitamin C, pH and Relative Water Content

Fruit soluble protein was measured according to the method described by Wei Jinmei with some modifications [49]. Vitamin C content was analyzed using the spectrophotometric method described [50]. The vitamin C content was determined using a standard curve generated from an ascorbic acid solution. The pH was measured using an acidometer (Lei, PHS-3E) (Shanghai LeiCi, Shanghai, China). The relative water content was determined by heating and drying method. The apple pulp was dried at 105 °C for 12 h and weighed every 2 h. Constant weight was considered achieved when the difference between two consecutive weighing results was $\leq$0.1 g.

4.8. Determination of Total Phenols, Tannins, Anthocyanins and Flavonoids

The total phenol was determined by Folin-Ciocalteu method [51], the tannin content was determined by Folin-Denis method [52], and the anthocyanin and flavonoid were determined by spectrophotometer metric method [53].

4.9. High-Performance Liquid Chromatography (HPLC) Analysis of Sugars and Organic Acids

The contents of major organic acids and sugars in fruit pulp were measured according to the method [54], by using high-performance liquid chromatography (HPLC) (Waters Corp., Wilford, MA, USA). Each experiment was carried out with three biological replicates.

4.10. Determination of Aroma Substances and Their Thresholds

Aroma substances were collected using headspace solid-phase microextraction (HS-SPME). Specifically, 5.0 g of fruit was weighed into a 50 mL sample vial, followed by the addition of 2.0 g NaCl, 20 μL of internal standard (2-octanol, 8.19 mg·L^{-1}), and a magnetic stir bar. The vial was then placed on a magnetic stirrer, where the sample was stirred at 500 r·min^{-1} and heated at 55 °C for 30 min to achieve equilibrium. After this conditioning step, the SPME fiber was inserted into the headspace of the vial, and extraction was carried out for 30 min under continuous stirring and heating at 55 °C. Then were determined by gas chromatography-mass spectrometry [55], VOCs were identified and quantified using an Agilent Model 8860 GC (Agilent Technologies Inc., Santa Clara, CA, USA) and a 5977B mass spectrometer (Agilent) (Agilent Technologies Inc., Santa Clara, CA, USA), both equipped with a 30 m × 0.25 mm × 0.25 μm HP-5MS capillary column. Helium was the carrier gas at a 1.0 mL·min^{-1} linear velocity. The injector temperature was kept at 230 °C, and the detector at 250 °C. The oven temperature was programmed from 40 °C (4 min), increasing at 8 °C·min^{-1} to 180 °C, at 20 °C·min^{-1} to 230 °C, at 5 °C·min^{-1} to 250 °C, and held for 5 min. Mass spectra were recorded in electron impact (EI) ionization mode at 70 eV. Identification of VOCs was achieved by comparing the mass spectra with the data system library (MWGC or NIST) and linear retention index (RI, determined by alkane C5–C40), referencing RI information to increase qualitative accuracy and eliminating the interference of false positive substances.

The following ROAV formula was developed based on the relative concentrations of various aroma components and their respective threshold values in water: $ROAV = (C_n/C_{max}) \times (T_{max}/T_n) \times 100$. Here, C_n represents the relative content of any VOC (μg·g^{-1}), C_{max} represents the component with the highest relative content (μg·g^{-1}), T_n represents the threshold of any VOC (μg·g^{-1}), and T_{max} represents the threshold of the component with the largest relative content (μg·g^{-1}) [56].

4.11. Data Processing

Data processing and figure generation were performed using Excel 2021 (Microsoft, Redmond, WA, USA). Graphs were created using Origin 2024 (Origin Inc., San Francisco, CA, USA). Statistical analyses, including significance testing and one-way ANOVA, were conducted using SPSS 26.0 (SPSS Inc., Chicago, IL, USA), with a significance level of $p < 0.05$.

5. Conclusions

This study demonstrated that foliar fertilization has the potential to improve fruit quality in 'Snick' apples (e.g., higher sugar-acid ratio and phenolic content) by enhancing physiological functions such as leaf photosynthetic performance (Figure 9). Although positive effects of foliar fertilizers, particularly mineral source potassium fulvic acid, were observed, it should be noted that fruit load and harvest date were identified as key co-variates influencing intrinsic fruit quality and maturity. A higher fruit load may lead to nutrient dilution, which could partially offset the benefits conferred by foliar fertilization. Furthermore, the timing of harvest directly affects fruit acidity, firmness, and metabolite accumulation. These factors may interact with foliar fertilizer treatments and potentially confound the results obtained in this study.

Therefore, the conclusions regarding the efficacy of foliar fertilizers require further validation in subsequent experiments that rigorously control fruit load and incorporate multiple harvest time points. Provided these variables are effectively managed, mineral source potassium fulvic acid can be recommended as a preferred foliar treatment for enhancing 'Snick' apple production efficiency. Its application should also be prioritized in further trials involving other apple varieties.

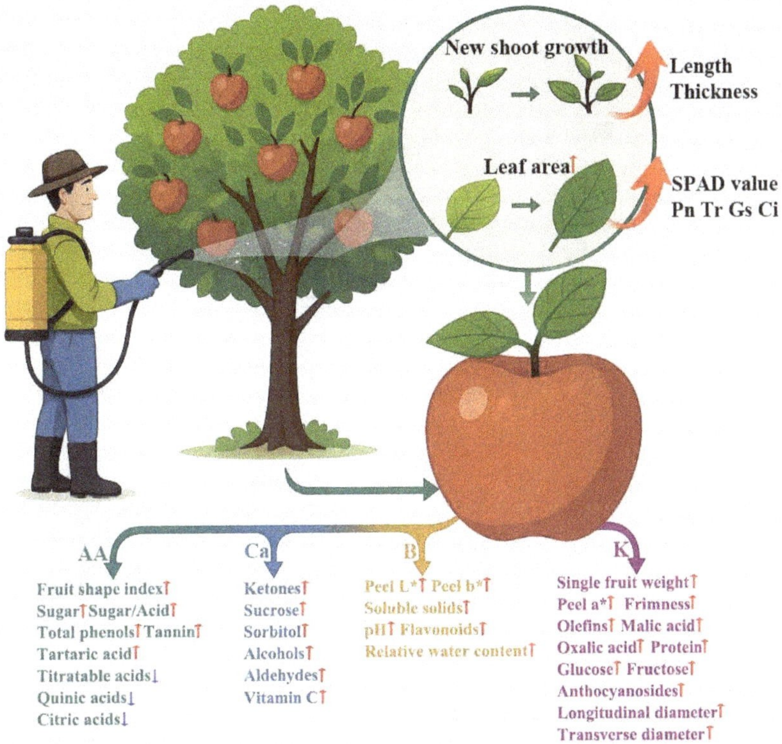

Figure 9. Effects of foliar fertilizer on leaf, fruit quality and aroma content. AA: amino acid foliar fertilizer. Ca: Calcium foliar fertilizer. B: Boron foliar fertilizer. K: Foliar potassium fertilizer. The red upward arrow represents an increase in its content, and the blue downward arrow represents a decrease in its content.

Author Contributions: Writing—original draft preparation, H.-F.X.; writing-review and editing, G.-P.L., Z.-Y.B. and J.M.; methodology, H.-F.X., S.-M.L. and W.-F.M.; formal analysis, H.-F.X., S.-M.L. and S.-X.L.; investigation, H.-F.X., W.-F.M. and S.-X.L.; data curation, H.-F.X. and S.-M.L.; supervision, J.M.; project administration, J.M.; funding acquisition, J.M. All authors have read and agreed to the published version of the manuscript.

Funding: This research was supported by the Technology Major Project of Gansu Province, China (23ZDNA001-4), Science and Industrial Support Plan of Higher Education of Gansu (2023CYZC-48), Science and Technology Major Project of Gansu Province, China (24ZDNL001). All the above projects are funded by Professor Juan Mao.

Data Availability Statement: Data are contained within the article.

Conflicts of Interest: The authors declare no conflicts of interest.

References

1. Zeng, C.; Mu, Y.; Yuan, J.; Zhang, H.; Song, J.; Kang, S. Apple cultivar-influenced differences in cider: A comprehensive analysis of physicochemical and aroma compounds. *Microchem. J.* **2025**, *208*, 112382. [CrossRef]
2. Wen, S.; Cui, N.; Wang, Y.; Gong, D.; Xing, L.; Wu, Z.; Zhang, Y.; Wang, Z. Deficit irrigation enhances yield and water productivity of apples by inhibiting excessive vegetative growth and improving photosynthetic performance. *Agric. Water Manag.* **2025**, *307*, 109220. [CrossRef]

3. Chen, Y.; Wen, X.; Sun, Y.; Zhang, J.; Wu, W.; Liao, Y. Mulching practices altered soil bacterial community structure and improved orchard productivity and apple quality after five growing seasons. *Sci. Hortic.* **2014**, *172*, 248–257. [CrossRef]
4. Gao, X.; Mei, X.; Zhang, J.; Cai, J.; Gu, F.; Hao, W.; Gong, D. Comparison of three modified models in evapotranspiration and its components over a rainfed spring maize cropland on the Loess Plateau, China. *Agric. For. Meteorol.* **2023**, *330*, 109322. [CrossRef]
5. Yu, J.; Liu, X.; Wang, W.; Zhang, L.; Wang, C.; Zhang, Q.; Wang, J.; Du, M.; Sheng, L.; Hu, D. *MdCIbHLH1* modulates sugar metabolism and accumulation in apple fruits by coordinating carbohydrate synthesis and allocation. *Hortic. Plant J.* **2025**, *11*, 578–592. [CrossRef]
6. Sredojevic, A.; Radivojevic, D.; Levic, S.M.; Aksic, M.F.; Milivojevic, J.; Djekic, I. Effects of the fruit harvest date and shelf-life nexus of apples on different quality perspectives. *Appl. Sci.* **2024**, *14*, 11737. [CrossRef]
7. Hassan, M.R.; Arakawa, O.; Nissato, K.; Ito, D. Changes in the harvesting window and quality of apple fruit cultivated under long-term high temperature and CO_2. *Sci. Hortic.* **2024**, *338*, 113611. [CrossRef]
8. Wang, Q.; Zhou, Q.; Feng, Y.; Yang, X. Foliar application protected vegetable against poisonous element cadmium and mitigated human health risks. *Sci. Total Environ.* **2024**, *926*, 171915. [CrossRef]
9. Zaben Habib, J.; Nazhan Ali, O.; Ahmed Alsajri, F.; Hatta, M. Effect of shade covers and foliar fertilizer on the growth of pepper seeds. *IOP Conf. Ser. Earth Environ. Sci.* **2021**, *807*, 042011.
10. Ishfaq, M.; Kiran, A.; Rehman, H.; Farooq, M.; Ijaz, N.H.; Nadeem, F.; Azeem, I.; Li, X.; Wakeel, A. Foliar nutrition: Potential and challenges under multifaceted agriculture. *Environ. Exp. Bot.* **2022**, *200*, 104909. [CrossRef]
11. Khomphet, T.; Promwee, A.; Islam, S.S. Effects of foliar fertilizer application on the growth and fruit quality of commercial melon varieties grown in a soilless culture system. *PeerJ* **2023**, *11*, e14900. [CrossRef]
12. Dong, H.; Li, F.; Xuan, X.; Ahiakpa, J.K.; Tao, J.; Zhang, X.; Ge, P.; Wang, Y.; Gai, W.; Zhang, Y. The genetic basis and improvement of photosynthesis in tomato. *Hortic. Plant J.* **2025**, *11*, 69–84. [CrossRef]
13. Hong, T.; Cai, Z.; Li, R.; Liu, J.; Li, J.; Wang, Z.; Zhang, Z. Effects of water and nitrogen coupling on watermelon growth, photosynthesis and yield under CO_2 enrichment. *Agric. Water Manag.* **2022**, *259*, 107229. [CrossRef]
14. Therby-Vale, R.; Lacombe, B.; Rhee, S.Y.; Nussaume, L.; Rouached, H. Mineral nutrient signaling controls photosynthesis: Focus on iron deficiency-induced chlorosis. *Trends Plant Sci.* **2022**, *27*, 502–509. [CrossRef] [PubMed]
15. Zhang, Y.; Peng, Y.; Liu, J.; Yan, J.; Zhu, K.; Sun, X.; Bu, X.; Wang, X.; Ahammed, G.J.; Liu, Y.; et al. Tetratricopeptide repeat protein SlREC2 positively regulates cold tolerance in tomato. *Plant Physiol.* **2023**, *192*, 648–665. [CrossRef] [PubMed]
16. Lu, J.; Pan, C.; Li, X.; Huang, Z.; Shu, J.; Wang, X.; Lu, X.; Pan, F.; Hu, J.; Zhang, H.; et al. *OBV* (obscure vein), a C_2H_2 zinc finger transcription factor, positively regulates chloroplast development and bundle sheath extension formation in tomato (*Solanum lycopersicum*) leaf veins. *Hortic. Res.* **2021**, *8*, 230. [CrossRef]
17. Nasar, J.; Wang, G.Y.; Ahmad, S.; Muhammad, I.; Zeeshan, M.; Gitari, H.; Adnan, M.; Fahad, S.; Khalid, M.H.; Zhou, X.B.; et al. Nitrogen fertilization coupled with iron foliar application improves the photosynthetic characteristics, photosynthetic nitrogen use efficiency, and the related enzymes of maize crops under different planting patterns. *Front. Plant Sci.* **2022**, *13*, 988055. [CrossRef]
18. El-Tanahy, A.M.M.; Mahmoud, S.H.; Elwahed, M.S.A.; Salama, D.M. Enhancing celery's growth, production, quality, and nutritional status using tryptophan and glycine amino acids. *Sci. Rep.* **2024**, *14*, 26571. [CrossRef]
19. Wang, T.; Tan, L.; Chen, Z.; Yang, Y.; Yuan, Y.; Zheng, Z.; Deng, L.; Zhang, M.; Sun, G.; He, S.; et al. Mitigating citrus fruit cracking: The efficacy of chelated calcium or silicon foliar fertilizers in 'Okitsu no. 58' citrus fruit. *Front. Plant Sci.* **2024**, *15*, 1402945. [CrossRef]
20. Gholamnejad, S.; Haghighi, M.; Etemadi, N.; Pessarakli, M. Effects of boron on nutrient partitioning, Ca movement, and fruit quality of tomatoes. *J. Plant Nutr.* **2023**, *46*, 697–713. [CrossRef]
21. Xu, W.; Wang, P.; Yuan, L.; Chen, X.; Hu, X. Effects of application methods of boron on tomato growth, fruit quality and flavor. *Horticulturae* **2021**, *7*, 223. [CrossRef]
22. Singh, G.; Gschwend, A.R.; Dami, I.E. Effect of foliar application of potassium fertilizer on yield, fruit quality, and cold hardiness of *Vitis* spp. 'Chambourcin'. *Int. J. Fruit Sci.* **2024**, *24*, 102–114. [CrossRef]
23. Makale, A.R.; Mourice, S.K.; Kapinga, F.A. Assessment of farmers' knowledge on the foliar fertilizer use for improved cashew (*Anacardium occidentale* L.) productivity in south-eastern Tanzania. *J. Agric. Food Res.* **2024**, *16*, 101092. [CrossRef]
24. Pimentel, C.; Pina, C.M.; Müller, N.; Lara, L.A.; Melo Rodriguez, G.; Orlando, F.; Schoelkopf, J.; Fernández, V. Mineral particles in foliar fertilizer formulations can improve the rate of foliar uptake. *Plants* **2023**, *13*, 71. [CrossRef]
25. Seo, H.J.; Sawant, S.S.; Lee, B.; Kim, K.; Song, J.; Choi, E.D. Mechanisms driving fruit cracking in 'Sinhwa' pears (*Pyrus pyrifolia* Nakai) and effect of foliar fertilizer application on fruit quality. *Sci. Hortic.* **2024**, *332*, 113232. [CrossRef]
26. Liu, W.; Tang, Y.; Xie, Z.; Zeng, G.; Wu, T.; Liu, J.; Lin, Z. Optimizing paclobutrazol application for regulating dwarfing in ougan (*Citrus reticulata* cv. Suavissima): Comprehensive insights from growth, photosynthesis, and physiological responses. *Plants* **2025**, *14*, 763. [CrossRef]

27. Zheng, H.J.; Wang, X.; Ma, W.F.; Gou, H.M.; Liang, G.P.; Mao, J. Temporal variations in photosynthesis and leaf element contents of '*Marselan*' grapevines in response to foliar fertilizer application. *Plants* **2025**, *14*, 946. [CrossRef] [PubMed]
28. Gli, J.Q.H.; Khalaf, J.M. Effect of foliar application of humic acid and nano-NPK on some growth characteristic of apple trees *Malus Domestica* Borkh. *IOP Conf. Ser. Earth Environ. Sci.* **2023**, *1158*, 042027. [CrossRef]
29. Hassan, A.; Margay, A.R.; Din, S. Effect of foliar sprays of phenylalanine, nano-potash and potassium sulphate on fruit quality attributes of apple (*Malus* × *domestica* Borkh.) cv. Ambri. *Int. J. Plant Soil Sci.* **2024**, *36*, 317–325. [CrossRef]
30. Kufesh, S.A.; Abdul, A.S.; Al-Issawi, A.S. The impact of stem site and foliar application of zinc and boron in growth, flowers and yield of apples (Brahimi Class). *IOP Conf. Ser. Earth Environ. Sci.* **2023**, *1158*, 042038. [CrossRef]
31. Seif El-Yazal, M.A.; Morsi, M.E. Metabolic changes in chemical constituents, flowering behavior, yield and fruit quality of 'Anna' apple (*Malus sylvestris* Mill) trees with foliar application some compounds. *IJAGST* **2022**, *2*, 30–39. [CrossRef]
32. ValizadehKaji, B.; Mohammaei, M. Foliar application of biostimulants and potassium silicate enhances yield and fruit quality of mandarin Cv. 'Page'. *Appl. Fruit Sci.* **2025**, *67*, 8. [CrossRef]
33. Khan, A.S.; Munir, M.; Shaheen, T.; Tassawar, T.; Rafiq, M.A.; Ali, S.; Anwar, R.; Rehman, R.N.U.; Hasan, M.U.; Malik, A.U. Supplemental foliar applied mixture of amino acids and seaweed extract improved vegetative growth, yield and quality of citrus fruit. *Sci. Hortic.* **2022**, *296*, 110903. [CrossRef]
34. Sajid, M.; Basit, A.; Shah, S.T.; Khan, A.; Ullah, I.; Bilal, M.; Khan, M.S.; Khan, W. Enhancing the quality and fruit yield of sweet cherry (*Prunus avium*) cultivars by foliar application of boron. *Appl. Fruit Sci.* **2024**, *66*, 485–494. [CrossRef]
35. Guo, X.; Zhan, N.; Wang, Z.; Ding, F.; Yang, Z.; Cui, X. Effects of the combination of *Bacillus aryabhattai* and calcium peroxide on soil silicon and potassium contents, the yield and quality of facility tomato. *Arch. Agron. Soil Sci.* **2024**, *70*, 1–12. [CrossRef]
36. Sediqui, N.; Amin, M.W.; Dawlatzai, N.; Gulab, G.; Poyesh, D.S.; Terada, N.; Sanada, A.; Kamata, A.; Koshio, K. Elucidation of shoot and root growth, physiological responses, and quality traits of tomato (*Solanum lycopersicon* L.) exposed to elevated calcium carbonate concentrations. *Horticulturae* **2024**, *10*, 573. [CrossRef]
37. Wu, Y.; Yang, X.; Wang, X.; Yan, L.; Hu, X.; Lian, M. Effect of foliar calcium fertilization on fruit quality, cell wall enzyme activity and expression of key genes in Chinese cherry. *Int. J. Fruit Sci.* **2023**, *23*, 200–216. [CrossRef]
38. Walli, S.; Hafiz, I.A.; Khan, R.I.; Bashir, M.A.; Uddin, S.; Khan, S.U.; Ajmal, U.; Hussain, S.; Khalid, M.F. Zinc and boron application at different phenological stages alleviates tree growth, fruit yield and quality of sweet orange cv. 'Blood Red'. *Gesunde Pflanzen.* **2022**, *74*, 385–396. [CrossRef]
39. Thu, A.M.; Alam, S.M.; Khan, M.A.; Han, H.; Liu, D.H.; Tahir, R.; Ateeq, M.; Liu, Y.Z. Foliar spraying of potassium sulfate during fruit development comprehensively improves the quality of citrus fruits. *Sci. Hortic.* **2024**, *338*, 113696. [CrossRef]
40. Malekzadeh, M.R.; Esmaeilizadeh, M.; Roosta, H.R. Optimizing strawberry growth and fruit quality through fertigation frequency and foliar application of potassium sulfate. *J. Soil Sci. Plant Nutr.* **2024**, *24*, 3042–3055. [CrossRef]
41. Ali, Q.; Kurubas, M.S.; Ustun, H.; Balkhi, M.; Erkan, M. Evaluation of foliar organic fertilizer, biofertilizer and biological fungicide on the antioxidant compounds and postharvest quality attributes of strawberry fruit. *Erwerbs-Obstbau* **2022**, *64*, 365–376. [CrossRef]
42. Yu, J.Q.; Ji, F.Y.; Yang, X.K.; Cheng, Y.; Gao, H.S.; Sheng, L.X. A genome-wide investigation of the mechanism underlying the effect of exogenous boron application on sugar content and overall quality of 'Benihoppe' strawberries. *Plant Physiol. Biochem.* **2024**, *216*, 109116. [CrossRef] [PubMed]
43. Ma, L.; Chi, C.; Lv, S.; Tong, Y.; Yang, L. Effects of different sources of potassium fertiliser on yield, fruit quality and nutrient absorption in 'Harward' kiwifruit (*Actinidia deliciosa*). *Open Life Sci.* **2025**, *20*, 20251114. [CrossRef] [PubMed]
44. Peng, J.J.; Du, J.J.; Ma, W.Q.; Chen, T.T.; Shui, X.; Zhou, K.B. Effect of calcium and magnesium foliar fertilizer on sugar contents and sugar metabolizing enzyme activities in 'Feizixiao' litchi pulp. *J. Trop. Biol.* **2024**, *15*, 217–223.
45. Cook, B.S.; Brown, M.G.; Lin, Y.; Kwasniewski, M.T.; Ac-Pangan, M.F.; Stewart, A.C.; Peck, G.M. Foliar urea applications to apple trees increase yeast assimilable nitrogen, amino acids, and flavor volatiles in fruit and hard cider. *J. Agric. Food Res.* **2024**, *16*, 101136. [CrossRef]
46. Henderson, B.C.R.; Sanderson, J.M.; Fowles, A. A review of the foliar application of individual amino acids as biostimulants in plants. *Discov. Agric.* **2025**, *3*, 69. [CrossRef]
47. Jiang, L.; Chen, X.; Gu, X.; Deng, M.; Li, X.; Zhou, A.; Suo, M.; Gao, W.; Lin, Y.; Wang, Y.; et al. Light quality and sucrose-regulated detached ripening of strawberry with possible involvement of abscisic acid and auxin signaling. *Int. J. Mol. Sci.* **2023**, *24*, 5681. [CrossRef]
48. Wang, J.; Lu, Y.; Zhang, X.; Hu, W.; Lin, L.; Deng, Q.; Xia, H.; Liang, D.; Lv, X. Effects of potassium-containing fertilizers on sugar and organic acid metabolism in grape fruits. *Int. J. Mol. Sci.* **2024**, *25*, 2828. [CrossRef]
49. Wei, J.; Zhang, Y.; Li, W. Comparison of physiochemical characteristics of Lanzhou bulb of lily during storage periods. *Postharvest Biol. Technol.* **2025**, *219*, 113271. [CrossRef]

50. Yang, T.; Samarakoon, U.; Altland, J.; Ling, P. Photosynthesis, biomass production, nutritional quality, and flavor-related phytochemical properties of hydroponic-grown arugula (*Eruca sativa* Mill.) 'Standard' under different electrical conductivities of nutrient solution. *Agronomy* **2021**, *11*, 1340. [CrossRef]
51. Stratil, P.; Klejdus, B.; Kubáň, V. Determination of total content of phenolic compounds and their antioxidant activity in vegetables—Evaluation of spectrophotometric methods. *J. Agric. Food Chem.* **2006**, *54*, 607–616. [CrossRef] [PubMed]
52. Khomdram, S.D.; Singh, P.K. Polyphenolic compounds and free radical scavenging activity in eight Lamiaceae herbs of Manipur. *Not. Sci. Biol.* **2011**, *3*, 108–113. [CrossRef]
53. Nastasi, J.R.; Fitzgerald, M.A.; Kontogiorgos, V. Tuning the mechanical properties of pectin films with polyphenol-rich plant extracts. *Int. J. Biol. Macromol.* **2023**, *253*, 127536. [CrossRef]
54. Li, X.; Zeng, Y.; Wang, T.; Jiang, B.; Liao, M.; Lv, Y.; Li, J.; Zhong, Y. Dynamic analysis of the fruit sugar-acid profile in a fresh-sweet mutant and wild type in 'Shatangju' (*Citrus reticulata* cv.). *Plants* **2024**, *13*, 2722. [CrossRef]
55. Ma, C.; Wang, Q.; Tian, D.; Yuan, W.; Tang, X.; Deng, X.; Liu, Y.; Gao, C.; Fan, G.; Xiao, X.; et al. HS-SPME-GC-MS combined with relative odor activity value identify the key aroma components of flowery and fruity aroma in different types of GABA tea. *Food Chem. X* **2024**, *24*, 101965. [CrossRef] [PubMed]
56. Wang, M.Q.; Ma, W.J.; Shi, J.; Zhu, Y.; Lin, Z.; Lv, H.P. Characterization of the key aroma compounds in Longjing tea using stir bar sorptive extraction (SBSE) combined with gas chromatography-mass spectrometry (GC–MS), gas chromatography-olfactometry (GC-O), odor activity value (OAV), and aroma recombination. *Food Res. Int.* **2020**, *130*, 108908. [CrossRef] [PubMed]

plants

MDPI

Article

Seaweed Foliar Biostimulants Improve Growth and Phytochemicals of Thai Basil (*Ocimum basilicum* L.) in a Plant Factory

Vu Phong Lam [1,2,†], Gwonjeong Bok [3,†], Dao Nhan Loi [2,4], Manh Cuong Do [5,6,*] and Jongseok Park [1,4,*]

1 Department of Horticultural Science, Chungnam National University, Daejeon 34134, Republic of Korea; phonglamdhtaybac@gmail.com
2 Faculty of Agriculture and Forestry, Tay Bac University, Son La 360000, Vietnam; daonhanloi@gmail.com
3 Glocal University Project Team, Sunchon National University, Suncheon 57922, Republic of Korea; bokc4578@scnu.ac.kr
4 Department of Bio-AI Convergence, Chungnam National University, Daejeon 34134, Republic of Korea
5 Institute of Theoretical and Applied Research, Duy Tan University, Hanoi 100000, Vietnam
6 College of Medicine and Pharmacy, Duy Tan University, Da Nang 550000, Vietnam
* Correspondence: domanhcuong2@duytan.edu.vn (M.C.D.); jongseok@cnu.ac.kr (J.P.); Tel.: +82-42-821-5737 (J.P.)
† These authors contributed equally to this work.

Abstract: This study aimed to identify the optimal concentration of seaweed extract (SE) for enhancing growth, photosynthetic traits, antioxidant activity, and bioactive compound accumulation in Thai basil (*Ocimum basilicum* L.) plants cultivated in a fully controlled plant factory. Basil plants were foliar-sprayed twice weekly with five SE concentrations (0.5, 1.0, 1.5, 2.0, and 2.5 mL·L^{-1}), while untreated plants served as controls. After 28 days of transplanting, plant growth parameters, photosynthetic parameters, chlorophyll pigments, antioxidant activity, and the concentrations of phenolic acids and rosmarinic acid (RA) were analyzed. Moderate SE concentrations (1.0–2.0 mL·L^{-1}) significantly enhanced plant growth, chlorophyll a, carotenoid levels, DPPH radical scavenging, and total flavonoid content relative to control. The 2.0 mL·L^{-1} treatment produced the highest total phenolic content (1.88-fold increase over the control) and was associated with elevated benzoic acid, rutin, quercetin, and kaempferol, along with reduced trans-cinnamic acid, indicating activation of the phenylpropanoid pathway. Moreover, all SE treatments significantly increased RA accumulation. These findings demonstrate that SE is an effective, sustainable biostimulant for Thai basil, with 2.0 mL·L^{-1} as the optimal concentration for maximizing growth and phytochemical production.

Keywords: antioxidant activity; chlorophyll; concentration; flavonoid; photosynthetic parameters

1. Introduction

Thai basil (*Ocimum basilicum* L.), a member of the Lamiaceae family, is a widely cultivated culinary and medicinal herb, especially popular in Southeast Asia [1,2]. Its distinctive aroma and flavor come from a rich composition of bioactive compounds, including phenolic acids, flavonoids, and essential oils such as linalool, methyl chavicol, and eugenol [3,4]. These compounds exhibit diverse biological activities, including antioxidant, antimicrobial, anti-inflammatory, and anticancer properties, making Thai basil valuable for the food, pharmaceutical, and cosmetic industries [5]. With the growing global demand for functional foods and natural health products, there is increasing interest in improving both the yield and phytochemical quality of basil [6,7]. However, the accumulation of these compounds is strongly affected by environmental factors and cultivation methods, highlighting the need for optimized production strategies to enhance its economic and nutritional value.

In recent years, plant factories with artificial lighting (PFALs) have become a cutting-edge approach for sustainable and high-quality herb production [8,9]. These systems create a fully controlled environment where key factors—such as temperature, humidity, light intensity, photoperiod, and nutrient supply—can be precisely regulated [9]. By eliminating external environmental fluctuations, PFALs enable consistent plant growth, improved uniformity, and year-round production of high-value crops [10]. To maximize the potential of PFALs, it is essential to complement environmental control with bio-based inputs that further stimulate plant performance.

Among such innovations, seaweed-based biostimulants have emerged as effective and eco-friendly solutions for improving plant growth, productivity, and stress resilience [11–13]. Derived primarily from brown algae such as *Ascophyllum nodosum* and *Sargassum* spp., these products contain diverse bioactive molecules—including phytohormones, polysaccharides, amino acids, and minerals—that regulate plant metabolism and enhance physiological efficiency [14]. Numerous studies have demonstrated that seaweed extracts (SE) can enhance photosynthetic activity, nutrient uptake, and root development while also stimulating the production of antioxidant enzymes and secondary metabolites [14–16]. Their application has shown positive effects across a wide range of crops and environments: enhanced yield and nutrient uptake in soybean under rainfed conditions [15], increased sucrose content and photosynthetic activity in sugarcane [17], and improved drought tolerance and flower development in *Hydrangea paniculata* [18]. In fruit crops such as peach and apricot, foliar applications of seaweed and yeast extracts have boosted fruit uniformity, polyphenol accumulation, and antioxidant capacity [16]. Similarly, leafy and microgreen vegetables like cabbage and arugula exhibited elevated phenolic and flavonoid contents after treatment with *Ascophyllum nodosum* extracts [19,20]. In various horticultural crops, foliar application of SE has led to measurable improvements in productivity and biochemical quality. For instance, treatments of 4.0–5.0 L ha^{-1} enhanced fruit uniformity and polyphenol accumulation in apricot [16], while similar formulations increased tomato yield by 4.6–6.9% through improved photosynthetic capacity [21]. In hot pepper, a 0.5% SE raised the yield to 4.23 kg m^{-2} and boosted total antioxidant activity by nearly 18% [22], and antioxidant values in tomato fruits rose by 38% (ABTS) and 11% (DPPH) after biostimulant treatment [23]. These findings underline the broad-spectrum potential of biostimulants to promote crop quality and sustainability in both open-field and controlled environments.

While seaweed-based biostimulants have been widely studied in field and greenhouse settings, their impact under fully controlled PFAL conditions—particularly for aromatic and medicinal plants such as Thai basil—remains limited. Therefore, this study aimed to evaluate how foliar application of seaweed-based biostimulants influences growth, antioxidant enzyme activities, and bioactive compound accumulation in Thai basil grown in a plant factory. The findings will provide valuable insights into sustainable, high-quality production systems for basil and other medicinal herbs.

2. Results

2.1. Growth Parameters

The results in Figures 1 and 2 reveal that SE treatments of 1.0, 1.5, and 2.0 mL·L^{-1} significantly promoted leaf number, leaf area, and shoot biomass, both fresh and dry, compared with untreated controls (Figures 1 and 2A–C,E). By contrast, there was no significant difference in root fresh and dry weights among control and SE treatments (Figure 2D,F). As shown in Figures 1 and 3 the whole plant dry weight was significantly enhanced under the 1.0, 1.5, 2.0, 2.5 mL·L^{-1} SE treatments (Figure 3A) compared to the control, whereas the shoot-to-root fresh and dry weight ratios did not differ significantly among treatments, indicating balanced biomass allocation between shoots and roots (Figure 3B–D).

Figure 1. Growth performance of Thai basil plants under different seaweed extract treatments (0, 0.5, 1.0, 1.5, 2.0, and 2.5 mL·L^{-1}) at four weeks after transplantation.

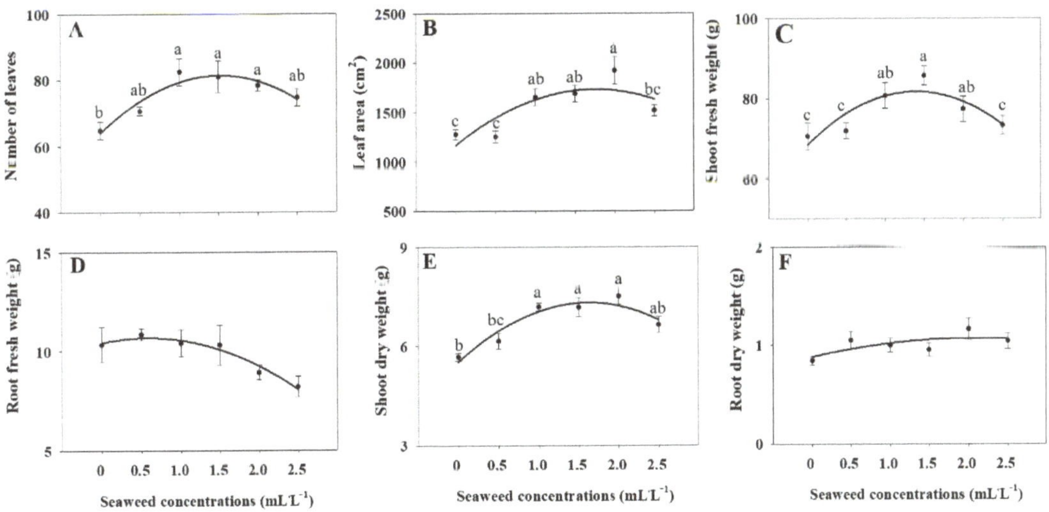

Figure 2. Number of leaves (**A**), leaf area (**B**), shoot fresh weight (**C**), root fresh weight (**D**), shoot dry weight (**E**), and root dry weight (**F**) of Thai basil under different seaweed extract treatments (0, 0.5, 1.0, 1.5, 2.0, and 2.5 mL·L^{-1}). Values are means ± SE (n = 6). Different letters indicate statistically significant differences ($p \leq 0.05$) based on ANOVA and Tukey's post hoc test.

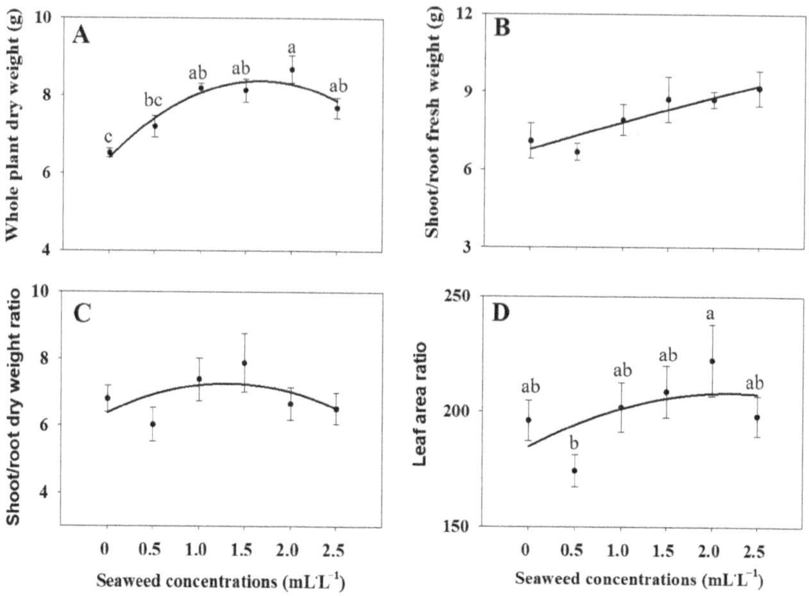

Figure 3. Whole plant dry weight (**A**), shoot/root fresh weight (**B**), shoot/root dry weight (**C**), and leaf area ratio (**D**) of Thai basil under different seaweed extract treatments (0, 0.5, 1.0, 1.5, 2.0, and 2.5 mL·L^{-1}). Values are means ± SE (n = 6). Different letters indicate significant differences at $p \leq 0.05$ based on ANOVA and Tukey's test.

2.2. Chlorophyll, Carotenoids, and Total Flavonoid

Chlorophyll a (Chl a) levels were significantly higher in the 1.0, 1.5, and 2.0 mL·L^{-1} treatments compared with the untreated group (Figure 4A), while chlorophyll b (Chl b) remained unchanged (Figure 4B). The higher values for the total Chl were observed in the 1.0 and 2.0 mL·L^{-1} SE treatments group compared to the untreated group (Figure 4D). Significant increases in flavonoid content were observed in plants subjected to 1.0, 1.5, and 2.0 mL·L^{-1} SE treatments, compared to the untreated group (Figure 4C). The highest total carotenoid content was observed in the 2.0 mL·L^{-1} SE treatment, which was 1.98-fold higher than that of the control group (Figure 4E). A significant increase in DPPH radical scavenging was observed in the 0.5, 1.0, 1.5, and 2.0 mL·L^{-1} SE treatments (Figure 4F). Flavonoid levels and DPPH radical scavenging activity also increased significantly at moderate SE concentrations (1.0–2.0 mL·L^{-1}), indicating enhanced antioxidant potential (Figure 4C,F).

2.3. Photosynthetic Characteristics

The net photosynthetic rate was significantly higher under the 2.0 mL·L^{-1} treatment than in the control, indicating enhanced photosynthetic efficiency (Figure 5A). However, stomatal conductance, intercellular CO_2 concentration, and transpiration rate were not significantly affected by SE treatments (Figure 5B–D), suggesting that improvements in photosynthesis were driven primarily by biochemical factors rather than changes in stomatal behavior.

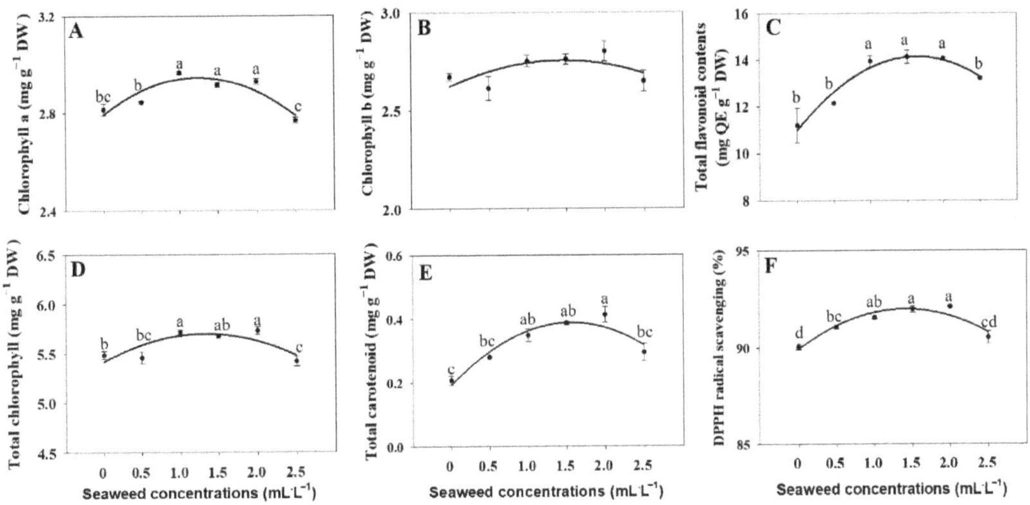

Figure 4. Chlorophyll a (**A**), chlorophyll b (**B**), total flavonoid contents (**C**), total chlorophyll (**D**), total carotenoid (**E**), and DPPH radical scavenging activity (**F**) of Thai basil under different seaweed extract treatments (0, 0.5, 1.0, 1.5, 2.0, and 2.5 mL·L^{-1}). Values are means ± SE (n = 3). Different letters indicate significant differences at $p \leq 0.05$.

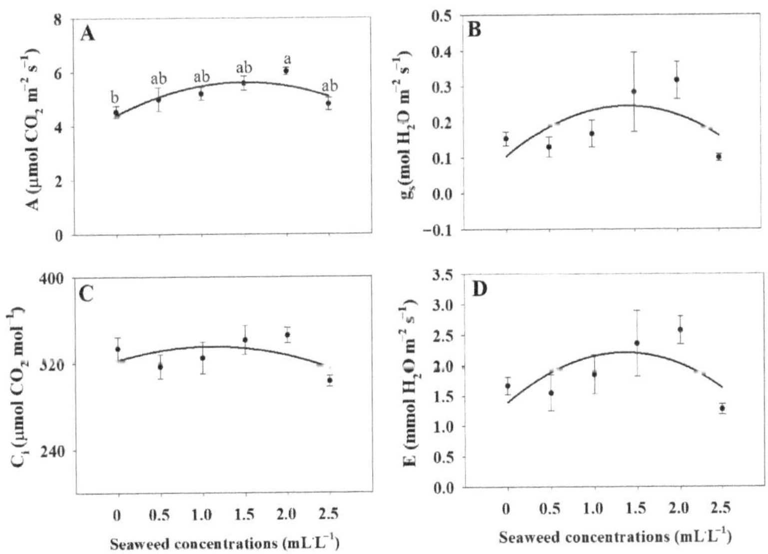

Figure 5. Net photosynthetic rate (A; **A**), stomatal conductance (gs; **B**), intercellular CO_2 concentration (Ci; **C**), and transpiration rate (E; **D**) of Thai basil under different seaweed extract treatments (0, 0.5, 1.0, 1.5, 2.0, and 2.5 mL·L^{-1}). Values are means ± SE (n = 3). Different letters indicate significant differences at $p \leq 0.05$ as determined by one-way ANOVA followed by Tukey's test.

2.4. Rosmarinic Acid and Total Phenolic Concentrations and Contents

The concentration of caffeic acid was slightly higher in the 2.5 mL·L^{-1} treatment compared to the control, though differences among treatments were not significant (Table 1). Benzoic acid content increased significantly in the 1.0, 1.5, and 2.0 mL·L^{-1} SE treatments relative to the control. Rutin concentrations were significantly higher in the 0.5, 1.0, 1.5, and 2.5 mL·L^{-1} SE treatments compared to the untreated group. In contrast, trans-cinnamic acid concentrations decreased significantly in all SE treatments compared to the untreated group. The highest quercetin concentration was recorded in the 1.5 mL·L^{-1} SE treatment, which was significantly greater than the control. Moreover, kaempferol levels increased significantly in the 1.5, 2.0, and 2.5 mL·L^{-1} SE treatments compared to the control (Table 1).

Table 1. Phenolic compounds in basil under different seaweed treatments (0-, 0.5-, 1.0-, 1.5-, 2.0-, and 2.5- mL·L^{-1}. Values represent means (n = 3).

^{w} SE Concentration Treatments (mL·L^{-1})	Phenolic Compounds in Basil (mg·g^{-1} DW)					
	Caffeic Acid (mg·g^{-1} DW)	Benzoic Acid (mg·g^{-1} DW)	Rutin (mg·g^{-1} DW)	Trans-Cinnamic Acid (mg·g^{-1} DW)	Quercetin (mg·g^{-1} DW)	Kaempferol (mg·g^{-1} DW)
0	0.42 ± 0.002 ab	6.04 ± 0.475 d	32.17 ± 0.266 d	0.0159 ± 0.00019 a	0.159 ± 0.0004 b	0.395 ± 0.0040 c
0.5	0.35 ± 0.005 c	5.75 ± 0.208 d	34.55 ± 0.020 c	0.0067 ± 0.00011 b	0.152 ± 0.00005 c	0.335 ± 0.0066 d
1.0	0.34 ± 0.008 c	9.45 ± 1.228 c	39.35 ± 0.531 a	0.0039 ± 0.00026 b	0.146 ± 0.0005 d	0.383 ± 0.0044 c
1.5	0.40 ± 0.496 b	13.78 ± 0.516 b	37.14 ± 0.355 b	0.0095 ± 0.00293 b	0.174 ± 0.0004 a	0.435 ± 0.0042 b
2.0	0.42 ± 0.005 ab	16.99 ± 0.100 a	33.30 ± 0.147 cd	0.0086 ± 0.00038 b	0.147 ± 0.0016 d	0.461 ± 0.0022 a
2.5	0.43 ± 0.002 a	6.09 ± 0.561 d	39.88 ± 0.401 a	0.0077 ± 0.00001 b	0.144 ± 0.0004 d	0.455 ± 0.0048 ab
Significance ^{z}	***	***	***	***	***	***

^{w} SE concentration; ^{z} significant at *** $p \leq 0.001$. Values are means ± SE (n = 3). Different letters in a column indicate significant differences among treatments at $p < 0.05$ (Tukey's test), DW = dry weight.

Significant increases in RA concentrations and contents were observed in plants subjected to all SE treatments, compared to the control (Figure 6A,B). Exposure to seaweed extract for 1.0, 1.5, 2.0, and 2.5 mL·L^{-1} resulted in a significant elevation of total phenolic concentrations and contents in shoot (Figure 6C,D). The higher levels of RA content were observed with the 1.0, 1.5, 2.0, and 2.5 mL·L^{-1} SE treatments compared to the untreated group. The highest levels of total phenolic content were observed with the 2 mL·L^{-1} SE treatment, showing increases of 1.88-fold, respectively, compared to the control.

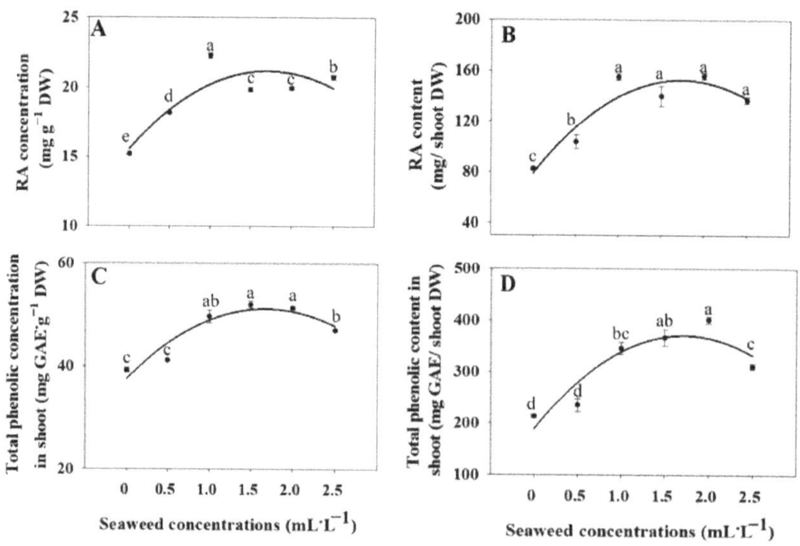

Figure 6. Rosmarinic acid (RA) concentration (**A**) and content (**B**), total phenolic concentration (**C**) and content (**D**) in basil shoot under different seaweed extract treatments (0, 0.5, 1.0, 1.5, 2.0, and 2.5 mL·L^{-1}). Values are means ± SE (n = 3). Different letters indicate significant differences at $p \leq 0.05$ using ANOVA followed by Tukey's test.

3. Discussion

3.1. Plant Growth Parameters

Seaweed extract (SE), especially at 1.0–2.0 mL·L^{-1}, significantly boosted basil growth by increasing leaf number, leaf area, and shoot biomass, likely due to natural hormones like auxins and cytokinins in the SE [14]. These compounds stimulate cell division and elongation, enhance bud initiation, and improve chloroplast development, which collectively boost photosynthetic efficiency and shoot biomass accumulation [24,25]. Additionally, the extract enhances nutrient uptake and activates antioxidant defense systems, allowing plants to better tolerate stress and allocate more resources to shoot growth [13,26]. Consistent with previous studies on other crops, the SE markedly promoted tomato plant growth by increasing flower and fruit numbers, enhancing flower cluster development, extending root length, boosting root and shoot dry weight, and raising SPAD values, while simultaneously improving fruit quality and overall yield [27]. SE application increased tomato yield by 4.6–6.9%, with the 60 kg hm^{-2} (=60 kg per hectare) treatment achieving the highest yield through enhanced photosynthesis [21]. Similar growth-promoting effects were observed in peppermint and purple basil, where SE enhanced leaf biomass and overall plant vigor [28]. In crops such as soybean and spinach, SE treatment improved vegetative growth, leaf chemical composition, and antioxidant activity, suggesting that the stimulatory effect of SE may be generalizable across multiple plant species [15,29]. The foliar application of 7.5 mL·L^{-1} SE resulted in optimal biomass yield, growth, and essential oil content and composition in holy basil [30]. Seaweed application significantly enhanced all measured traits, except for flower length and stem diameter [31]. Biostimulant treatments increased plant fresh weight, leaf number, and plant height in basil compared to the control [32]. In our study, although the whole-plant dry weight significantly increased at concentrations of 1.0 to 2.5 mL·L^{-1}, the shoot-to-root fresh and dry weight ratios did not change significantly among treatments. This indicates that both shoot and root biomasses increased proportionally, reflecting balanced whole-plant growth rather than preferential biomass allocation to a specific organ. The foliar application of seaweed extract likely exerted its main effects on above-ground tissues, while root growth was already optimized under hydroponic conditions. Similar observations were reported by Sleight et al. (2023) [33], who found that although the distribution of below-ground biomass varied among willow cultivars and sites, the root-to-shoot ratio did not differ significantly across sites (p = 0.8970), cultivars (p = 0.2834), or their interaction (p = 0.8481). At the highest concentration (2.5 mL·L^{-1}), growth did not increase further, possibly due to hormonal imbalance or mild phytotoxic effects. Vargas-Hernández et al. (2017) reported that excessive biostimulant concentrations can cause hormonal or metabolic disturbances that suppress plant growth [34]. Similarly, Arun et al. (2014) observed that increasing seaweed extract concentrations beyond optimal levels inhibited seed germination and growth in *Solanum lycopersicum* due to toxicity effects [35].

Overall, these findings demonstrate that moderate SE concentrations, especially 1.5–2.0 mL·L^{-1}, effectively enhance basil growth by promoting shoot development and overall plant vigor without negatively impacting root growth. Thus, moderate foliar application of SE can maximize shoot development without compromising root growth in hydroponic systems.

3.2. Chlorophyll and Total Carotenoid Levels

The results showed that seaweed extract had a strong positive effect on basil growth and physiology at moderate concentrations. Treatments with 1.0, 1.5, and 2.0 mL·L^{-1} SE significantly increased chlorophyll a (Chl a) content compared to the control, likely due to the presence of natural plant hormones such as cytokinins and auxins reported by the manufacturer to be contained in the commercial SE product, which stimulate chloroplast development and chlorophyll biosynthesis, thereby enhancing photosynthetic efficiency [13,36]. In contrast, chlorophyll b (Chl b) levels did not change significantly, as this pigment mainly functions in light harvesting and remains relatively stable under optimal growing conditions. Total chlorophyll was highest in the 1.0 and 2.0 $mL \cdot L^{-1}$ treatments, primarily driven by the increase in Chl a. The total carotenoid content reached its peak at 2.0 $mL \cdot L^{-1}$, nearly doubled relative to untreated plants, indicating improved photoprotection and antioxidant defense. Similarly, SE application increased chlorophyll and carotenoid levels in peppermint, arugula, and spinach, suggesting that the enhancement of photosynthetic pigments by algae-based biostimulants is a broadly conserved response [20,28,29]. Similarly, flavonoid content was significantly elevated at 1.0, 1.5, and 2.0 $mL \cdot L^{-1}$, which may be attributed to the elicitor-like activity of seaweed extracts that can enhance the biosynthesis of antioxidant compounds such as flavonoids [13,37]. In agreement with previous studies, flavonoid content in cabbage increased under SE treatment compared to the control [19]. This was supported by the marked increase in DPPH radical scavenging activity observed in the 0.5–2.0 $mL \cdot L^{-1}$ treatments, reflecting higher antioxidant capacity in the treated plants. Similar increases in flavonoid accumulation and antioxidant activity have been reported in sweet basil, mint, and spinach following SE application, suggesting that SE consistently enhances secondary metabolite production across plant species [29,38]. Consistent with earlier reports, the highest rise in flavonoid content (89%) and DPPH antioxidant activity (82%) in arugula occurred at a 2.5% biostimulant concentration compared to the control [20]. Overall, these findings suggest that SE acts as an effective biostimulant, enhancing photosynthesis, secondary metabolite production, and antioxidant defense in basil, with 1.0–2.0 $mL \cdot L^{-1}$ being the optimal range for maximum benefits, while higher concentrations may not provide additional improvements due to nutrient saturation or slight phytotoxic effects.

3.3. Photosynthetic Parameters

The application of SE at 2.0 $mL \cdot L^{-1}$ significantly enhanced the net photosynthetic rate (A) in basil compared to the control, likely due to the presence of natural plant growth regulators such as cytokinins and auxins, as well as essential nutrients found in SE. These compounds promote chloroplast development, stimulate chlorophyll biosynthesis, and improve the efficiency of the photosynthetic machinery, leading to greater CO_2 fixation and higher photosynthetic capacity [28,39,40]. Studies in sugarcane and tomato also reported increased photosynthetic rates following SE application, confirming that the biochemical stimulation of photosynthesis by algae-derived biostimulants is a common mechanism across different plant species [17,21]. In contrast, stomatal conductance (gs), intercellular CO_2 concentration (Ci), and transpiration rate (E) did not show significant differences among treatments. This suggests that the improvement in photosynthesis was primarily due to biochemical and physiological enhancements within the leaf tissue, rather than changes in stomatal behavior or water regulation. Since SE was applied foliarly and the plants were grown under controlled hydroponic conditions with optimal water availability, stomatal activity remained stable, while internal biochemical processes drove the observed increase in photosynthetic performance [18]. In controlled hydroponic systems, stable stomatal behavior helps maintain efficient photosynthesis and water use, as plants are continuously supplied with adequate water and nutrients.

3.4. Rosmarinic Acid and Total Phenolic Concentrations and Contents

The increase in rosmarinic acid (RA) and total phenolic content in basil under seaweed extract (SE) treatments can be attributed to the bioactive compounds present in SE, such as natural plant growth regulators (auxins, cytokinins, and gibberellins), amino acids, trace minerals, and vitamins [30,38]. These compounds stimulate the phenylpropanoid pathway by enhancing the activity of key enzymes like phenylalanine ammonia-lyase, which promotes the biosynthesis of RA and other phenolic compounds [13,41]. Similar effects have been observed in arugula, spinach, and yarrow, where SE application enhanced phenolic accumulation, flavonoid content, and antioxidant activity, indicating a conserved mechanism for the stimulation of secondary metabolism by SE across different plant species [20,29,42]. As a result, plants treated with higher SE concentrations, particularly 2.0 mL·L^{-1}, showed a substantial increase total phenolics, reaching 1.88-fold higher levels than the control. The increase in benzoic acid and rutin suggests an upregulation of early and intermediate steps of the phenylpropanoid pathway, while the decrease in trans-cinnamic acid indicates its rapid conversion into downstream metabolites [43]. Additionally, the significant rise in quercetin and kaempferol concentrations reflects an enhanced flavonoid biosynthesis and antioxidant defense. Lower concentrations of SE (0.5 and 1.0 mL·L^{-1}) provided insufficient bioactive compounds to fully activate these metabolic pathways, whereas very high concentrations (2.5 mL·L^{-1}) may have caused feedback regulation or metabolic diversion. Overall, the results demonstrate that foliar application of SE stimulates secondary metabolism and phenolic accumulation in basil, with 2.0 mL·L^{-1} being the optimal concentration for maximizing total phenolic production. These findings align with previous studies showed that foliar application of seaweed extract significantly enhanced vegetative growth, seed yield components, and leaf chemical composition (including sugars, proteins, amino acids, antioxidant activity, and volatile oil), as well as leaf and stem structure in basil. The most pronounced effects were observed at 1.5 mL·L^{-1} [44]. Applying 9 mL·L^{-1} seaweed extract boosted leaf phenolics and antioxidant activity by up to 30.44% over the control, improving yarrow's medicinal quality [42]. SE treatments enhanced total phenols and ascorbic acid, thereby improving the functional quality of spinach [29]. Overall, these findings support a general mechanism in which SE foliar application stimulates phenylpropanoid metabolism, flavonoid biosynthesis, and antioxidant defense in diverse plant species.

4. Materials and Methods

4.1. Plant Material and Seedling Conditions

Thai basil (*Ocimum basilicum* Linn.) seeds sourced from Namheung Seed Co., Ltd. (Seoul, Republic of Korea) were sown in rockwool cubes and cultivated in a controlled environment chamber maintained at 21.2 ± 2 °C with 71 ± 10% relative humidity. The plants were exposed to LED lighting (TL5 14W/865, Philips, Amsterdam, The Netherlands) delivering a photosynthetic photon flux density (PPFD) of 200 ± 10 μmol·m^{-2}·s^{-1} on a 16 h light/8 h dark cycle. Fourteen days after sowing, Thai basil seedlings were soaked in Hoagland solution with a pH of 6.6 and an EC of 1.04 dS·m^{-1}.

4.2. Treatments and Experimental Design

At 28 days after sowing, basil seedlings were transferred to a plant factory and grown using a deep flow technique hydroponic system, with environmental conditions maintained as during the seedling stage. Each treatment consisted of 10 plants per replicate. The plants were cultivated for 28 days using Hoagland solution maintained at a pH of approximately 6.7 and an EC of 2.0 dS·m^{-1}. To evaluate the effects of seaweed extract (SE), five concentrations (0.5, 1.0, 1.5, 2.0, and 2.5 mL·L^{-1}) were applied twice per week via foliar spraying, while distilled water was used as the control. Each spraying was performed until all leaf surfaces were thoroughly wetted. The SE used in this study was a commercial product (Living Farm, K-INGS Co., Ltd., Busan, Republic of Korea) and was not prepared by the authors. According to the manufacturer's label, this product is a liquid compound fertilizer containing SE derived from *Ascophyllum nodosum*, with 5% nitrogen (N), 3% available phosphoric acid (P_2O_5), 2% soluble potassium (K_2O), 0.05% boron (B), and 0.0005% molybdenum (Mo). The SE also contains natural plant hormones such as auxins and cytokinins, which promote root growth, nutrient uptake, photosynthesis, and stress tolerance, while activating antioxidant enzymes. The experiment was arranged using a completely randomized design with two replicates.

4.3. Growth Parameter Measurements

Twenty-eight days after transplanting, leaf area, leaf number, and the fresh weights of shoots and roots were recorded with an electronic scale (EW220-3NM, Kern & Sohn GmbH., Balingen, Germany). Afterward, the samples were oven-dried at 70 °C for seven days to determine the dry weights (DW). Leaf area was measured with a leaf area meter (LI-3000A, Li-Cor, Lincoln, NE, USA). The leaf area ratio, shoot-to-root DW ratio, and shoot-to-root FW ratio were then calculated, with six plants used per replicate (n = 6).

4.4. Photosynthetic Characteristics

Photosynthetic parameters were measured on three plants (n = 3) per replicate using a LI-COR 6400 system (LI-COR Inc., Lincoln, NE, USA) under fixed chamber settings (400 μmol mol^{-1} CO_2, 300 μmol m^{-2} s^{-1} PPFD, 25 °C, 500 cm^3 s^{-1} airflow). Recorded variables included photosynthetic rate, stomatal conductance, CO_2 concentration, and transpiration rate.

4.5. Photosynthetic Pigment Analysis

After harvest, basil shoots were rapidly frozen in liquid nitrogen and transferred within four days to a –50 °C dry freezer (TFD5503, IL Shinbiobase Co., Ltd., Dongducheon, Republic of Korea). Samples were ground with a porcelain mortar and pestle, sieved, and analyzed for chlorophyll (Chl) a, Chl b, and total cartotenoid, using an Epoch microplate spectrophotometer (EMS) (EPOCH-SN; Agilent Technologies, Inc., Santa Clara, CA 95051, USA). Powdered samples (20 mg DW) were extracted with 2 mL of 90% MeOH, centrifuged at 1308× *g* for 20 min, and absorbance was measured at 665.2, 652.4, and 470 nm for Chl a, Chl b, and Car, respectively, following Lichtenthaler, 1987 and Lam et al., 2023a [45,46]. Pigment concentrations were calculated as follows:

$$\text{Chl a } (\text{mg}\cdot\text{g}^{-1}) = (16.82 \times A_{665.2} - 9.28 \times A_{652.4})/10$$

$$\text{Chl b } (\text{mg}\cdot\text{g}^{-1}) = (36.92 \times A_{652.4} - 16.54 \times A_{665.2})/10$$

$$\text{Total carotenoid } (\text{mg}\cdot\text{g}^{-1}) = ([1000 \times A_{470} - 1.91 \times \text{Chl a} - 95.15 \text{ Chl b}]/225)/10$$

$$\text{Total Chl a + b } (\text{mg}\cdot\text{g}^{-1}) = \text{Chl a} + \text{Chl b}$$

where A is the absorbance at wavelength.

4.6. DPPH Radical Scavenging Activity Analysis

DPPH radical scavenging activity was measured at 517 nm using an EMS, following Rahman Et Al., 2015 and Lam Et Al., 2023 [46,47]. DPPH activitiy was calculated as:

$$\text{DPPH } (\%) = (A_{\text{blank}} - A_{\text{sample}})/A_{\text{blank}} \times 100$$

where A is the absorbance at 517 nm.

4.7. Total Flavonoid Content

Total flavonoid content in basil plants was determined using the aluminum chloride ($AlCl_3$) colorimetric method of Lin and Tang (2007) [48] method, with 20 mg of dry sample extracted in 2 mL of 90% methanol and sonicated at 20 °C for 20 min. The mixture was then centrifuged at 1308× *g* for 20 min at 4 °C. A 100 µL aliquot of the sample was mixed with 300 µL of 95% ethanol, 20 µL of 10% $AlCl_3$, 20 µL of 10% potassium acetate, and 600 µL of triple-distilled water. The absorbance was measured at 415 nm after 40 min of incubation at room temperature at 22 °C using an EMS. Total flavonoid content was determined in triplicate (n = 3) using quercetin (QE) as the standard, with a calibration curve prepared from quercetin solutions of 0, 50, 75, 125, 250, and 500 µg/mL in methanol, and the results expressed as mg QE/g DW.

4.8. Phenolic Compound Analysis by HPLC

Phenolic constituents in basil were quantified using a 1260 Infinity Series High-performance liquid chromatography (HPLC) system (1260 Infinity; Agilent Technologies Inc., Santa Clara, CA, USA) [49,50]. Dry powder (100 mg) was extracted with 2 mL of 80% methanol, sonicated for 1 h with intermittent vortexing, centrifuged (12,000× *g*, 10 min), and filtered (0.45 µm). Analysis was performed by HPLC on a Symmetry C18 column (30 °C) using 0.15% acetic acid in water (solvent A) and methanol (solvent B) at 1.0 mL·min^{-1}, with UV detection at 280 nm. The 98 min gradient of solvent B was: 5% (0–1 min), 15% (1–9 min), 20% (9–24 min), 30% (24–54 min), 45% (54–66 min), 56% (66–76 min), 60% (76–80 min), 80% (80–91 min), and 5% (91–98 min). External standards (100–500 µg·mL^{-1}; Sigma-Aldrich, Co., Ltd., Seoul, Republic of Korea) of caffeic acid, benzoic acid, rutin, trans-cinnamic acid, quercetin, and kaempferol were used to construct calibration curves ($R^2 \geq 0.9999$). Retention times were 27.59, 56.81, 61.75, 72.34, 75.13, and 82.08 min, respectively. Each sample was analyzed in triplicate with an injection volume of 20 µL.

4.9. Determination of Rosmarinic Acid

Rosmarinic acid (RA) content was determined following Lam Et Al. (2021) [51]. Samples were frozen in liquid nitrogen, freeze-dried at −50 °C for four days, ground, and sieved. Dry powder (200 mg) was extracted with 10 mL methanol, sonicated for 30 min, centrifuged (1358× *g*, 20 min), and filtered (0.45 µm). HPLC was performed using 0.1% formic acid in water (solvent A) and acetonitrile (solvent B) under the following gradient: 20% B (0–5 min), 50% B (5–20 min), and 100% B (20–22 min). The flow rate was 0.8 mL·min^{-1} with 10 µL injection and detection at 330 nm. Standard RA (Sigma-Aldrich, Co., Ltd., Seoul, Republic of Korea) showed a retention time of 11.655 min, with calibration linearity ($R^2 = 0.9999$) across 0.1–100 µg·mL^{-1} (LOD = 0.06 µg·mL^{-1}; LOQ = 0.20 µg·mL^{-1}). The RA content of shoots was determined by multiplying RA concentration (mg g^{-1} DW) by shoot DW (g).

4.10. Statistical Analysis

Growth parameters were measured in six plants (n = 6), and three plants (n = 3) per replicate were used for physiological and biochemical analyses, plants were randomly selected from each replicate. The experiment used a completely randomized design with two replicates. Data were analyzed by one-way ANOVA and Tukey's test ($p < 0.05$) using SPSS version 20.0 (IBM Corp., Armonk, NY, USA).

5. Conclusions

Foliar application of Ascophyllum nodosum-derived SE significantly improved the growth, photosynthetic performance, and phytochemical quality of Thai basil cultivated in a fully controlled plant factory. Moderate concentrations (1.0–2.0 mL·L^{-1}) consistently enhanced leaf area, shoot biomass, chlorophyll a, carotenoids, flavonoids, and antioxidant capacity. The 2.0 mL·L^{-1} treatment yielded the greatest increases in total phenolics and rosmarinic acid, approximately 1.9-fold higher than in untreated plants. The concurrent rise in key phenolics (benzoic acid, rutin, quercetin, kaempferol) and the reduction in trans-cinnamic acid indicate activation of the phenylpropanoid pathway. Higher concentrations provided no additional benefit, establishing 2.0 mL·L^{-1} as the practical optimum for maximizing both biomass and phytochemical quality without affecting root allocation. These findings demonstrate that A. nodosum-based SE is a sustainable and effective biostimulant for premium basil production in plant factory systems. Future research should refine application strategies across cultivars and growth stages and employ metabolomic and transcriptomic approaches to better link biochemical improvements with sensory attributes and postharvest quality.

Author Contributions: V.P.L. experimental design, performed the experiments, data collection and analysis, writing—original manuscript, writing—review and editing. G.B. experimental design, performed the experiments, writing—review and editing. D.N.L. performed the experiments, preparation for the manuscript, writing—review and editing. M.C.D. experimental design, performed the experiments, writing—review and editing. J.P. Project administration, supervision, constructing idea, experimental design, data analysis, writing—original manuscript, writing—review and editing. All authors have read and agreed to the published version of the manuscript.

Funding: This work was supported by BK21 FOUR Program by Chungnam National University Research Grant, 2025.

Data Availability Statement: The original contributions are included in the article. Further queries can be directed to the corresponding author.

Acknowledgments: This research was supported by the National Research Foundation (NRF) funded by the Korean government (MSIT) (No. RS-2024-00459463).

Conflicts of Interest: There are no conflicts of interest to declare.

References

1. Jimayu, G. Review on production and importance of basil (*Ocimum basilicum* L.) and roles of fertilizer on basil yield. *J. Biol. Agric. Healthc.* **2021**, *11*, 39–47.
2. Spence, C. Sweet basil: An increasingly popular culinary herb. *Int. J. Gastron. Food Sci.* **2024**, *36*, 100927. [CrossRef]
3. Romano, R.; De Luca, L.; Aiello, A.; Pagano, R.; Di Pierro, P.; Pizzolongo, F.; Masi, P. Basil (*Ocimum basilicum* L.) leaves as a source of bioactive compounds. *Foods* **2022**, *11*, 3212. [CrossRef]
4. Bhatti, M.; Kaur, D.; Thakur, M. A research article: Comparison on phytochemical analysis and antioxidant profile of two species of *Ocimum* (*Ocimum basilicum* and *Ocimum sanctum*). *Der Pharma Chem.* **2020**, *12*, 15–20.
5. Shiam, M.A.H.; Alam, A.; Biswas, M.; Alam, M.; Zahid, M.A.; Alam, S.M.S.; Akhtaruzzaman, M.; Ahmed, S. A comprehensive review on basil seeds as a source of nutrients and functional ingredients with health benefit properties. *Appl. Food Res.* **2025**, *5*, 100859. [CrossRef]
6. Eghbal, E.; Aliniaeifard, S.; Mehrjerdi, M.Z.; Abdi, S.; Hassani, S.B.; Rassaie, T.; Gruda, N.S. Growth, phytochemical, and phytohormonal responses of basil to different light durations and intensities under constant daily light integral. *BMC Plant Biol.* **2024**, *24*, 935. [CrossRef]
7. Eskandarzade, P.; Mehrjerdi, M.Z.; Gruda, N.S.; Aliniaeifard, S. Phytochemical compositions and antioxidant activity of green and purple basils altered by light intensity and harvesting time. *Heliyon* **2024**, *10*, e30931. [CrossRef] [PubMed]
8. Lozano-Castellanos, L.F.; Navas-Gracia, L.M.; Lozano-Castellanos, J.C.; Correa-Guimaraes, A. Technologies applied to artificial lighting in indoor agriculture: A review. *Sustainability* **2025**, *17*, 3196. [CrossRef]

9. Guo, M.; Jin, Z.; Ma, L.; Ou, S. Application of plant factory with artificial lighting in horticultural production: Current progress and future trends. *Hortic. Plant J.* 2025; *in press*. [CrossRef]
10. Yoshida, H.; Shimada, K.; Hikosaka, S.; Goto, E. Effect of UV-B irradiation on bioactive compounds of red perilla (*Perilla frutescens* (L.) Britton) cultivated in a plant factory with artificial light. *Horticulturae* **2022**, *8*, 725. [CrossRef]
11. Deolu-Ajayi, A.O.; van der Meer, I.M.; van der Werf, A.; Karlova, R. The power of seaweeds as plant biostimulants to boost crop production under abiotic stress. *Plant Cell Environ.* **2022**, *45*, 2537–2553. [CrossRef]
12. Mandal, S.; Anand, U.; López-Bucio, J.; Radha; Kumar, M.; Lal, M.K.; Tiwari, R.K.; Dey, A. Biostimulants and environmental stress mitigation in crops: A novel and emerging approach for agricultural sustainability under climate change. *Environ. Res.* **2023**, *233*, 116357. [CrossRef] [PubMed]
13. Ali, O.; Ramsubhag, A.; Jayaraman, J. Biostimulant properties of seaweed extracts in plants: Implications towards sustainable crop production. *Plants* **2021**, *10*, 531. [CrossRef]
14. Mughunth, R.J.; Velmurugan, S.; Mohanalakshmi, M.; Vanitha, K. A review of seaweed extract's potential as a biostimulant to enhance growth and mitigate stress in horticulture crops. *Sci. Hortic.* **2024**, *334*, 113312. [CrossRef]
15. Rathore, S.S.; Chaudhary, D.R.; Boricha, G.N.; Ghosh, A.; Bhatt, B.P.; Zodape, S.T.; Patolia, J.S. Effect of seaweed extract on the growth, yield and nutrient uptake of soybean (*Glycine max*) under rainfed conditions. *S. Afr. J. Bot.* **2009**, *75*, 351–355. [CrossRef]
16. Gatti, N.; Maghrebi, M.; Serio, G.; Gentile, C.; Bunea, V.V.; Vigliante, I.; Boitte, C.; Garabello, C.; Contartese, V.; Bertea, C.M.; et al. Seaweed and yeast extracts as sustainable phytostimulant to boost secondary metabolism of apricot fruits. *Front. Plant Sci.* **2025**, *15*, 1455156. [CrossRef]
17. Chen, D.W.; Zhou, W.L.; Yang, J.; Ao, J.H.; Huang, Y.; Shen, D.C.; Jiang, Y.; Huang, Z.R.; Shen, H. Effects of seaweed extracts on the growth, physiological activity, cane yield and sucrose content of sugarcane in China. *Front. Plant Sci.* **2021**, *12*, 659130. [CrossRef]
18. De Clercq, P.; Pauwels, E.; Top, S.; Steppe, K.; Van Labeke, M.C. Effect of seaweed-based biostimulants on growth and development of *Hydrangea paniculata* under continuous or periodic drought stress. *Horticulturae* **2023**, *9*, 509. [CrossRef]
19. Lola-Luz, T.; Hennequart, F.; Gaffney, M. Enhancement of phenolic and flavonoid compounds in cabbage (*Brassica oleraceae*) following application of commercial seaweed extracts of the brown seaweed, (*Ascophyllum nodosum*). *Agric. Food Sci.* **2013**, *22*, 288–295. [CrossRef]
20. Drygas, B.; Piechowiak, T.; Balawejder, M.; Matlok, N.; Kreczko, J.; Puchalski, C. The eliciting effect of aqueous extracts from *Ascophyllum nodosum* algae on the cultivation of arugula (*Eruca sativa* Mill.) microgreens. *Sustainability* **2024**, *16*, 7436. [CrossRef]
21. Yao, Y.Y.; Wang, X.Q.; Chen, B.C.; Zhang, M.; Ma, J.Z. Seaweed extract improved yields, leaf photosynthesis, ripening time, and net returns of tomato (*Solanum lycopersicum* Mill.). *ACS Omega* **2020**, *5*, 4242–4249. [CrossRef] [PubMed]
22. Ashour, M.; Hassan, S.M.; Elshobary, M.E.; Ammar, G.A.G.; Gaber, A.; Alsanie, W.F.; Mansour, A.T.; El-Shenody, R. Impact of commercial seaweed liquid extract (TAM®) biostimulant and its bioactive molecules on growth and antioxidant activities of hot pepper (*Capsicum annuum*). *Plant* **2021**, *10*, 1045. [CrossRef]
23. Mannino, G.; Campobenedetto, C.; Vigliante, I.; Contartese, V.; Gentile, C.; Bertea, C.M. The application of a plant biostimulant based on seaweed and yeast extract improved tomato fruit development and quality. *Biomolecules* **2020**, *10*, 1662. [CrossRef] [PubMed]
24. Mueller, D.; Leyser, O. Auxin, cytokinin and the control of shoot branching. *Ann. Bot.* **2011**, *107*, 1203–1212. [CrossRef]
25. Sosnowski, J.; Truba, M.; Vasileva, V. The impact of auxin and cytokinin on the growth and development of selected crops. *Agriculture* **2023**, *13*, 724. [CrossRef]
26. Mukherjee, A.; Patel, J.S. Seaweed extract: Biostimulator of plant defense and plant productivity. *Int. J. Environ. Sci. Technol.* **2020**, *17*, 553–558. [CrossRef]
27. Hussain, H.I.; Kasinadhuni, N.; Arioli, T. The effect of seaweed extract on tomato plant growth, productivity and soil. *J. Appl. Phycol.* **2021**, *33*, 1305–1314. [CrossRef]
28. Laribi, B.; Annabi, H.A.; Bettaieb, T. Effects of ulva intestinalis linnaeus seaweed liquid extract on plant growth, photosynthetic performance and water status of two hydroponically grown lamiaceae species: Peppermint (*Mentha* × *piperita* L.) and purple basil (*Ocimum Basilicum* var. *purpurascens* Benth.). *S. Afr. J. Bot.* **2023**, *158*, 63–72. [CrossRef]
29. Rouphael, Y.; Giordano, M.; Cardarelli, M.; Cozzolino, E.; Mori, M.; Kyriacou, M.C.; Bonini, P.; Colla, G. Plant- and seaweed-based extracts increase yield but differentially modulate nutritional quality of greenhouse spinach through biostimulant action. *Agronomy* **2018**, *8*, 126. [CrossRef]
30. Jamwal, S.; Kumari, A.; Veeragurunathan, V.; Prasad, K.; Ghosh, A.; Kumar, R. Enhancing growth, yield, essential oil content, and composition of holy basil (*Ocimum tenuiflorum* L.) using red algae-based bio-stimulant under acidic conditions of the Western Himalayas. *BMC Plant Biol.* **2025**, *25*, 84. [CrossRef]
31. Pirani, H.; Ebadi, M.-T.; Rezaei, A. Effect of seaweed fertilizer foliar application on growth parameters, yield, essential oil content and composition of hyssop (*Hyssopus officinalis* L.). *Iran. J. Med. Aromat. Plants Res.* **2020**, *36*, 376–389. [CrossRef]

32. Alibakhshi, M.; Asadi-Gharneh, H.A. Growth and biochemical properties of green basil (*Ocimum basilicum* L.) affected by foliar application of biostimulants. *J. Crop Nutr. Sci.* **2021**, *7*, 34–45.
33. Sleight, N.J.; Volk, T.A.; Eisenbies, M. Belowground biomass and root:Shoot ratios of three willow cultivars at two sites. *Forests* **2023**, *14*, 525. [CrossRef]
34. Vargas-Hernandez, M.; Macias-Bobadilla, I.; Guevara-Gonzalez, R.G.; Romero-Gomez, S.D.; Rico-Garcia, E.; Ocampo-Velazquez, R.V.; Alvarez-Arquieta, L.D.; Torres-Pacheco, I. Plant hormesis management with biostimulants of biotic origin in agriculture. *Front. Plant Sci.* **2017**, *8*, 1762. [CrossRef] [PubMed]
35. Arun, D.; Kothandaraman, G.; Chandran, M.; Dinakarkumar, Y. Studies on effect of seaweed extracts on crop plants and microbes. *Int. J. ChemTech Res.* **2014**, *6*, 4235–4240.
36. Stirk, W.A.; VanStaden, J. Comparison of cytokinin- and auxin-like activity in some commercially used seaweed extracts. *J. Appl. Phycol.* **1996**, *8*, 503–508. [CrossRef]
37. Shukla, P.S.; Mantin, E.G.; Adil, M.; Bajpai, S.; Critchley, A.T.; Prithiviraj, B. *Ascophyllum nodosum*-Based biostimulants: Sustainable applications in agriculture for the stimulation of plant growth, stress tolerance, and disease management. *Front. Plant Sci.* **2019**, *10*, 655. [CrossRef]
38. Elansary, H.O.; Yessoufou, K.; Shokralla, S.; Mahmoud, E.A.; Skaicka-Wozniak, K. Enhancing mint and basil oil composition and antibacterial activity using seaweed extracts. *Ind. Crops Prod.* **2016**, *92*, 50–56. [CrossRef]
39. Shehata, A.M.; Walid, S.E.N. Response of sweet basil plants (*Ocimum basilicum*, L.) grown under salinity stress to spraying seaweed extract. *Future J. Biol.* **2019**, *2*, 16–28.
40. EL Boukhari, M.E.; Barakate, M.; Bouhia, Y.; Lyamlouli, K. Trends in seaweed extract based biostimulants: Manufacturing process and beneficial effect on soil-plant systems. *Plants* **2020**, *9*, 359. [CrossRef]
41. Ertani, A.; Francioso, O.; Tinti, A.; Schiavon, M.; Pizzeghello, D.; Nardi, S. Evaluation of seaweed extracts from *Laminaria* and *Ascophyllum nodosum* spp. as biostimulants in *Zea mays* L. Using a combination of chemical, biochemical and morphological approaches. *Front. Plant Sci.* **2018**, *9*, 428. [CrossRef]
42. Pacheco, A.C.; Sobral, L.A.; Gorni, P.H.; Carvalho, M.E.A. Ascophyllum nodosum extract improves phenolic compound content and antioxidant activity of medicinal and functional food plant *Achillea millefolium* L. *Aust. J. Crop Sci.* **2019**, *13*, 418–423. [CrossRef]
43. Kisa, D.; Ceylan, Y.; Imamoglu, R. Accumulation of phenolic compounds and expression of phenylpropanoid biosynthesis-related genes in leaves of basil transformed with *A. rhizogenes* strains. *Physiol. Mol. Biol. Plants* **2023**, *29*, 629–640. [CrossRef]
44. Azza, S.M.; Yousef, R.S. Response of basil plant (*Ocimum sanctum* L.) to foliar spray with amino acids or seaweed extract. *J. Hortic. Sci. Ornam. Plants* **2015**, *7*, 94–106.
45. Lichtenthaler, H.K. Chlorophylls and carotenoids—Pigments of photosynthetic biomembranes. *Methods Enzymol.* **1987**, *148*, 350–382. [CrossRef]
46. Lam, V.P.; Beomseon, L.; Anh, V.K.; Loi, D.N.; Kim, S.; Kwang-Ya, L.; Park, J. Effectiveness of silver nitrate application on plant growth and bioactive compounds in *Agastache rugosa* (Fisch. & C.A.Mey.) kuntze. *Heliyon* **2023**, *9*, e20205. [CrossRef]
47. Rahman, M.M.; Islam, M.B.; Biswas, M.; Khurshid Alam, A.H. In vitro antioxidant and free radical scavenging activity of different parts of *Tabebuia pallida* growing in Bangladesh. *BMC Res. Notes* **2015**, *8*, 621. [CrossRef]
48. Lin, J.Y.; Tang, C.Y. Determination of total phenolic and flavonoid contents in selected fruits and vegetables, as well as their stimulatory effects on mouse splenocyte proliferation. *Food Chem.* **2007**, *101*, 140–147. [CrossRef]
49. Yeo, H.J.; Baek, S.A.; Sathasivam, R.; Kim, J.K.; Park, S.U. Metabolomic analysis reveals the interaction of primary and secondary metabolism in white, pale green, and green pak choi (*Brassica rapa* subsp. *chinensis*). *Appl. Biol. Chem.* **2021**, *64*, 3. [CrossRef]
50. Meyer, P.; Förster, N.; Huyskens-Keil, S.; Ulrichs, C.; Geilfus, C.M. Phenolic compound abundance in Pak choi leaves is controlled by salinity and dependent on pH of the leaf apoplast. *Plant Environ. Interact.* **2021**, *2*, 36–44. [CrossRef]
51. Lam, V.P.; Loi, D.N.; Shin, J.; Mi, L.K.; Park, J. Optimization of salicylic acid concentrations for increasing antioxidant enzymes and bioactive compounds of *Agastache rugosa* in a plant factory. *PLoS ONE* **2024**, *19*, e0306340. [CrossRef]

plants

MDPI

Article

Hemp Inflorescence as a Sustainable Biostimulant Tool to Boost Growth and Antioxidant Capacity in Oilseed Pumpkin

Ivana Varga [1,*], Manda Antunović [1], Monika Tkalec Kojić [1], Antonela Markulj Kulundžić [2], Dario Iljkić [1], Renata Baličević [3] and Marija Ravlić [3]

[1] Department of Plant Production and Biotechnology, Faculty of Agrobiotechnical Sciences Osijek, Josip Juraj Strossmayer University of Osijek, Vladimira Preloga 1, 31000 Osijek, Croatia; manda.antunovic@fazos.hr (M.A.); monikat@fazos.hr (M.T.K.); dario.iljkic@fazos.hr (D.I.)

[2] Department of Industrial Plants Breeding and Genetics, Agricultural Institute Osijek, 31000 Osijek, Croatia; antonela.markulj@poljinos.hr

[3] Department of Phytomedicine, Faculty of Agrobiotechnical Sciences Osijek, Josip Juraj Strossmayer University of Osijek, Vladimira Preloga 1, 31000 Osijek, Croatia; renata.balicevic@fazos.hr (R.B.); marija.ravlic@fazos.hr (M.R.)

* Correspondence: ivana.varga@fazos.hr

Abstract: The study investigates whether water extracts from industrial hemp inflorescences influence the germination and early growth of hull-less oilseed pumpkin (*Cucurbita pepo* L.), with the hypothesis that industrial hemp extracts may act as a biostimulant, enhancing growth, biomass, and bioactive compound accumulation in pumpkin seedlings. Fully developed and healthy inflorescences of industrial hemp were harvested, dried, ground into powder, filtered, and diluted to concentrations of control (water), 1.0%, 2.5%, and 5.0% for the seed germination bioassay. Morphological, growth parameters, and bioactive compounds of the hull-less oilseed pumpkin sprouts were determined. Total germination rate was not affected with industrial hemp inflorescent water extracts, while sprout vigor index and biomass increased at 2.5 and 5.0% of the extract applied. The average root length of hull-less oilseed pumpkin sprouts was 14.19 cm, the stem length was 5.45 cm, and the fresh mass of the sprouts was 14.10 g per plant. Water extracts of 2.5 and 5.0% significantly ($p \leq 0.001$) increased stem length by more than double, and the sprouts' fresh mass by about 35% compared to the control. The average *Chl a* (chlorophyll), *Chl b*, *Chl a + b*, and *Car* (carotenoids) content was on average 0.161, 0.115, 0.268, and 0.136 mg g^{-1} FW, respectively, and were significantly affected compared to the control. The highest total phenol (TPC) and flavonoid content (TFC) were determined for hull-less oilseed pumpkin sprouts at 1.0% of water extract (100.21 µg QC/1 g tissue and 0.02 µg GA/1 g tissue, respectively). Low absolute values are consistent with the early seedling stage, where secondary metabolism is underdeveloped. The antioxidant activity was determined with the FRAP (Ferric Reducing Antioxidant Power) method and a significant influence ($p \leq 0.05$) of industrial hemp inflorescence water extracts on antioxidant activity of pumpkin sprouts was observed, which significantly ($p \leq 0.05$) increased on all treatments compared to the control, by 36% on average, with no significant differences among different concentrations of water extracts. Overall, industrial hemp inflorescence water extracts have a positive influence on the observed parameters, supporting the potential use of industrial hemp inflorescence water extracts as a biostimulant for hull-less oilseed pumpkin.

Keywords: organic agriculture; phytoextracts; *Cannabis sativa* L. inflorescences extract; root; stem; growth parameters; *Cucurbita pepo* var. hull-less; antioxidant activity

1. Introduction

Many plants have been part of folk medicine for centuries, especially due to their medicinal purposes. There is an increasing use of plant extracts in agriculture, mainly in organic farming and sustainable alternative systems to traditional agriculture, which use many medicinal, aromatic, and herbal plants for multiple roles, e.g., in the regulation of plant diseases, insects, herbicides, allelopathic use, biostimulants, growth retardant, etc. [1–8]. Godlewska et al. [9] state that different plant parts can have varying influences, and seeds, fruits, flowers, stems, leaves, and roots can be used to produce plant-based extracts. Nowadays, chemical fertilizers play a major role in increasing crop yields and food production. However, their excessive and long-term use has led to several issues such as loss of fertility, structure, and microbial life, as well as runoff of nitrogen and phosphorus into the soil and water [10–12]. Biostimulants have become increasingly important in agricultural production, especially because they can reduce the amount of chemical fertilizers needed, thereby lowering input costs [13–15]. Biostimulants are substances of various organic or inorganic origins and chemical compositions that exert a biostimulator effect on plants. Application of biostimulants such as amino acids, humic extracts, and seaweed extracts, helps mitigate the impact of stressful conditions and enhances plant resistance to both abiotic and biotic stress factors [16–20]. Special benefits can be found in the use of natural sources, as the plant extracts are eco-friendly and most often cost-effective. The use of natural sources can bring special benefits, improve crop resilience and help plants cope with drought, salinity, and extreme temperatures [21–23].

Industrial hemp (*Cannabis sativa* L.) is increasingly attracting attention due to its multipurpose applications and environmental sustainability [24–27]. As a fast-growing crop with low pesticide and water requirements, hemp is used to produce fibers, seeds, oil, bioplastics, construction materials, and bioactive compounds such as cannabidiol (CBD). It is mostly permitted to cultivate industrial hemp with a total delta-9-tetrahydrocannabinol (THC) content of 0.2% to 1.0% [28–30]. Industrial hemp has significant development potential within the framework of sustainable agriculture and bioeconomy.

Hull-less oilseed pumpkin (*Cucurbita pepo* subsp. *pepo*) seeds are usually used as a raw material for cold-pressed oil, because seeds contain about 40–50% of oil [31–33]. Except for oil, there is a great advantage to human consumption, since the seeds do not require de-hulling, they are used as a snack and in the baking industries [34]. Many farmers grow it in Europe, especially in Austria, Poland, the Czech Republic, Hungary, Slovenia, Serbia, Romania, and Slovakia [35–41]. The seeds are dark green and have a high content of free fatty acids and oil and are often used in salads. The seeds are rich in vitamins, healthy fatty acids, phytosterols, magnesium, zinc, selenium, and carotenoids [42–46]. Seeds also contain cucurbitacins—compounds with antitumor and antimicrobial properties [46,47].

Both seed germination and early seedling development are influenced by a range of factors. The speed of the sprouts' growth and development depends on several factors: quality, temperature of the substrate and air, water supply, and lighting conditions [48–52]. First, seeds need to absorb water and swell, and under the influence of moisture, the seed coat softens, and the conditions necessary for germination and emergence are activated. The success of germination in vegetables, flowers, and other plant species is most influenced by the ability to absorb water, i.e., swelling and heat [53]. There are several environmental factors such as heat, cold, disease attack, drought, acidity, salinity, etc., all of which can significantly affect the speed and uniformity of germination and early seedling growth [54]. Seedling or sprout vigor reflects the health status and growth rate, both of which are critical for successful establishment and subsequent crop performance. During this early stage, seedlings rely heavily on the nutrient reserves stored within the seeds, as they are not yet fully photosynthetically active and have not developed a sufficiently extensive root system to absorb nutrients and water from the soil.

Wazeer et al. [55] stated that protein-based biostimulants are typically applied via foliar sprays or soil irrigation; however, their use in seed priming has recently emerged as a promising and sustainable approach to enhance seed quality and germination. Use of different types of extracts as biostimulants in organic agriculture is increasing in order to achieve sustainable and environmentally friendly food production [56,57]. Plant-based biostimulants enhance plant metabolism, nutrient uptake, and resistance to abiotic stress without introducing synthetic chemicals into the agroecosystem. Skowronek et al. [58] reported that industrial hemp extracts significantly increased the activity of antioxidant enzymes, extending the life of bees (*Apis mellifera*). Jolayemi et al. [59] highlight the growing scientific importance of protein-based biostimulants, which activate natural physiological processes to enhance plant growth. These biostimulants are typically derived from plant or animal sources, often as by-products or waste materials from agro-industrial processes, an approach that is particularly significant in the context of the circular bioeconomy and sustainability [55–60].

Plant extracts can have a two-way effect; they can inhibit plant growth and development, but they can also promote plant growth or growth at certain phases of development. Although certain effects of plant species on each other are known, there are still many unknowns when it comes to the interaction of different plant species and their influence on plant growth and development. The effects of plant extracts varied depending on whether fresh or dry biomass was used and whether the extracts exhibited stimulatory or inhibitory activity. Since higher concentrations (generally $\geq$5%) often showed inhibitory effects, concentration proved to be a key factor influencing the response. According to the literature, plant extracts are usually used in small portions generally $\geq$5%, and not higher than 10% [5,8,8,23,61–63].

Industrial hemp is still, in many cases, an underutilized crop with great potential for organic farming. Its inflorescences are rich in bioactive compounds such as terpenes, phenolics, flavonoids, and phytocannabinoids, many of which are known to have antioxidant, antimicrobial, allelopathic, or biostimulant effects on plants [64–67]. Moreover, hemp is a sustainable crop that grows quickly, requires minimal inputs, and improves soil structure, making it well-suited for use in ecological and regenerative agriculture. Its low THC content of industrial hemp genotypes also makes it safe and legally acceptable for research and field application.

Hull-less oilseed pumpkin is a crop of increasing economic and nutritional value, particularly in organic systems where its seeds are used for oil production and as a functional food. In professional and organic farming, pumpkins are often grown from seedlings, rather than direct seeding, to achieve better early establishment, uniform growth, and earlier harvest [68]. However, young seedlings are more vulnerable to environmental stress (e.g., drought, salinity, pathogens), and the use of natural biostimulants during this sensitive stage can significantly support plant development and resilience.

Based on the above, this study aimed to determine the effect of the industrial hemp inflorescence water extract of different concentrations on germination and early growth of the hull-less oilseed pumpkin during the sprouts phase.

2. Results

2.1. One-Way ANOVA

According to the results of the one-way ANOVA of the present study, the industrial hemp inflorescence water extracts significantly influenced ($p \leq 0.05$) oilseed pumpkin root growth (Table 1). The stem and total sprout length, as well as the fresh mass per plant of oilseed pumpkin sprouts, were very significantly influenced ($p \leq 0.001$) by the industrial hemp inflorescence water extracts. There was a significant influence of industrial hemp inflorescence water extracts on *Car* content.

Table 1. The one-way ANOVA and mean values of the total germination rate, morphological parameters, sprout growth parameters, bioactive compounds, and antioxidant activity of the hull-less oilseed pumpkin sprouts.

Parameter	Unit	Mean Square	F Value	$p > F$	$LSD_{0.05}$	Mean
Root length	cm	5.682	3.10	*	2.09	14.19
Stem length		18.376	23.27	***	0.91	5.45
Total length		38.41	16.34	***	2.36	19.64
Fresh mass	g per plant	0.195	9.09	***	0.23	14.10
Germination rate	%	24.759	0.21	ns	-	78.08
VI		18.994	2.99	*	3.88	15.30
RSR		4.347	12.79	***	0.898	2.93
SBI		315.67	18.99	***	6.282	28.21
Chl a	mg/g FW	0.003	39.06	***	0.017	0.161
Chl b		0.004	20.77	***	0.029	0.115
Chl a + b		0.009	8.36	*	0.062	0.268
Car		0.008	28.38	***	0.032	0.136
FRAP	mM $FeSO_4$	1067.38	5.26	*	24.833	131.57
TPC	µg GA/1 g tissue	0.001	1.39	*	0.006	0.017
TFC	µg QC/1 g tissue	992.25	13.24	**	16.297	81.65

VI—Vigor index, SBI—seedlings biomass index; *Chl*—chlorophyll, FRAP—Ferric Reducing Antioxidant Power, TPC—total phenols content, TFC—total flavonoids content. Significance: ns—not significant * $p \leq 0.05$, ** $p \leq 0.01$, *** $p \leq 0.001$.

The total phenol content of the hull-less oilseed pumpkin sprouts was significantly ($p \leq 0.05$) influenced by the industrial hemp water extracts (Table 1), whereas total flavonoid content in the hull-less oilseed pumpkin sprouts was very significantly ($p \leq 0.01$) affected by the industrial hemp inflorescence water extracts. The antioxidant activity measured by the FRAP method differed significantly ($p \leq 0.05$) in the hull-less oilseed pumpkin sprouts under the influence of different water extracts.

For sprout growth parameters of the hull-less oilseed pumpkin, there was a significant influence ($p \leq 0.05$) of industrial hemp inflorescence water extracts determined for the vigor index (VI), and a significant ($p \leq 0.05$) influence for root-to-shoot ratio (RSR) and sprout biomass index (SBI) (Table 1).

2.2. Morphological and Growth Parameters of the Hull-Less Oilseed Pumpkin Sprouts

The results concerning the morphological parameters of oilseed pumpkin sprouts are presented in Figures 1 and 2. The highest average root length (Figures 1 and 2a) was observed at the 2.5% treatment (15.92 cm), which was significantly greater than the 5.0% treatment (13.17 cm). However, it was not significantly different from the 0% (13.83 cm) and 1.0% (13.85 cm) treatments. The 5.0% treatment showed the lowest root length and was the only one statistically different from the highest value, suggesting that higher concentrations may negatively affect root growth.

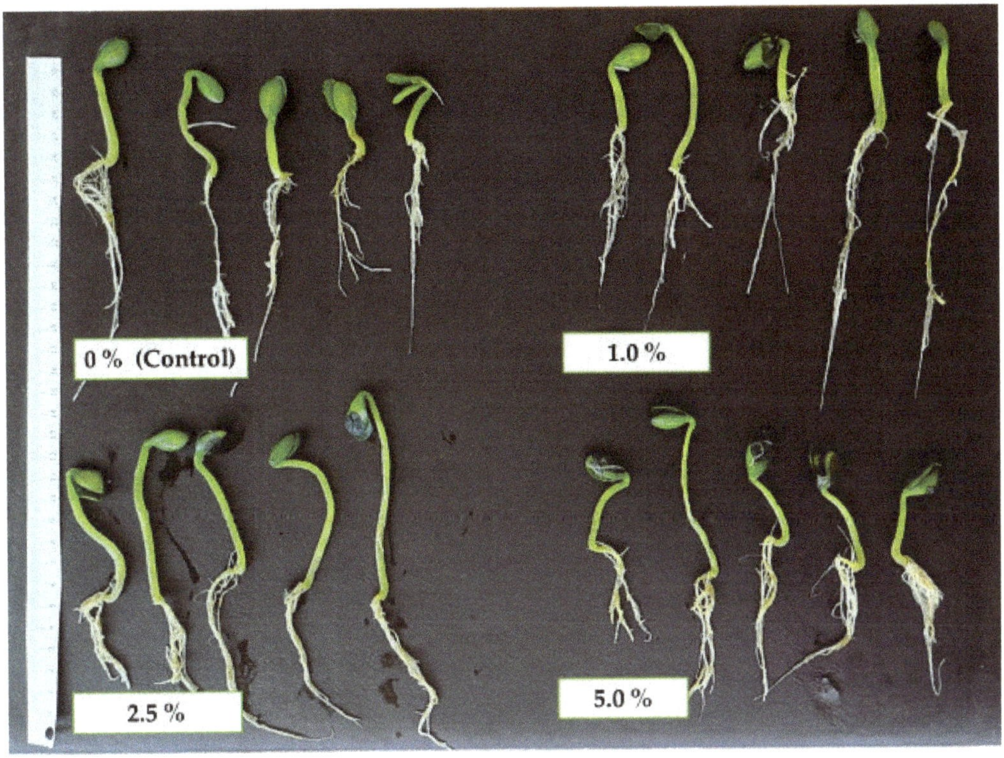

Figure 1. Hull-less oilseed pumpkin sprouts regarding different concentrations of industrial hemp inflorescence water extracts.

The longest stem was measured at the 2.5% treatment (8.22 cm), which was significantly higher than all other treatments (Figure 2b). The 5.0% treatment (6.01 cm) was significantly lower than 2.5%, but higher than both the control, 0% (3.42 cm), and 1.0% industrial hemp inflorescence water extracts (4.16 cm), which did not differ significantly from each other. This suggests that moderate treatment levels may enhance the measured outcome, while higher or lower concentrations reduce the effect.

The total length of oilseed pumpkin sprouts differs significantly ($p \leq 0.001$) from the industrial hemp inflorescence water extracts (Table 1). The 2.5% treatment resulted in the longest sprouts (24.14 cm), which was significantly greater than all other treatments (Figure 2c). The industrial hemp inflorescence water extracts 0%, 1.0%, and 5.0% showed no significant differences among each other.

The fresh mass of the oilseed pumpkin sprouts was also significantly different among the different concentrations of industrial hemp inflorescence water extracts (Table 1). The longest sprouts at the 2.5% treatment also showed the highest fresh mass (16.88 g/plant), which was significantly greater than all other treatments (Figure 2d). The 0% treatment had the lowest value of the fresh mass per plant (11.58), while the sprouts' fresh mass at 1.0% and 5.0% were intermediate and not significantly different from each other.

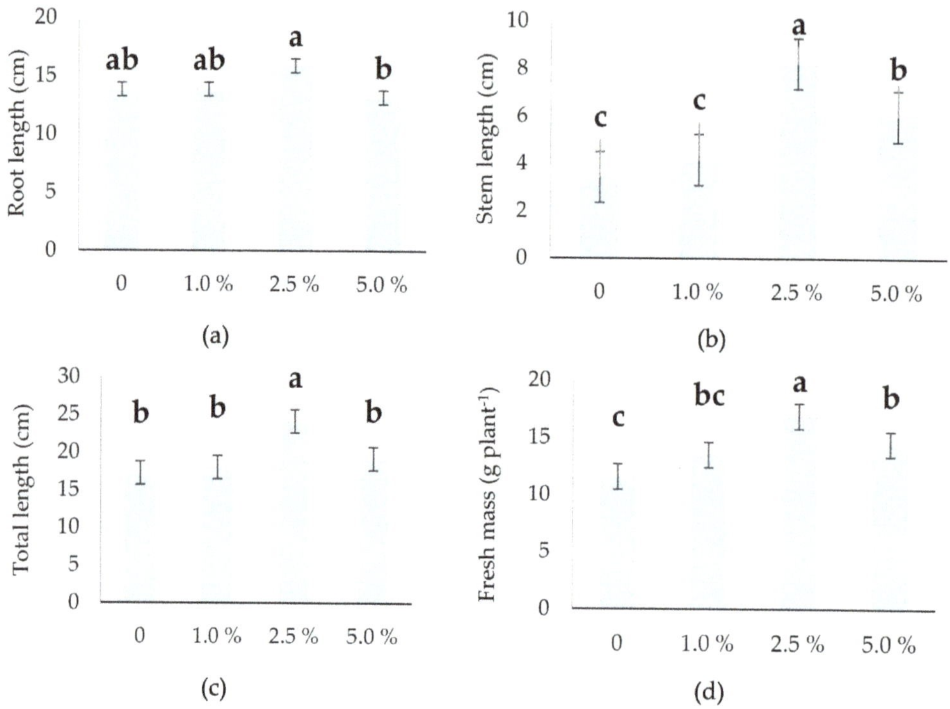

Figure 2. Morphological parameters of oilseed pumpkin sprouts regarding different concentrations of industrial hemp inflorescence water extracts: (**a**) stem length (cm), (**b**) root length (cm), (**c**) total length (cm), and (**d**) fresh mass (g plant $^{-1}$). Different letters above the bars indicate statistically significant differences at $p \leq 0.05$ (one-way ANOVA). Vertical error bars represent the standard error (SE) of the mean at 5%.

2.3. Total Germination Rate and Growth Parameters of the Hull-Less Oilseed Pumpkin Sprouts

There was no significant difference determined for total germination rate (Table 1), even though industrial hemp inflorescence water extracts had a mild inhibitory effect on seed germination across all tested concentrations (1–5%) in comparison to the control (Figure 3a).

The greatest reduction was recorded at 1%, but differences between the treated groups were relatively small. The highest germination rate, 82%, was recorded at the control treatment (0%), while all treatments with the extract showed a reduction in germination. At 1.0%, the germination rate dropped to 76%, which was the lowest value observed.

In the present study, at the 2.5% industrial hemp inflorescence water extract, the VI was the highest (~19%), which was significantly greater than the control and 1.0% treatment (Figure 3b). The RSR of the hull-less oilseed pumpkin sprouts was 2.93 (Table 1) and was very significantly ($p \leq 0.001$) affected by the industrial hemp inflorescence water extract. The highest RSR was found at the control (2.16), and the lowest at 2.5% industrial hemp inflorescence water extract (1.94) (Figure 3c). The SBI was also significantly ($p \leq 0.001$) affected by the industrial hemp inflorescence water extract (Table 1). The lowest SBI was found at the control (19.99), and the highest at the 2.5% industrial hemp inflorescence water extract (40.64) (Figure 3d).

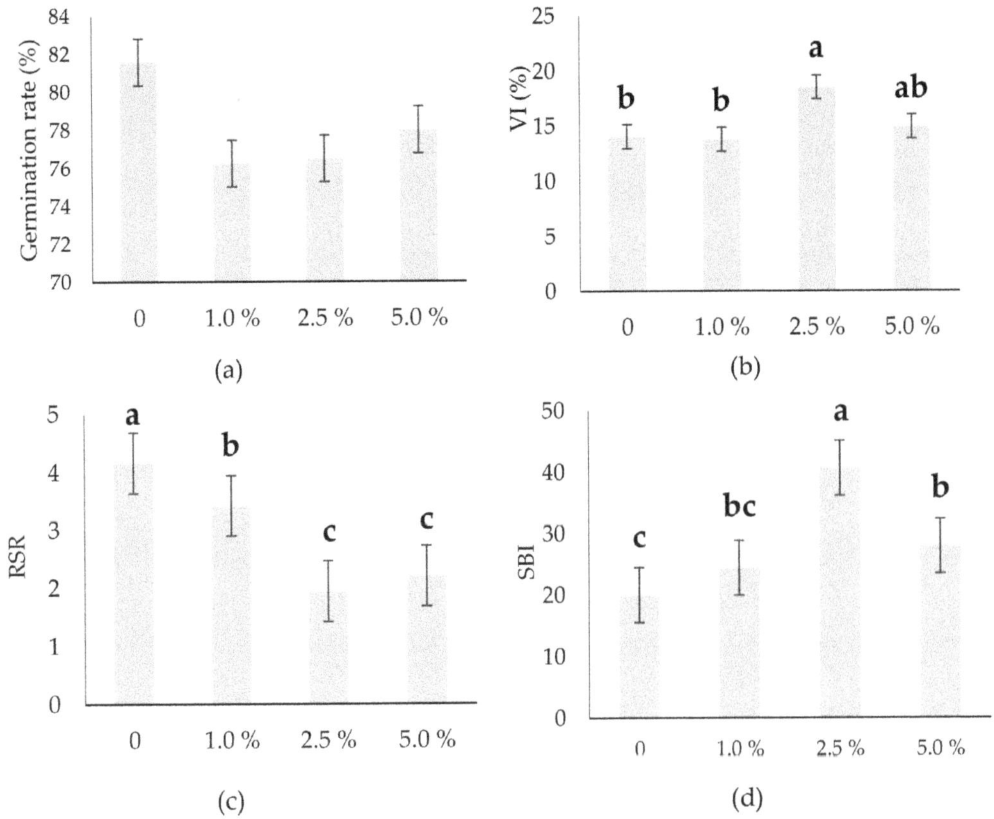

Figure 3. Germination rate and growth parameters of hull-less oilseed pumpkin sprouts regarding different concentrations of industrial hemp inflorescence water extracts: (**a**) germination rate (%), (**b**) vigor index (VI) (%), (**c**) root-to-shoot ratio, and (**d**) sprout biomass index. Different letters above the bars indicate statistically significant differences at $p \leq 0.05$ (one-way ANOVA). Vertical error bars represent the standard error (SE) of the mean at 5%.

2.4. The Pigment Status of the Hull-Less Oilseed Pumpkin Sprouts

Industrial hemp inflorescence water extracts had a very significant ($p \leq 0.001$) influence on the average *Chl a* content of oilseed pumpkin sprouts (Table 1).

The average *Chl a* content of oilseed pumpkin sprouts was 0.161 mg g $^{-1}$ FW. Industrial hemp inflorescence water extracts had a positive influence on increasing *Chl a* of the oilseed pumpkin sprouts (Figure 4a). Control (0%) had the lowest *Chl a* content (~0.125 mg g^{-1} FW), 1.0% extract had a moderate increase in *Chl a* (~0.145 mg g^{-1} FW), while the highest content was determined at 2.5% and 5% (0.187 mg g^{-1} FW on average), but with no significant difference between the latter two concentrations.

A very significant ($p \leq 0.001$) influence of industrial hemp inflorescence water extracts was found for *Chl b* content (Table 1). The average *Chl b* content was 0.115 mg/g FW (Figure 4b). Industrial hemp inflorescence water extracts positively influenced the *Chl b* content in oilseed pumpkin sprouts, which was the highest at 5% (0.174 mg/g FW).

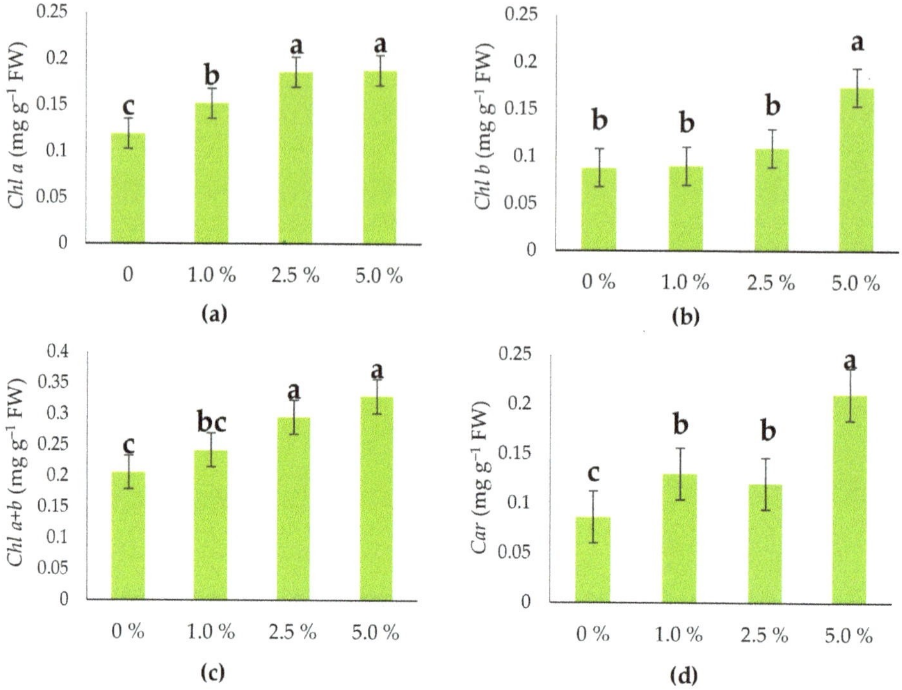

Figure 4. Pigment content of hull-less oilseed pumpkin sprouts (FW) at different levels of industrial hemp inflorescence water extracts: (**a**) *Chl a* (mg g^{-1} FW), (**b**) *Chl b* (mg g^{-1} FW), (**c**) *Chl a* + *b* (mg g^{-1} FW), and (**d**) *Car* (mg g^{-1} FW). Different letters above the bars indicate statistically significant differences at $p \leq 0.05$ (one-way ANOVA). Vertical error bars represent the standard error (SE) of the mean at 5%.

The content of *Chl a* + *b* has also been significantly ($p \leq 0.05$) affected by the industrial hemp water extracts (Table 1). The average content was 0.268 mg/g FW (Figure 4c). The lowest value was recorded in the control (0%), at 0.206 mg g^{-1} FW, and increased up to 0.329 mg g^{-1} FW at 5% of industrial hemp inflorescence water extracts.

The average *Car* content was 0.136 mg/g FW (Table 1). The highest *Car* content was observed at 5.0% (0.210 mg/g FW), suggesting a strong stimulatory effect at higher concentrations (Figure 4d).

2.5. Total Phenols, Flavonoids, and Antioxidant Activity of the Hull-Less Oilseed Pumpkin Sprouts

The TPC was significantly ($p \leq 0.05$) influenced by the industrial hemp water extracts (Table 1). An average of 0.017 µg GA/1 g tissue content in oilseed pumpkin sprouts was found for the TPC (Table 1). Also, the TFC in the oilseed pumpkin sprouts was significantly influenced ($p \leq 0.01$) by the industrial hemp inflorescences' water extracts (Table 1).

The treatment with 1.0% resulted in the highest TPC (Figure 5a), approximately 0.020 µg GA/g tissue, which is significantly higher than all other groups. In contrast, the control group (0%) and the treatments with 2.5% and 5.0% showed lower and statistically similar values, indicating no significant difference among them.

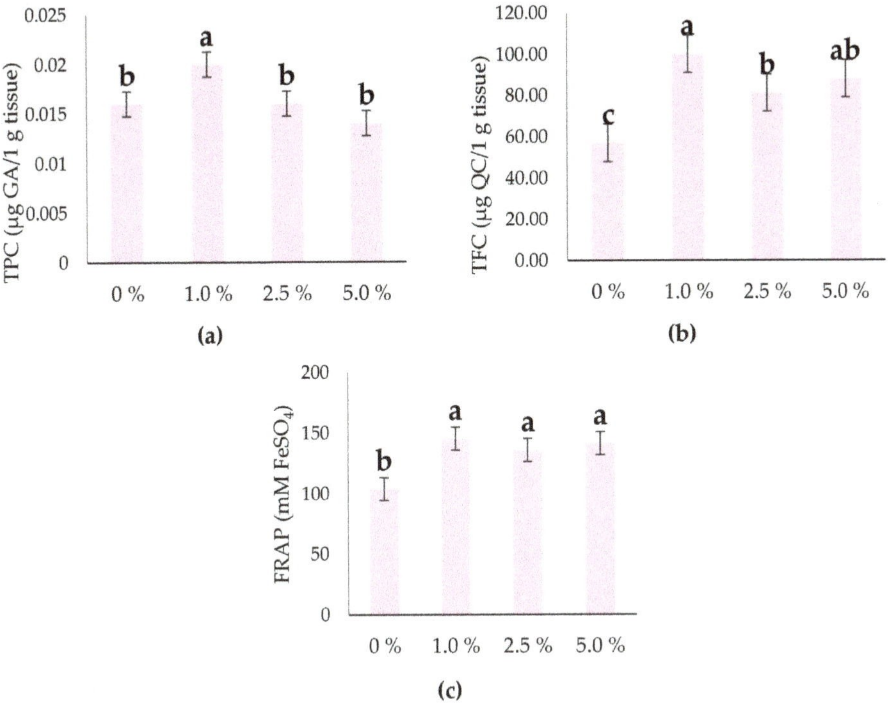

Figure 5. Total phenols (TPC), flavonoids (TFC), and antioxidant activity of hull-less oilseed pumpkin sprouts (FW) at different levels of industrial hemp inflorescence water extracts: (**a**) TPC (μg GA/1 g tissue), (**b**) TFC (μg QC/1 g tissue), and (**c**) antioxidant activity—FRAP (mM $FeSO_4$). Different letters above the bars indicate statistically significant differences at $p \leq 0.05$ (one-way ANOVA). Vertical error bars represent the standard error (SE) of the mean at 5%.

The 1.0% treatment significantly increased QC concentration compared to all other treatments (Figure 5b). At the 2.5% and 5.0% industrial hemp inflorescence water extract treatments were not significantly different from each other for TFC, but the 2.5% was significantly lower than the 1.0% (Figure 5a).

Figure 5b shows the TFC under different treatments. The treatment with 1.0% resulted in the highest TFC level, approximately 100 μg QC/g tissue, which is statistically different from the other groups. In contrast, the control group (0%) exhibited the lowest TFC (57.03 μg QC/g tissue). The 2.5% treatment showed a significantly higher TFC compared to the control but lower than the 1.0% treatment, suggesting that increasing the concentration beyond 1.0% does not further enhance TFC levels and may even result in a slight decrease.

Antioxidant activity was significantly affected by the industrial hemp inflorescence water extracts (Table 1), being the lowest in oilseed pumpkin sprouts in the control (103.886 mM $FeSO_4$), while all treatments with industrial hemp inflorescence water extracts increased the antioxidant activity of oilseed pumpkin sprouts (Figure 5c).

2.6. Correlation Analysis

To determine the relationship among all measured parameters, the Pearson correlation coefficient was determined (Table S1). To visually identify which parameters responded similarly or oppositely under industrial hemp inflorescence water extract treatment, a heat map was created (Figure 6). The heat map illustrates the interconnection between morphological growth, pigment synthesis, and antioxidant activity in seedlings exposed to industrial hemp extract.

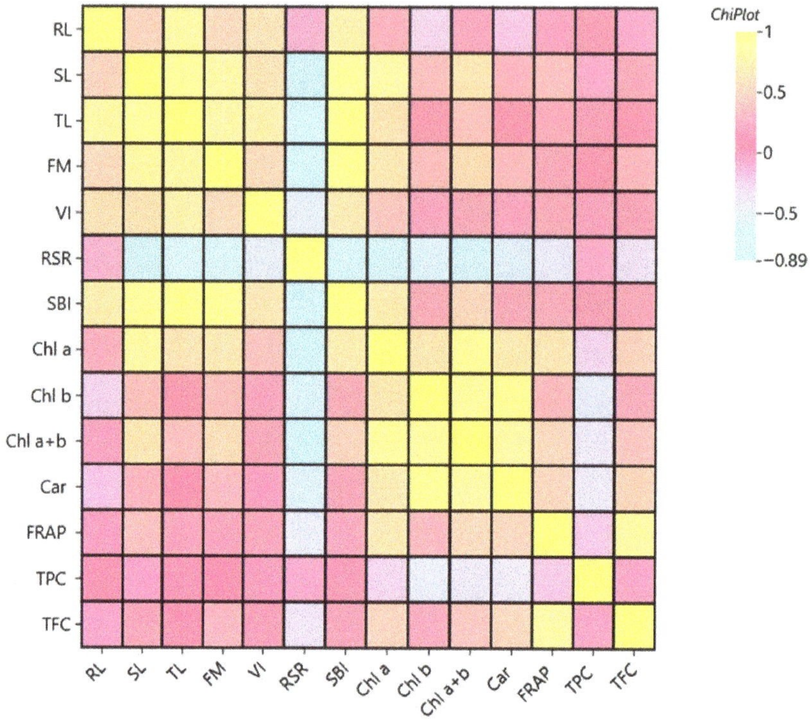

Figure 6. The heatmap of Pearson correlation coefficient for total gemination, morphological parameters, bioactive compounds, and antioxidant activity of the hull-less oilseed pumpkin sprouts (TG—total germination rate, RL—root length, SL—stem length, TL—total sprout length, FM—fresh mass, VI—vigor index, RSR—root to stem ratio, SBI—sprouts biomass index, FRAP—antioxidant activity FRAP (mM $FeSO_4$), TPC—total phenol content, TFC—total flavonoid content). Each cell shows the Pearson correlation coefficient (r) between two variables, which ranges from: +1.0 (perfect positive correlation), 0 (no correlation), and −1.0 (perfect negative correlation) (N = 16).

Total germination (TG) showed no significant correlations with most other parameters in the present study (Figure 6), except for a weak positive correlation with fresh mass (FM) (r = 0.576; $p \leq 0.05$) (Table S1). Shoot length (SL) showed strong positive correlations with TL (r = 0.887 **), FM (r = 0.797 **), SBI (r = 0.888 **), and VI (r = 0.626 **). This indicates that shoot elongation is a central trait reflecting seedling performance under extract influence.

The shoot length (SL) also showed a strong negative correlation with root to shoot ratio (RSR) (r = −0.890, $p \leq 0.01$), indicating that as shoot growth increases, the proportion of biomass allocated to roots decreases (Table S1). The root length (RL) was positively correlated with hull-less oilseed pumpkin sprouts' total length (TL) (r = 0.823; $p \leq 0.001$), fresh mass (r = 0.541; $p \leq 0.05$), and the vigor index (VI) (r = 0.632; $p \leq 0.001$), confirming its importance in overall seedling growth.

Chl a was strongly correlated with *Chl b* (r = 0.668, $p \leq 0.01$), *Chl a + b* (r = 0.877, $p \leq 0.01$), and carotenoids (r = 0.689, $p \leq 0.01$), as expected due to their shared role in photosynthesis (Table S1). *Chl b* and *Car* had a particularly strong correlation (r = 0.916, $p \leq 0.001$), suggesting coordinated pigment synthesis.

The antioxidant capacity of the hull-less oilseed pumpkin sprouts determined with the FRAP method was significantly positively correlated with total TFC ($r = 0.840$, $p \leq 0.001$), and moderately with TPC, confirming the major role of flavonoids in antioxidant activity (Table S1, Figure 6).

3. Discussion

3.1. Morphological Parameters and Growth Analysis of the Hull-Less Oilseed Pumpkin Sprouts

There is still a lack of studies involving industrial hemp as a plant-based biostimulant in agricultural production. Du Jardin [68] states that biostimulants, including plant-based biostimulants, are increasingly used worldwide. With the present study, the influence on hull-less pumpkin sprouts is evident and statistically confirmed. Even though the total germination rate was not significantly influenced by the industrial hemp water extracts' inflorescence, other growth parameters were significantly affected by their application. Similarly, total germination did not show significant correlation with the majority of the parameters (Figure 6, Table S1). This suggests that although germination rate itself was not directly associated with shoot/root length or biochemical traits, seedlings that did germinate tended to produce more biomass under certain conditions. Thus, the results of the total germination rate of the present study suggest that certain bioactive compounds present in the industrial hemp inflorescence water extract may moderately interfere with physiological processes involved in seed germination. The VI responded non-linearly to increasing concentrations of hemp extract. Interestingly, the 2.5% extract concentration significantly enhanced seedling vigor, suggesting a possible stimulatory effect at moderate doses. In contrast, both lower (1.0%) and higher (5.0%) concentrations showed reduced vigor, possibly indicating inhibitory or phytotoxic effects at suboptimal and elevated extract levels. Islam et al. [69] found that industrial hemp was used as seed priming media, and among the 14 hemp varieties, Han FNQ (China) showed the highest seed germination and seedling vigor, while low-concentration ClO_2 pre-treatment (500 $mg \cdot L^{-1}$) improved early seedling growth despite a general decline in germination across all pre-treatments.

According to Andrić [70], seed aging is one of the main causes of reduced vigor and poor field emergence, which becomes especially evident under adverse environmental conditions such as drought, low temperatures, or soil compaction, ultimately leading to decreased crop yield and quality. For the cultivar of oilseed pumpkin with the hull, in media with various pH values (2.5–8.5), Varga et al. [71] found an average germination rate of 92%, with differences among pH values, reflecting the plant's stress response to acidic and alkaline media. The highest germination rate was observed at pH 8.5 (94%), followed by a total of 93% at pH 2.5, 4.5, and 7.5. Lynch et al. [72] state that root architecture reflects the most stressful conditions in the soil systems. Additionally, authors stated that root-to-shoot ratio usually varies widely among plant species, and it is highly modified by external factors. In the present study, the significantly lower RSR of hull-less oilseed pumpkin sprouts was at 2.5% of industrial hemp inflorescence water extract, which may also align with the earlier observation of increased VI at that same concentration, potentially reflecting balanced or prioritized shoot elongation in favorable conditions. A decrease in the RSR at the concentration of 2.5% and 5.0% (Figure 2c) suggests that these treatments led to relatively greater shoot growth or inhibition of root development, thereby lowering the root-to-shoot proportion. This shift could indicate a physiological stress response at specific extract concentrations.

3.2. Leaf Pigments, Phenols, Flavonoid Content, and Antioxidant Capacity of the Hull-Less Oilseed Pumpkin Sprouts

The average *Chl a* content in the hull-less oilseed pumpkin sprouts of the present study was 0.161 mg g^{-1} FW. The application of industrial hemp inflorescence water extracts significantly enhanced *Chl a* level, with maximum values observed at concentrations of 2.5% and 5.0%. These findings indicate a positive and concentration-dependent stimulatory effect, particularly up to 2.5%. The results suggest that industrial hemp inflorescence water extracts significantly enhance *Chl a* level in oilseed pumpkin sprouts. The most significant increase in *Chl b* was observed at the 5.0% concentration, suggesting a concentration-dependent stimulation of *Chl a* biosynthesis. There are several studies that also confirm the increase in leaf pigment status with the application of biostimulants. Francesca et al. [73] stated that application of biostimulant CycoFlow (Agriges, Benevento, Italy) on different tomato genotypes significantly influences the changes of Chl *a*, *Chl b*, *Car*, and ascorbic acid in a way that *Chl a* content is consistently higher than *Chl b* across both control and with biostimulant application, while *Car* levels remained relatively stable across treatments. In potato (*Solanum tuberosum*) leaves, Mystkowska [74] found that the application of biostimulants increased the SPAD (Soil–Plant Analysis Development) leaf greenness index and plant height in all three potato cultivars compared to the control. Di Mola et al. [75] found that in baby rocket leaf (*Diplotaxis erucoides* L., cv. 'Reset') treated with foliar applications of biostimulators, one a legume-derived protein hydrolysate and the other a tropical plant extract, there was a significant increase of 112.0% and 79.2%, respectively, in total ascorbic acid, with no significant difference on lipophilic antioxidant activity. The authors stated that *Chl a* and *Chl b* content increased with both biostimulants, as compared to the control. Carotenoids, a class of terpenoids functioning as antioxidants and photosynthetic pigments, are positively affected by various environmental and biochemical factors [76,77]. In our study, the highest carotenoid content (0.210 mg g^{-1} FW) was recorded in hull-less oilseed pumpkin sprouts at the 5.0% of the industrial hemp water extract concentration, indicating a strong stimulatory effect of higher biostimulant levels on carotenoid accumulation in the sprouts.

According to the correlation analysis of the present study, the heatmap illustrates the strength and direction of relationships (correlations) between various morphological, physiological, and biochemical parameters of seedlings treated with industrial hemp extract (Figure 6). Moderate to strong positive correlations were found between growth indices (e.g., FM, TL, VI) and pigment or antioxidant traits (e.g., *Chl a*, Car, FRAP), possibly reflecting stress adaptation or industrial hemp inflorescence water extract influence. No strong correlation was observed between FRAP and total *Chl* content, indicating that pigment synthesis and antioxidant mechanisms may be independently regulated under the tested conditions. In our study, 1.0% concentration was optimal for enhancing TPC (Figure 4a), while higher concentrations (2.5% and 5.0%) did not provide additional benefit and resulted in a return to baseline levels observed in the control. Moreover, in the present study, the 1.0% treatment appears to be the most effective concentration for enhancing TFC (Figure 5b), suggesting that higher concentrations of industrial hemp inflorescence (2.5% and 5.0%) do not provide additional benefits and may indicate a saturation point or possible inhibitory effects at elevated levels.

The TPC in plants plays a key role in plant defense, growth regulation, and stress response. Kalinová et al. [78] stated that the levels of flavonoids such as rutin, orientin, and vitexin in the cookies and pasta made from wheat and common buckwheat sprouts or microgreens are influenced by multiple factors, including the food matrix, heating method, temperature, processing duration, interactions with other compounds, and the chemical properties of the flavonoids themselves. Total Flavonoid Content (TFC) represents the combined amount of all different flavonoid compounds found in the plant tissue.

In the present study, the antioxidant activity of hull-less oilseed pumpkin (Figure 4c) was increased in all treatments, which was significantly higher ($p \leq 0.05$) compared to the control. These findings highlight the positive impact of biostimulants on enhancing the antioxidant potential in plants. Relatively low concentrations of TPC and TFC observed in this study are due to the early developmental stage of the seedlings. At this stage, secondary metabolism is still underdeveloped, as the plants prioritize primary growth processes. In contrast, older seedlings, transplants, or mature plants typically accumulate higher levels of these compounds, which explains the differences reported in other studies. Posmyk and Szafrańska [79] state that biostimulants can enhance plant tolerance to adverse environmental conditions, including chilling, drought, salinity, chemical pollutants, and heavy metal stress. For tomato plants (*Solanum lycopersicum* L., cv.'Coronel') Cozzolino et al. [80] found that the application of plant-based biostimulants (an extract of brown seaweed *Ecklonia maxima* (Osbeck), legume derived protein hydrolysate, and a tropical plant extract), significantly affected lipophilic antioxidant activity, hydrophilic antioxidant activity, and ascorbic acid content, compared with an untreated control, while no statistically significant differences were observed in total phenol content. To evaluate the biostimulant effects of biological preparation based on rosemary (*Salvia rosmarinus* L.) and eucalyptus (*E. globulus* L.) essential oils, Chrysargyris et al. [81] used it for tomato seedlings (cv. Brillande Grade A). The authors stated that the ascorbic acid content of the fruits was markedly enhanced by application of biostimulant (147.02 mg 100 g^{-1} FW), more than doubling the control value, and that phenol content was highest in foliar spray with the *Razormin* commercial product treated (2.59 μmol GAE g^{-1} FW). On the other hand, plant extracts can have an inhibitory influence on plant germination and bioactive compounds. Plant extracts, particularly at higher concentrations, may release allelopathic compounds that interfere with seed germination, and could potentially be used for natural weed control or bioherbicide development, though with caution regarding crop safety. Teacă et al. [82] tested different concentrations (0–0.2 g/L) of ethanol extracts from *Melissa officinalis* (lemon balm, V1 and V2) and *Lavandula angustifolia* (lavender, V3) on the germination percentage of various crop seeds, including wheat, rye, triticale, soybean, and bean. Authors stated that all three plant extracts exhibit allelochemical effects, reducing seed germination in a dose-dependent manner, in a way that the wheat was consistently the most resilient species across all treatments, while legumes (soybean and bean) and triticale appeared most susceptible to the inhibitory effects. Authors stated that lemon balm extract has the strongest inhibitory effect, likely due to higher concentrations or different types of active compounds (e.g., phenolics, terpenoids).

In the present study, a concentration of 2.5% industrial hemp inflorescence water extract improved growth parameters such as root, stem, and total length, as well as fresh mass. However, for leaf pigments, the 5.0% concentration showed the best results for *Chl a*, *Chl b*, *Chl* (*a* + *b*), and *Car*. The FRAP antioxidant activity increased in all treatments compared to the control, but the highest TPC and total flavonoid content TFC were observed at the 1.0% concentration. In the context of sustainable agriculture and the growing demand for environmentally friendly production systems, there is increasing interest in identifying natural biostimulants that can promote plant growth, enhance stress tolerance, and improve crop quality—all without the use of synthetic agrochemicals.

For other plant species and water extracts, Mutlu-Durak et al. [83] stated that water extracts of weeping willow leaves and bark, rich in bioactive compounds such as salicylates and phenolics, were shown to act as effective plant-based biostimulants for maize, enhancing shoot fresh weight, root area, and leaf protein content. Yadav et al. [84] examine twenty-five herbal extracts for their biostimulant activity on wheat through seed priming, assessing morphological parameters such as germination, seedling growth, and biomass, as well as biochemical traits including phenolics, flavonoids, enzyme activity, and antioxidant capacity. Extracts from *Phyllanthus emblica*, *Plumbago zeylanica*, *Catharanthus roseus*, and *Baccopa monnieri* were particularly effective, with principal component analysis grouping the samples based on their growth-promoting activity. Although it remains difficult to predict how findings of our study would translate into field conditions for pumpkin cultivation, it opens up new opportunities for further research, especially due to many growers producing pumpkins as seedlings in controlled environments, where the biostimulant effect could be more effective.

4. Materials and Methods

4.1. Preparation of Water Extracts

The full, expanded, and healthy inflorescences of industrial hemp (cv. Finola) were collected at the full flowering stage on 5 July 2024, from the experimental field of the Faculty of Agrobiotechnical Sciences Osijek, located in Tenja, Croatia. The inflorescence was used for water extract preparation in the present study. The harvested inflorescences were oven-dried (Kambič, SP-440,Semič, Slovenia) for 48 h. Once completely dried, the biomass was chopped and ground using an electric mill (Retsch, Haan, Germany) into a fine powder that passed through a 1 mm sieve. The preparation of *C. sativa* water extracts was based on the method described by Norsworthy [85], with slight modifications. Specifically, 50 g of the dried industrial hemp inflorescence powder were soaked in 1000 mL of distilled water (1:20 w/v) at room temperature (21 $\pm$ 2 °C) for 24 h. The mixture was initially filtered through a fine cloth to remove coarse particles, followed by two rounds of filtration through filter paper (Munktell, Bärenstein, Germany), to obtain a stock solution with a 5% concentration, reflecting the ratio of plant material to solvent used in preparation. This stock extract was then diluted with distilled water to achieve final working concentrations of 0% (control), 1.0%, 2.5%, and 5.0% for the seed germination bioassay in Petri dishes. All extracts were freshly prepared one day before the start of the experiment and stored at 4 °C to prevent microbial growth and preserve the integrity of bioactive compounds.

4.2. Seed Germination Bioassay in Petri Dishes

The experiment was conducted under controlled conditions at the Faculty of Agrobiotechnical Sciences Osijek (Croatia). The experiment was carried out using a completely randomized design with four replications per treatment. Each replicate consisted of a separate sterilized Petri dish (90 mm diameter) lined with a double layer of filter paper (Munktell 80 g/qm, Germany). For the experiment, the seeds of hull-less oilseed pumpkin (*Cucurbita pepo* L.), variety "Štrigovec" (Croatian plant genetic resources database Accession number IND00071 [86]), collected from the project National program for the conservation and sustainable use of plant genetic sources for food and agriculture in the Republic of Croatia, were used. For each replicate, 10 seeds of hull-less oilseed pumpkin were evenly distributed on the filter paper.

The filter paper was moistened with 10 mL of industrial hemp (*Cannabis sativa* L.) inflorescence water extract, applied in three concentrations: 1.0%, 2.5%, and 5.0%. The 0% treatment (distilled water) served as the solvent control. Seeds were incubated at 21 $\pm$ 2 °C under a 12 h light/12 h dark photoperiod.

4.3. Total Germination and Morphological Parameters of Hull-Less Oilseed Pumpkin Sprouts

On the final day (14th day) of the experiment, the total germination rate percentage for each treatment was calculated using the equation: G (%) = (Number of germinated seeds/Total number of seeds) × 100. The root and shoot lengths of seedlings were measured manually using a ruler, and the fresh mass of seedlings was recorded using a precision balance. To assess early seedling performance, the Vigor Index (VI) was calculated according to the following equation: VI = (Average seedling length (cm) × Germination rate percentage)/100. The Root-to-Shoot Ratio (RSR) was determined according to the equation: Mean root length (cm)/Mean stem length (cm). Sprouts Biomass Index (SBI) was calculated using the equation: SBI = Fresh weight (g per plant) × Total seedling length (cm).

Fresh hull-less oilseed sprouts were put in the laboratory freezer at −80 °C for further analysis of the bioactive compounds.

4.4. Analyses of the Pigment Content of the Hull-Less Oilseed Pumpkin Sprouts

To determine photosynthetic pigment content, 0.05 g of the powdered hull-less oilseed pumpkin sprouts extract was accurately weighed into a 15 mL test tube, followed by the addition of 10 mL of acetone. The mixture was homogenized using a vortex mixer and subsequently centrifuged at 4000 rpm for 10 min at 4 °C. From the resulting supernatant, 2 mL was transferred into a cuvette, and absorbance was measured at A662 nm, A644 nm, and A440 nm using a spectrophotometer (Shimadzu, UV-1800, Kyoto, Japan).

The concentrations of chlorophyll a (*Chl a*), chlorophyll b (*Chl b*), total chlorophyll (*Chl a + b*), and carotenoids (*Car*) were calculated using the Holm–Wettstein [87,88] equations, expressed in mg dm^{-3}:

$$\text{Chlorophyll } a\ (Chl\ a) = 9.784 \times A662 - 0.990 \times A644$$

$$\text{Chlorophyll } b\ (Chl\ b) = 21.426 \times A644 - 4.650 \times A662$$

$$\text{Total chlorophyll } (a + b)\ (Chl\ a + b) = 5.134 \times A662 + 20.436 \times A644$$

$$\text{Carotenoids } (Car) = 4.695 \times A440 - 0.268 \times (Chl\ a + b)$$

To express the results per gram of fresh weight (mg/g FW), the concentration in the extract was multiplied by the extract volume (dm^3) and divided by the fresh weight of the plant material (g) used for extraction. This provides the pigment content relative to the fresh mass of the sample.

4.5. Analyses of the Total Phenolic and Flavonoid Content of the Industrial Hemp Water Extract and Hull-Less Oilseed Pumpkin Sprouts

The TPC in the industrial hemp water extract and hull-less oilseed pumpkin sprouts was determined using the Folin–Ciocalteu method as described by Singleton and Rossi [89], with minor modifications in three technical replications. The reaction mixture consisted of the sample extract, distilled water, Folin–Ciocalteu reagent, and sodium carbonate (Na_2CO_3). Following incubation at 37 °C for 60 min, absorbance was measured at A765 nm using a spectrophotometer (Shimadzu, UV-1800, Japan) in three technical repetitions. A standard calibration curve ($y = 2.939x + 0.0677$, $R^2 = 0.9953$) was generated using increasing concentrations of gallic acid, and results were expressed as gallic acid equivalents (mg GAE/g FW).

The total flavonoid content was determined in the ethanol extract after the addition of aluminum chloride ($AlCl_3$) and 96% ethanol in three technical replications. The reaction mixture was homogenized and incubated at room temperature for 60 min, after which absorbance was measured at 415 nm. A calibration curve ($y = 0.0076x + 0.0304$, R^2 =

0.9988) was prepared using standard solutions of quercetin, and results were expressed as quercetin equivalents (mg QC/g FW).

For the industrial hemp inflorescence water extracts, the total phenolic content was 0.0128 μg GA/1 g tissue, and total flavonoid content was 30.52 μg QC/g tissue. The results were recalculated from mg GAE/g FW and mg QC/g FW to μg GA/g and μg QC/g for consistency in the tables.

4.6. Antioxidant Activity of the Hull-Less Oilseed Pumpkin Sprouts

The antioxidant capacity of the hemp extract was assessed using the Ferric Reducing Antioxidant Power (FRAP) assay, following the method described by Keutgen and Pawelzik [90], with slight modifications. The FRAP reagent was added to the ethanol extract of the sample, and the reaction mixture was incubated at 37 °C for 4 min. Subsequently, the absorbance was measured at A593 nm using a spectrophotometer (Shimadzu, UV-1800, Japan).

A standard calibration curve ($y = 1.2461x + 0.0505$, $R^2 = 0.9969$) was prepared using increasing concentrations of $FeSO_4$, and the antioxidant capacity was expressed as mM $FeSO_4$ per gram of fresh weight (mM $FeSO_4$ g^{-1} FW).

4.7. Biometric Approach

All obtained data were transferred into MS Excel (MS Office 2011), and afterward, the SAS Enterprise Guide, 7.1, was used for biometric analysis. The one-way ANOVA was used to determine the differences in total germination, morphological parameters, and bioactive compounds. The normality of data distribution was tested using the Shapiro–Wilk test, while Bartlett's test was applied to verify the homogeneity of variances. The influence of industrial hemp inflorescence water extracts on hull-less oilseed pumpkin early growth was expressed as the significance: ns—not significant, * $p \leq 0.05$, ** $p \leq 0.01$, *** $p \leq 0.001$. In the case of the significant F value, the post hoc Student t-test (Least Significant Difference, LSD test) was used to determine the differences between the means. There were three degrees of freedom (DF) for all parameters, and eight error degrees of freedom (DF).

The Pearson correlation coefficient was used to determine correlations. To visualize the relationships among all parameters in the present study, a correlation heat map was drawn using *ChiPlot* [91] (https://www.chiplot.online/) (accessed on 30 July 2025).

5. Conclusions

The use of biostimulants derived from plant extracts offers a sustainable and environmentally friendly approach to crop management, particularly within the context of organic agriculture, which is experiencing significant global growth. Hull-less oilseed pumpkin is a crop often cultivated under organic practices, and the use of plant-based biostimulants can play an important role in the growing process. Even though antioxidant activity was primarily driven by flavonoid content, not directly by seedling morphology or pigment levels, the water extract of industrial hemp inflorescence significantly influenced all bioactive compound content in the hull-less oilseed pumpkin sprouts. The promising results also highlight the need for follow-up greenhouse and field experiments to evaluate practical applicability.

Although different concentrations had different strengths, the 2.5% industrial hemp water extract can be considered the most effective overall, as it significantly enhanced plant growth and biomass, which are key indicators of vitality. Meanwhile, the 5.0% concentration was most beneficial for enhancing photosynthetic pigments, and 1.0% was best for increasing certain antioxidant compounds.

Industrial hemp shows considerable potential in this regard, and future research should further explore its applications, including for other crop species. Further research is recommended to include other arable crops, as well as various parts of the industrial hemp plant. Such studies could contribute to a more sustainable and efficient use of agricultural resources and potentially reveal new functionalities and value-added uses for hemp and other crop by-products.

Supplementary Materials: The following supporting information can be downloaded at: https://www.mdpi.com/article/10.3390/plants14223473/s1, Table S1. The correlation analysis of the hull-less oilseed pumpkin sprouts' morphological and growth parameters, bioactive compounds, and antioxidant activ-ity (N = 16).

Author Contributions: Conceptualization, I.V. and M.R.; methodology, M.R., M.T.K. and R.B.; software, I.V.; validation, A.M.K.; formal analysis, M.T.K.; investigation, M.A.; resources, M.A.; data curation, D.I.; writing—original draft preparation, I.V., M.R., A.M.K. and D.I.; writing—review and editing, M.A. and R.B.; visualization, I.V. and M.R.; supervision, M.A.; project administration, M.A.; funding acquisition, M.A.. All authors have read and agreed to the published version of the manuscript.

Funding: This study was a part of the project National program for the conservation and sustainable use of plant genetic sources for food and agriculture in the Republic of Croatia for the period from 2021 to 2027.

Data Availability Statement: Data are contained within the article.

Conflicts of Interest: The authors declare no conflicts of interest.

Abbreviations

TG	total germination rate
RL	root length
SL	stem length
TL	total sprout length
FM	fresh mass
FW	Fresh weight
Chl a	Chlorophyll *a*
Chl b	Chlorophyll *b*
Car	Carotenoids
VI	Vigor index
RSR	Root to shoot ratio
SBI	Sprouts biomass index
SE	Standard Error
FRAP	ferric reducing antioxidant power
TPC	total phenol content
TFC	total flavonoid content

References

1. Đurić, M.; Mladenović, J.; Bošković-Rakočević, L.; Šekularac, G.; Brković, D.; Pavlović, N. Effect of Mineral Fertilizers on the Yield and Quality of Industrial Hemp (*Cannabis sativa* L.). *Acta Agric. Serbica* **2019**, *24*, 27–39. [CrossRef]
2. Durán-Lara, E.F.; Valderrama, A.; Marican, A. Natural Organic Compounds for Application in Organic Farming. *Agriculture* **2020**, *10*, 41. [CrossRef]
3. Suteu, D.; Rusu, L.; Zaharia, C.; Badeanu, M.; Daraban, G.M. Challenge of Utilization Vegetal Extracts as Natural Plant Protection Products. *Appl. Sci.* **2020**, *10*, 8913. [CrossRef]
4. Cenobio-Galindo, A.d.J.; Hernández-Fuentes, A.D.; González-Lemus, U.; Zaldívar-Ortega, A.K.; González-Montiel, L.; Madariaga-Navarrete, A.; Hernández-Soto, I. Biofungicides Based on Plant Extracts: On the Road to Organic Farming. *Int. J. Mol. Sci.* **2024**, *25*, 6879. [CrossRef]

5. Ravlić, M.; Baličević, R.; Vinković, Ž.; Brozović, B.; Sarajić, A.; Kranjac, D. The Allelopathic Potential of Ruderal Plant Species on Tomato and Lettuce. *Poljoprivreda* **2024**, *30*, 3–9. [CrossRef]
6. Povero, G.; Mejia, J.F.; Di Tommaso, D.; Piaggesi, A.; Warrior, P. A Systematic Approach to Discover and Characterize Natural Plant Biostimulants. *Front. Plant Sci.* **2016**, *7*, 435. [CrossRef]
7. Perić, K.; Čupić, T.; Krizmanić, G.; Tokić, B.; Andrić, L.; Ravlić, M.; Meglič, V.; Tucak, M. The Role of Crop Wild Relatives and Landraces of Forage Legumes in Pre-Breeding as a Response to Climate Change. *Agronomy* **2024**, *14*, 1385. [CrossRef]
8. Ravlić, M.; Baličević, R.; Lisjak, M.; Vinković, Ž.; Ravlić, J.; Županić, A.; Svitlica, B. Allelopathic Effect of *Salvia pratensis* L. on Germination and Growth of Crops. *Crops* **2025**, *5*, 45. [CrossRef]
9. Godlewska, K.; Ronga, D.; Michalak, I. Plant Extracts—Importance in Sustainable Agriculture. *Ital. J. Agron.* **2021**, *16*, 1851. [CrossRef]
10. Parađiković, N.; Vinković, T.; Vinković Vrček, I.; Žuntar, I.; Bojić, M.; Medić-Šarić, M. Effect of Natural Biostimulants on Yield and Nutritional Quality: An Example of Sweet Yellow Pepper (*Capsicum annuum* L.) Plants. *J. Sci. Food Agric.* **2011**, *91*, 2146–2152. [CrossRef]
11. Brozović, B.; Jug, I.; Jug, D.; Stipešević, B.; Ravlić, M.; Đurđević, B. Biochar and Fertilization Effects on Weed Incidence in Winter Wheat. *Agronomy* **2021**, *11*, 2028. [CrossRef]
12. Han, M.; Kasim, S.; Yang, Z.; Deng, X.; Saidi, N.; Uddin, M.; Shuib, E. Plant Extracts as Biostimulant Agents: A Promising Strategy for Managing Environmental Stress in Sustainable Agriculture. *Phyton* **2024**, *93*, 2149. [CrossRef]
13. Szparaga, A. From biostimulant to possible plant bioprotectant agents. *Agric. Eng.* **2023**, *27*, 87–98. [CrossRef]
14. Ali, O.; Ramsubhag, A.; Jayaraman, J. Biostimulant Properties of Seaweed Extracts in Plants: Implications towards Sustainable Crop Production. *Plants* **2021**, *10*, 531. [CrossRef] [PubMed]
15. Akram, N.A.; Saleem, M.H.; Shafiq, S.; Naz, H.; Farid-ul-Haq, M.; Ali, B.; Shafiq, F.; Iqbal, M.; Jaremko, M.; Qureshi, K.A. Phytoextracts as Crop Biostimulants and Natural Protective Agents—A Critical Review. *Sustainability* **2022**, *14*, 14498. [CrossRef]
16. Iljkić, D.; Vuković, M.; Dvojković, K.; Horvat, D.; Szpunar-Krok, E.; Jańczak-Pieniążek, M.; Rastija, M. Variety, Chemical Protection and Biostimulator Effect on Winter Wheat Status. *Poljoprivreda* **2024**, *30*, 28–35. [CrossRef]
17. Gluhić, D. Primjena biostimulatora na bazi aminokiselina u poljoprivrednoj proizvodnji. *Glas. Zašt. Bilja* **2020**, *43*, 38–46. [CrossRef]
18. Varga, I.; Kristić, M.; Lisjak, M.; Tkalec Kojić, M.; Iljkić, D.; Jović, J.; Kristek, S.; Markulj Kulundžić, A.; Antunović, M. Antioxidative Response and Phenolic Content of Young Industrial Hemp Leaves at Different Light and Mycorrhiza. *Plants* **2024**, *13*, 840. [CrossRef]
19. Saa, S.; Olivos-Del Rio, A.; Castro, S.; Brown, P.H. Foliar Application of Microbial and Plant Based Biostimulants Increases Growth and Potassium Uptake in Almond (*Prunus dulcis* [Mill.] D.A. Webb). *Front. Plant Sci.* **2015**, *6*, 87. [CrossRef]
20. Ravlić, M.; Markulj Kulundžić, A.; Baličević, R.; Marković, M.; Viljevac Vuletić, M.; Kranjac, D.; Sarajlić, A. Allelopathic Potential of Sunflower Genotypes at Different Growth Stages on Lettuce. *Appl. Sci.* **2022**, *12*, 12568. [CrossRef]
21. Ali, Q.; Shehzad, F.; Waseem, M.; Shahid, S.; Hussain, A.I.; Haider, M.Z.; Habib, N.; Hussain, S.M.; Javed, M.T.; Perveen, R. Plant-Based Biostimulants and Plant Stress Responses. In *Plant Ecophysiology and Adaptation Under Climate Change: Mechanisms and Perspectives I*; Hasanuzzaman, M., Ed.; Springer: Singapore, 2020. [CrossRef]
22. Graziani, G.; Cirillo, A.; Giannini, P.; Conti, S.; El-Nakhel, C.; Rouphael, Y.; Ritieni, A.; Di Vaio, C. Biostimulants Improve Plant Growth and Bioactive Compounds of Young Olive Trees under Abiotic Stress Conditions. *Agriculture* **2022**, *12*, 227. [CrossRef]
23. Baličević, R.; Ravlić, M.; Lucić, K.; Tatarević, M.; Lucić, P.; Marković, M. Allelopathic Effect of *Aloe vera* (L.) Burm. F. on Seed Germination and Seedlings Growth of Cereals, Industrial Crops and Vegetables. *Poljoprivreda* **2018**, *24*, 13–19. [CrossRef]
24. Varga, I.; Kraus, I.; Iljkić, D.; Jonjić, A.; Antunović, M. Tradicija proizvodnje industrijske konoplje u Hrvatskoj. *Sjemenarstvo* **2022**, *33*, 25–40. [CrossRef]
25. Visković, J.; Dunđerski, D.; Adamović, B.; Jaćimović, G.; Latković, D.; Vojnović, Đ. Toward an Environmentally Friendly Future: An Overview of Biofuels from Corn and Potential Alternatives in Hemp and Cucurbits. *Agronomy* **2024**, *14*, 1195. [CrossRef]
26. Klir, Ž.; Novoselec, J.; Antunović, Z. An Overview on the Use of Hemp (*Cannabis sativa* L.) in Animal Nutrition. *Poljoprivreda* **2019**, *25*, 52–61. [CrossRef]
27. Matassa, S.; Esposito, G.; Pirozzi, F.; Papirio, S. Exploring the Biomethane Potential of Different Industrial Hemp (*Cannabis sativa* L.) Biomass Residues. *Energies* **2020**, *13*, 3361. [CrossRef]
28. Varga, I.; Markulj Kulundžić, A.; Krolo, P.; Iljkić, D.; Tišma, M.; Kraus, I. Industrial Hemp Finola Variety Photosynthetic, Morphometric, Biomechanical, and Yield Responses to K Fertilization Across Different Growth Stages. *Agronomy* **2025**, *15*, 496. [CrossRef]
29. Varga, I.; Iljkić, D.; Krolo, P.; Perić Fekete, A.; Kraus, I. The Source of K Fertilizer for Industrial Hemp (*Cannabis sativa* L.): Mechanical and Chemical Properties of Stem for Rammed Earth Walls. *Agriculture* **2021**, *11*, 2196. [CrossRef]
30. Ranogajec, L.; Antunović, M.; Stipešević, B.; Varga, I. Stanje i potencijal proizvodnje industrijske konoplje u Hrvatskoj na osnovi SWOT analize. *Poljoprivreda* **2024**, *30*, 56–63. [CrossRef]

31. Murkovic, M.; Hillebrand, A.; Winkler, J.; Leitner, E.; Pfannhauser, W. Variability of fatty acid content in pumpkin seeds (*Cucurbita pepo* L.). *Z. Lebensm.-Unters. Forsch.* **1996**, *203*, 216–219. [CrossRef]
32. Sinkovič, L.; Verbič, J.; Kolmanič, A. Agronomical Traits of Oil Seed Pumpkin Cultivars (*Cucurbita pepo* subsp. *pepo*) and Nutritional Characteristics of Seeds, Oil Cakes and Pumpkin Oils. In *ISHS Acta Horticulturae 1326, Proceedings of the VII South-Eastern Europe Symposium on Vegetables and Potatoes, Maribor, Slovenia, June 20–23 2017*; International Society for Horticultural Science: Korbeek-Lo, Belgium, 2017; Volume 1326, pp. 215–222. [CrossRef]
33. Klir, Z.; Castro-Montoya, J.M.; Novoselec, J.; Molkentin, J.; Domacinovic, M.; Mioc, B.; Antunovic, Z. Influence of Pumpkin Seed Cake and Extruded Linseed on Milk Production and Milk Fatty Acid Profile in Alpine Goats. *Animals* **2017**, *11*, 1772–1778. [CrossRef]
34. Kaur, B.; Garcha, K.S.; Bhatia, D.; Khosa, J.S.; Sharma, M.; Mittal, A.; Dhatt, A.S. Identification of Single Major QTL and Candidate Gene(s) Governing Hull-Less Seed Trait in Pumpkin. *Front. Plant Sci.* **2022**, *13*, 948106. [CrossRef]
35. Křístková, E.; Lebeda, A.; Sedláková, B. Species Spectra, Distribution and Host Range of Cucurbit Powdery Mildews in the Czech Republic, and in Some Other European and Middle Eastern Countries. *Phytoparasitica* **2009**, *37*, 337–350. [CrossRef]
36. Jakop, M.; Grobelnik Mlakar, S.; Bavec, M.; Robačer, M.; Vukmanič, T.; Lisec, U.; Bavec, F. Yield Performance and Agronomic Efficiency in Oil Pumpkins (*Cucurbita pepo* L. Group *Pepo*) Depending on Production Systems and Varieties. *Agricultura* **2017**, *14*, 25–36. [CrossRef]
37. Pinke, G.; Karácsony, P.; Czúcz, B.; Botta-Dukát, Z. When Herbicides Don't Really Matter: Weed Species Composition of Oil Pumpkin (*Cucurbita pepo* L.) Fields in Hungary. *Crop Prot.* **2018**, *110*, 236–244. [CrossRef]
38. Godina Golija, M. Production and Consumption of Pumpkin Seed Oil in Goričko: Traditional and Contemporary Practices. *Traditiones* **2019**, *48*, 117–135. [CrossRef]
39. Tomić, D.; Marjanović, M.; Radovanović, M.; Đurović, V.; Lazarević, Đ.; Stevović, V.; Pavlović, N. Preliminary Report: Interdependence of Seed Yield Components of Pumpkin (*Cucurbita pepo* L.) Genotypes. In *AgroReS 2022, Proceedings of the XI International Symposium on Agricultural Sciences, Trebinje, Bosnia and Herzegovina, 26–28 May 2022*; University of Banja Luka: Banja Luka, Bosnia and Herzegovina, 2022; p. 31.
40. Tańska, M.; Ogrodowska, D.; Bartoszewski, G.; Korzeniewska, A.; Konopka, I. Seed Lipid Composition of New Hybrids of Styrian Oil Pumpkin Grown in Poland. *Agronomy* **2020**, *10*, 1104. [CrossRef]
41. Csipkés, M. The Importance of Pumpkins and Oil Gourds in Romania. *Ann. Univ. Oradea Econ. Sci. Ser.* **2023**, *32*, 13. [CrossRef]
42. Lelley, T.; Loy, B.; Murkovic, M. Hull-Less Oil Seed Pumpkin. In *Oil Crops*; Vollmann, J., Rajcan, I., Eds.; Springer: New York, NY, USA, 2009; pp. 469–492. [CrossRef]
43. Boye, J.; Brault, D.; Couture, I.; Estevez, B.; Grenier, M.; Lefrançois, E.; Leblanc, M.; Lefebvre, M.; Moreau, G.; Ribéreau, S. Hulless Pumpkin Seed: A New Crop for Organic Production in Québec. Available online: https://irda.qc.ca/media/id5pdbxs/irda-ullesspumpkinseednewcroporganicproductioncultivars-fichessyntheses-2014.pdf (accessed on 4 August 2025).
44. Meru, G.; Fu, Y.; Shrestha, S.; Michael, V.N.; Dorval, M.; Mainviel, R. Genomic Position and Markers Associated with the Hull-Less Seed Trait in Pumpkin. *Plants* **2022**, *11*, 1238. [CrossRef]
45. Wargala, E.; Chrzanowska, A.; Bernatek-Samoraj, W.; Kot, I. Pumpkin (*Cucurbita pepo* L.) Seed Oil—Cosmetic, Food and Medical Raw Material. *Herba Pol.* **2023**, *69*, 7–14. [CrossRef]
46. Salehi, B.; Capanoglu, E.; Adrar, N.; Catalkaya, G.; Shaheen, S.; Jaffer, M.; Giri, L.; Suyal, R.; Jugran, A.K.; Calina, D.; et al. *Cucurbits* Plants: A Key Emphasis to Its Pharmacological Potential. *Molecules* **2019**, *24*, 1854. [CrossRef] [PubMed]
47. Almeida, A.; Dong, L.; Thorsen, T.H.; Raadam, M.H.; Khakimov, B.; Carreño-Quintero, N.; Madsen, C.K.; Borkhardt, B.; Larsen, F.H.; Motawia, M.S.; et al. Metabolic Engineering of Cucurbitacins in *Cucurbita pepo* Hairy Roots. *Front. Plant Sci.* **2022**, *13*, 1021907. [CrossRef] [PubMed]
48. Šerá, B.; Hnilička, F. Genetic and Environmental Factors Affecting Seed Germination. *Plants* **2023**, *12*, 4106. [CrossRef]
49. Hirano, H.; Watanabe, T.; Fukuda, M.; Fukao, T. The Impact of Carbohydrate Management on Coleoptile Elongation in Anaerobically Germinating Seeds of Rice (*Oryza sativa* L.) under Light and Dark Cycles. *Plants* **2023**, *12*, 1565. [CrossRef]
50. Shah, S.S.H.; Latif, S.; Qureshi, R.; Ilyas, N.; Ahmad, M.S.; Rehman, S.; Khan, N.; Abdel-Maksoud, M.A.; El-Tayeb, M.A.; Saleh, I.A.; et al. Optimizing germination dynamics in seven key industrial and medicinal hemp varieties through seed priming techniques: An initial study for hemp cultivation in Pakistan. *Ind. Crops Prod.* **2024**, *222*, 119739. [CrossRef]
51. Lisjak, M.; Ocvirk, D.; Špoljarević, M.; Teklić, T.; Liović, I.; Špoljarić Marković, S.; Volenik, M.; Mijić, A. Učinak primiranja sjemena sumporovodikom na klijanje i biokemijske pokazatelje sušnog stresa kod klijanaca suncokreta. *Poljoprivreda* **2025**, *31*, 1–12. [CrossRef]
52. Jović, J.; Ivezić, V.; Popović, B.; Guberac, V.; Rastija, M.; Dujmović, L.; Orkić, V.; Kristek, S. Utjecaj primjene mikrobioloških preparata na odabrana svojstva kiselih tala. *Poljoprivreda* **2025**, *31*, 41–50. [CrossRef]
53. Song, J.; Wang, H.; Chu, R.; Zhao, L.; Li, X.; An, S.; Qiang, M.; Du, W.; Li, Q. Differences in Physiological Characteristics, Seed Germination, and Seedling Establishment in Response to Salt Stress between Dimorphic Seeds in the Halophyte *Suaeda liaotungensis*. *Plants* **2023**, *12*, 1408. [CrossRef]

54. Orkić, V.; Grubišić–Šestanj, S.; Ravnjak, B.; Rebekić, A.; Petrović, S.; Vila, S.; Kujundžić, S. Priježetveno Proklijavanje Kod Različitih Kultivara Pšenice. *Poljoprivreda* **2024**, *30*, 21–27. [CrossRef]
55. Wazeer, H.; Shridhar Gaonkar, S.; Doria, E.; Pagano, A.; Balestrazzi, A.; Macovei, A. Plant-Based Biostimulants for Seeds in the Context of Circular Economy and Sustainability. *Plants* **2024**, *13*, 1004. [CrossRef]
56. Petrović, E.; Vrandečić, K.; Ćosić, J.; Godena, S. Chemical Control of Olive Fungal Diseases: Strategies and Risks. *Poljoprivreda* **2024**, *30*, 44–53. [CrossRef]
57. Pugelnik, I.; Rebekić, A.; Jelić Milković, S.; Lončarić, R. Stavovi Ekoloških Proizvođača o Ekološkoj Poljoprivredi u Republici Hrvatskoj. *Poljoprivreda* **2024**, *30*, 91–99. [CrossRef]
58. Skowronek, P.; Wójcik, Ł.; Strachecka, A. Impressive Impact of Hemp Extract on Antioxidant System in Honey Bee (*Apis mellifera*) Organism. *Antioxidants* **2022**, *11*, 707. [CrossRef] [PubMed]
59. Jolayemi, O.L.; Malik, A.H.; Ekblad, T.; Fredlund, K.; Olsson, M.E.; Johansson, E. Protein-Based Biostimulants to Enhance Plant Growth—State-of-the-Art and Future Direction with Sugar Beet as an Example. *Agronomy* **2022**, *12*, 3211. [CrossRef]
60. Kovačić, Đ.; Radočaj, D.; Samac, D.; Jurišić, M. Influence of Thermal Pretreatment on Lignin Destabilization in Harvest Residues: An Ensemble Machine Learning Approach. *AgriEngineering* **2024**, *6*, 171–184. [CrossRef]
61. Ravlić, M.; Baličević, R.; Svalina, T.; Posavac, D.; Ravlić, J. Herbicidal Potential of Meadow Sage (*Salvia pratensis* L.) against Velvetleaf (*Abutilon theophrasti* Med.) and Common Corn-Cockle (*Agrostemma githago* L.). *Glas. Zaštite Bilja* **2023**, *67*, 116–121. [CrossRef]
62. Erhatić, R.; Horvat, D.; Zorić, Z.; Repajić, M.; Jović, T.; Herceg, M.; Habuš, M.; Srečec, S. Aqueous Extracts of Four Medicinal Plants and Their Allelopathic Effects on Germination and Seedlings: Their Morphometric Characteristics of Three Horticultural Plant Species. *Appl. Sci.* **2023**, *13*, 2258. [CrossRef]
63. Janusauskaite, D. The Allelopathic Activity of Aqueous Extracts of *Helianthus annuus* L., Grown in Boreal Conditions, on Germination, Development, and Physiological Indices of *Pisum sativum* L. *Plants* **2023**, *12*, 1920. [CrossRef]
64. Spano, M.; Di Matteo, G.; Ingallina, C.; Sobolev, A.P.; Giusti, A.M.; Vinci, G.; Cammarone, S.; Tortora, C.; Lamelza, L.; Prencipe, S.A.; et al. Industrial Hemp (*Cannabis sativa* L.) Inflorescences as Novel Food: The Effect of Different Agronomical Practices on Chemical Profile. *Foods* **2022**, *11*, 3658. [CrossRef]
65. Visković, J.; Sikora, V.; Latković, D.; Zeremski, T.; Dunđerski, D.; Astatkie, T.; Noller, J.S.; Zheljazkov, V.D. Optimization of Hemp Production Technology for Fiber and Seed. *Ind. Crops Prod.* **2024**, *219*, 119127. [CrossRef]
66. Habschied, K.; Jokić, S.; Aladić, K.; Šplajt, I.; Krstanović, V.; Mastanjević, K. Addition of Industrial Hemp (*Cannabis sativa* L.) Dry Inflorescence in Beer Production. *Appl. Sci.* **2025**, *15*, 624. [CrossRef]
67. Bavec, F.; Gril, L.; Grobelnik-Mlakar, S.; Bavec, M. Seedlings of oil pumpkins as an alternative to seed sowing: Yield and production costs. *Bodenkult.-Wien Munch.* **2002**, *53*, 39–44.
68. du Jardin, P. Plant Biostimulants: Definition, Concept, Main Categories and Regulation. *Sci. Hortic.* **2015**, *196*, 3–14. [CrossRef]
69. Islam, M.M.; Rengel, Z.; Storer, P.; Siddique, K.H.M.; Solaiman, Z.M. Industrial Hemp (*Cannabis sativa* L.) Varieties and Seed Pre-Treatments Affect Seed Germination and Early Growth of Seedlings. *Agronomy* **2022**, *12*, 6. [CrossRef]
70. Andrić, L. Ispitivanje vigora sjemena nekih domaćih kultivara soje (*Glycine max* (L.) Merrill). *Poljoprivreda* **2004**, *10*, 59–60.
71. Varga, I.; Iljkić, D.; Šimunović, M.; Agić, D.; Herman, G.; Antunović, M. Morphological parameters of oil-pumpkin sprouts in different water-solution pH values. *Sjemenarstvo* **2024**, *35*, 27–38. [CrossRef]
72. Lynch, J.; Marschner, P.; Rengel, Z. Effect of Internal and External Factors on Root Growth and Development. In *Marschner's Mineral Nutrition of Higher Plants*, 3rd ed.; Marschner, P., Ed.; Academic Press: London, UK, 2012; pp. 331–346. [CrossRef]
73. Francesca, S.; Arena, C.; Hay Mele, B.; Schettini, C.; Ambrosino, P.; Barone, A.; Rigano, M.M. The Use of a Plant-Based Biostimulant Improves Plant Performances and Fruit Quality in Tomato Plants Grown at Elevated Temperatures. *Agronomy* **2020**, *10*, 363. [CrossRef]
74. Mystkowska, I. The Effect of Biostimulants on the Chlorophyll Content and Height of *Solanum tuberosum* L. Plants. *J. Ecol. Eng.* **2022**, *23*, 72–77. [CrossRef]
75. Di Mola, I.; Ottaiano, L.; Cozzolino, E.; Senatore, M.; Giordano, M.; El-Nakhel, C.; Sacco, A.; Rouphael, Y.; Colla, G.; Mori, M. Plant-Based Biostimulants Influence the Agronomical, Physiological, and Qualitative Responses of Baby Rocket Leaves under Diverse Nitrogen Conditions. *Plants* **2019**, *8*, 522. [CrossRef]
76. Martínez-Lorente, S.E.; Martí-Guillén, J.M.; Pedreño, M.Á.; Almagro, L.; Sabater-Jara, A.B. Higher Plant-Derived Biostimulants: Mechanisms of Action and Their Role in Mitigating Plant Abiotic Stress. *Antioxidants* **2024**, *13*, 318. [CrossRef]
77. Park, Y.R.; Kwon, S.-J.; Kim, J.H.; Duan, S.; Eom, S.H. Light-Induced Antioxidant Phenolic Changes among the Sprouts of Lentil Cultivar. *Antioxidants* **2024**, *13*, 399. [CrossRef] [PubMed]
78. Kalinová, J.P.; Vrchotová, N.; Tříska, J. Flavonoids Profile in Pasta and Cookies Fortified with Common Buckwheat Sprouts or Microgreens Flour. *J. Cereal Sci.* **2025**, *121*, 104092. [CrossRef]
79. Posmyk, M.M.; Szafrańska, K. Biostimulators: A New Trend towards Solving an Old Problem. *Front. Plant Sci.* **2016**, *7*, 748. [CrossRef] [PubMed]

80. Cozzolino, E.; Di Mola, I.; Ottaiano, L.; El-Nakhel, C.; Rouphael, Y.; Mori, M. Foliar Application of Plant-Based Biostimulants Improves Yield and Upgrades Qualitative Characteristics of Processing Tomato. *Ital. J. Agron.* **2021**, *16*, 1825. [CrossRef]
81. Chrysargyris, A.; Charalambous, S.; Xylia, P.; Litskas, V.; Stavrinides, M.; Tzortzakis, N. Assessing the Biostimulant Effects of a Novel Plant-Based Formulation on Tomato Crop. *Sustainability* **2020**, *12*, 8432. [CrossRef]
82. Teacă, C.-A.; Bodîrlău, R.; Oprea, A.; Tănase, C.; Colceru, S. Influence of Plant Extracts on Germination and Postgermination Development of Different Species. *Cellulose Chem. Technol.* **2008**, *42*, 47–53.
83. Mutlu-Durak, H.; Yildiz Kutman, B. Seed Treatment with Biostimulants Extracted from Weeping Willow (*Salix babylonica*) Enhances Early Maize Growth. *Plants* **2021**, *10*, 1449. [CrossRef]
84. Yadav, A.; Singh, S.; Yadav, V. Screening Herbal Extracts as Biostimulant to Increase Germination, Plant Growth and Secondary Metabolite Production in Wheatgrass. *Sci. Rep.* **2024**, *14*, 607. [CrossRef]
85. Norsworthy, J.K. Allelopathic potential of wild radish (*Raphanus raphanistrum*). *Weed Technol.* **2003**, *17*, 307–313. [CrossRef]
86. Croatian Plant Genetic Resources Database, CPGRD Database. Available online: https://cpgrd.hapih.hr (accessed on 11 April 2025).
87. Holm, G. Chlorophyll mutations in barley. *Acta Agric. Scand.* **1954**, *4*, 457–461. [CrossRef]
88. Wettstein, D. Chlorophyll–letale und der submikroskopische Formwechsel der Plastiden. *Exp. Cell Res.* **1957**, *12*, 427–487. [CrossRef]
89. Singleton, V.L.; Rossi, J.A. Colorimetry of Total Phenolics with Phosphomolybdic-Phosphotungstic Acid Reagent. *AJEV* **1965**, *16*, 144–158. [CrossRef]
90. Keutgen, A.J.; Pawelzik, E. Quality and nutritional value of strawberry fruit under long term salt stress. *Food Chem.* **2008**, *107*, 1413–1420. [CrossRef]
91. Available online: https://www.chiplot.online/correlation_heatmap.html (accessed on 30 July 2025).

plants

MDPI

Article

Effect of Foliar Biostimulant Application on Bioactive Compounds and Antioxidant Capacity in Blueberry (*Vaccinium corymbosum* L.)

Tiago Lopes [1,2,3,4,*], Ana Paula Silva [1,2,5], Alfredo Aires [1,2,5], Rosa Carvalho [5], Maria Ferreira [5], António A. Vicente [3,4] and Berta Gonçalves [1,2,6]

1 Centre for the Research and Technology of Agro-Environmental and Biological Sciences (CITAB), University of Trás-os-Montes e Alto Douro (UTAD), 5000-801 Vila Real, Portugal; asilva@utad.pt (A.P.S.); alfredoa@utad.pt (A.A.); bertag@utad.pt (B.G.)
2 Institute for Innovation, Capacity Building and Sustainability of Agri-Food Production (Inov4Agro), University of Trás-os-Montes e Alto Douro (UTAD), 5000-801 Vila Real, Portugal
3 Centre of Biological Engineering (CEB), University of Minho (UM), Campus de Gualtar, 4710-057 Braga, Portugal; avicente@deb.uminho.pt
4 LABBELS—Associate Laboratory, 4710-057 Braga, Portugal
5 Department of Agronomy, University of Trás-os-Montes e Alto Douro (UTAD), 5000-801 Vila Real, Portugal; rpaula@utad.pt (R.C.); al66431@alunos.utad.pt (M.F.)
6 Department of Biology and Environment, University of Trás-os-Montes e Alto Douro (UTAD), 5000-801 Vila Real, Portugal
* Correspondence: tiagolopes97@hotmail.com

Abstract

Blueberries are rich in bioactive compounds and antioxidants whose contents can be significantly affected by pre-harvest agronomic practices. Thus, using natural biostimulants like *Ecklonia maxima* (EM) extract and glycine betaine (GB) is a promising strategy to improve blueberry quality. The effect of foliar pre-harvest application of EM and GB on bioactive compounds and antioxidant capacity (AC) of 'Duke' and 'Draper' blueberries was investigated in 2022 and 2023. Blueberries treated with GB had higher polyphenol contents and AC by the $ABTS^{\bullet+}$ and CUPRAC methods, particularly at higher doses. Vitamin C in 'Duke' was decreased by both doses of GB and low-dose EM + GB by 10–30% over two years, whereas in 'Draper', both doses of GB increased vitamin C in 2023 by 40–80%. High GB and EM + GB doses improved the AC of 'Duke' blueberries by 7–12% (CUPRAC) in both seasons. In 2022, the high dose of EM increased the levels of polyphenols in both cultivars by 14–21% and AC by $ABTS^{\bullet+}$ and CUPRAC in cv. 'Draper'. As a result, blueberries' nutritional value and AC can be enhanced by biostimulants, whose application must be adjusted according to cultivar and dose to optimize their use.

Keywords: glycine betaine; *Ecklonia maxima*; phenolic compounds; *ortho*-diphenols; flavonoids; anthocyanins

1. Introduction

Over the last decade, blueberries have become more popular mostly because of their excellent sensory attributes and beneficial health effects, including anti-inflammatory, antioxidant, neuroprotective, vision-enhancing, and anticancer properties, linked to their composition in bioactive compounds [1–4]. However, the qualitative parameters of blueberries depend on several factors, such as growing location, environmental conditions, cultivar, and fertilization strategy, among others [5–8].

Because of their complex chemical profile, blueberries are naturally rich in a variety of phenolic compounds, including flavonoids (e.g., anthocyanins, which are responsible for the characteristic pigmentation), *ortho*-diphenols, and other non-flavonoids, as well as vitamins and organic acids [3,4]. The antioxidant capacity (AC) of the fruit, which is an important factor determining its ability to scavenge free radicals, is mainly attributed to these compounds. Thus, elevated levels of these phytochemicals are directly responsible for the nutritional and functional improvements.

Regarding fertilization practices, the idea of using biostimulants to enhance tolerance to abiotic stresses, improve plant nutritional efficiency, and/or improve crop quality has been explored in recent decades across different plant species. Among these, biostimulants derived from seaweed extracts and glycine betaine (GB) have been used in several studies [9]. Previous work has reported that seaweed extracts may be useful for mitigating abiotic stresses, such as drought in blueberries [10]. Studies have also shown that seaweed-based biostimulants improved the total amount of phenolic compounds, anthocyanins, and antioxidant capacity (AC) of 'Bluecrop' blueberries [7]. Koort et al. [11] also reported an increase in the anthocyanin content of the 'Northblue' cultivar when treated with brown seaweed extracts, compared with mineral fertilization. Brown seaweeds, particularly from *Ecklonia maxima* (EM) species, contain several plant hormones, brassinosteroids, betaines, polyamines, phlorotannins, and alginate oligosaccharides, which can be responsible for the positive effects observed on bioactive compound contents and plant adaptation to environmental stresses [12]. On the other hand, GB, a non-toxic and highly water-soluble compound, has a fundamental role in osmotic adjustment while also protecting the photosynthetic apparatus, specifically photosystem II (PSII), from abiotic stress. GB stabilizes membrane integrity and various complex protein structures while preventing the accumulation of excess reactive oxygen species (ROS) [13]. The increase in yield and quality of different fruit crops was recently observed after the exogenous application of this compound [14–16]. In addition, the foliar application of GB has contributed to reducing enzyme activity crucial for energy metabolism and increasing the AC of 'Brightwell' blueberries, thus helping to maintain their post-harvest quality [17]. Recently, Lopes et al. [16] demonstrated that pre-harvest foliar application of EM and GB can enhance fruit size, firmness, and organic acid levels in blueberries of the 'Duke' and 'Draper' cultivars.

Although these positive effects have been observed on the physical and sensory attributes of blueberries, studies on the effects of foliar applications of EM- and GB-based biostimulants on phytochemical quality remain underexplored. Considering that the efficacy of biostimulants is influenced by cultivar, application method, and growing season, the aim of this research was to investigate whether the bioactive compound content and AC of blueberries would be significantly enhanced across different years by repeated foliar applications of different biostimulants, and whether the effects were synergistic, cultivar-dependent, and stable across different growing seasons. Therefore, the objective of this study was to evaluate the effect of pre-harvest foliar applications of EM- and GB-based biostimulants on the bioactive compounds and AC of 'Duke' and 'Draper' blueberry cultivars.

2. Results and Discussion

2.1. Total Phenolic Compounds, Flavonoids, and Ortho-Diphenols

The contents of total phenolic compounds (TPC), flavonoids, and *ortho*-diphenols are shown in Figure 1. In both years, the foliar treatment ($0.001 < p < 0.01$), cultivar ($p < 0.001$), and the interaction treatment × cultivar ($p < 0.001$) significantly affected the TPC. GB-treated (T4) 'Duke' blueberries had significantly 22% higher TPC in 2022 ($p < 0.001$) and 33% in 2023 ($p < 0.01$) compared with the control. In 2022, 4 L ha^{-1} EM (T1) treatment

https://doi.org/10.3390/plants15010092

significantly increased ($p < 0.001$) the TPC of 'Duke' fruits by 21%. For cv. 'Draper', the application of 4 kg ha^{-1} GB (T3) also increased ($p < 0.01$) TPC by 25%, compared with T0. Flavonoids were affected by treatment ($p < 0.01$) and cultivar ($p < 0.001$) in both years. However, the pattern was less consistent than for TPC, since the interaction treatment × cultivar was only significant in 2022 ($p < 0.05$). In that season, the use of EM + GB (T6) decreased ($p < 0.001$) the flavonoid content of 'Draper' blueberries by 16%, and in 2023, the same treatment also reduced ($p < 0.05$) flavonoids of 'Duke' fruits by 13%. This indicates a possible cultivar-specific sensitivity to combined treatments. The concentration of *ortho*-diphenols was affected by treatment ($p < 0.001$), cultivar ($p < 0.001$), and the interaction between these factors ($p < 0.05$) in the second year. For cv. 'Draper', the use of T6 also decreased the *ortho*-diphenol content of blueberries by 13% in 2022.

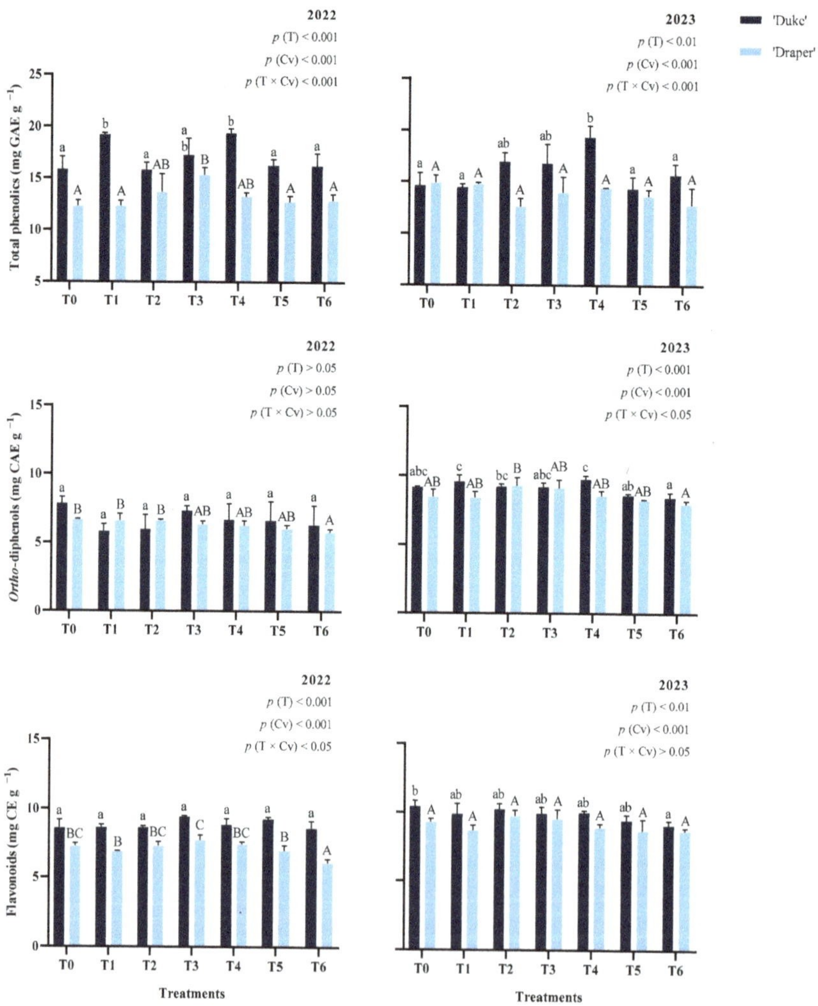

Figure 1. Total phenolics, flavonoids, and *ortho*-diphenols contents of fruits from 'Duke' and 'Draper' blueberry cultivars (Cv) depending on the treatments (T) used in the years 2022 and 2023. Lowercase letters indicate significant differences ($p < 0.05$) between the 'Duke' treatments and uppercase letters indicate significant differences ($p < 0.05$) between the 'Draper' treatments by Tukey's test; means ($n = 3$) ± SD.

https://doi.org/10.3390/plants15010092

Interestingly, across both cultivars, flavonoid and *ortho*-diphenol contents were higher ($p < 0.001$) in 2023 than in the previous year, which can be explained by the higher average temperatures and solar radiation reported during the months of fruit development. The interaction between treatment, cultivar, and year was also significant for TPC ($p < 0.001$) and flavonoids ($p < 0.05$) (Table A1, Appendix A).

Previous research has shown that foliar spraying of biostimulants improves the levels of bioactive compounds in fruits, although the effect depends on cultivar and edapho-climatic conditions. The increase in bioactive compounds after the use of biostimulants is consistent with studies in sweet cherries, grapes, and blueberries [10,14,15,18,19]. The higher concentration of phenolic compounds enhances plant tolerance against several biotic and abiotic stresses [20,21]. Indeed, these compounds can modulate gene expression related to the plant response to stress through the shikimate pathway, using phenylalanine and shikimic acid as precursors [22].

At the physiological level, the increase in TPC observed in the present work after GB treatment may be related to the stimulation of phenylpropanoid metabolism and the increase in associated enzyme activity [23]. Furthermore, GB can help prevent the oxidative degradation of phenolic compounds and the accumulation of ROS by improving enzymatic and non-enzymatic antioxidant systems [24]. As observed in this work, EM application can influence the levels of various phenolic compounds. These extracts contain several bioactive compounds, including phlorotannins, which can enhance phenolic synthesis by affecting key enzymes such as phenylalanine ammonia-lyase, peroxidase, and polyphenol oxidase [12,25]. In addition, an earlier study established that seaweed-based biostimulants can enhance the activity of genes involved in phenolic pathways that have revealed the potential to stimulate the biosynthesis of flavones [26].

Nonetheless, the accumulation of TPC and related compounds in treated plants can be affected by multiple factors. In a study on strawberries under water-stress conditions, Kapur et al. [27] found that some phenolic compounds were reduced while others increased following biostimulant treatment, reflecting the high complexity of the factors influencing these compounds. The increase in TPC observed in the present study is supported by previous findings on the effects of seaweed- and GB-based biostimulants in blueberries and other fruits like strawberries, sweet cherries, and grapes [7,15,18,19,28]. Post-harvest GB application has also improved TPC in blueberries [17]. On the other hand, decreased flavonoid and *ortho*-diphenol content in 'Draper' fruits treated with T6 in 2022 and in 'Duke' fruits in 2023 may suggest a cultivar-specific sensitivity or the stress-mitigating effect of the biostimulants. Based on this, although our results showed that foliar application of biostimulants generally enhanced TPC in blueberries, their effects on flavonoids and *ortho*-diphenols were more variable, which reflects a significant influence of cultivar, treatment, and environmental conditions.

2.2. *Individual Phenolic Compounds Identified by HPLC*

The individual polyphenol content, as determined by HPLC analysis, is presented in Tables 1 and 2. Blueberries are rich in anthocyanins, which are known for their important role in preventing chronic diseases and several other beneficial health effects [29,30]. The main anthocyanins identified in our study were delphinidin-3-*O*-galactoside, delphinidin-3-*O*-glucoside, delphinidin-3-*O*-arabinoside, petunidin-3-*O*-galactoside, malvidin-3-*O*-galactoside, and malvidin-3-*O*-arabinoside, while chlorogenic acid (phenolic acid), rutin, and hyperoside (flavonols) were also detected. Consistent with our work, Yang et al. [31] recorded that, among 17 blueberry cultivars studied, more than 90% of identified anthocyanins were malvidin, delphinidin, and petunidin glycosides.

https://doi.org/10.3390/plants15010092

Table 1. Average levels of individual polyphenols (μg g^{-1}) (Part 1) of fruits from 'Duke' and 'Draper' blueberry cultivars (Cv) depending on the treatments (T) used in the years 2022 and 2023.

Treatments	Years	Chlorogenic Acid 'Duke'	Chlorogenic Acid 'Draper'	Delphinidin-3-O-galactoside 'Duke'	Delphinidin-3-O-galactoside 'Draper'	Delphinidin-3-O-glucoside 'Duke'	Delphinidin-3-O-glucoside 'Draper'	Delphinidin-3-O-arabinoside 'Duke'	Delphinidin-3-O-arabinoside 'Draper'	Petunidin-3-O-galactoside 'Duke'	Petunidin-3-O-galactoside 'Draper'	Malvidin-3-O-galactoside 'Duke'	Malvidin-3-O-galactoside 'Draper'
T0	2022	312.92 ± 7.02 ^{ab}	305.97 ± 11.38 ^{AB}	1059.47 ± 115.62 ^{b}	625.71 ± 12.18 ^{A}	858.09 ± 99.31 ^{b}	513.18 ± 10.08 ^{A}	535.06 ± 61.89 ^{b}	378.26 ± 8.25 ^{A}	626.64 ± 68.75 ^{ab}	392.64 ± 8.99 ^{AB}	347.77 ± 40.20 ^{a}	247.45 ± 8.44 ^{A}
	2023	503.61 ± 5.82 ^{a}	547.27 ± 12.39 ^{A}	975.48 ± 29.61 ^{b}	702.77 ± 25.78 ^{A}	783.27 ± 11.44 ^{c}	599.50 ± 17.43 ^{A}	472.41 ± 22.41 ^{b}	341.57 ± 8.78 ^{ABC}	438.03 ± 22.69 ^{bc}	397.31 ± 16.55 ^{AB}	341.83 ± 9.16 ^{bc}	235.31 ± 9.63 ^{A}
T1	2022	356.32 ± 7.29 ^{bc}	376.13 ± 41.08 ^{BC}	889.01 ± 2.23 ^{ab}	737.12 ± 46.81 ^{ABC}	744.40 ± 0.80 ^{ab}	610.48 ± 39.47 ^{AB}	466.80 ± 1.77 ^{ab}	452.35 ± 25.07 ^{BC}	609.36 ± 15.60 ^{ab}	508.63 ± 28.68 ^{B}	311.82 ± 3.03 ^{a}	285.87 ± 10.62 ^{AB}
	2023	621.34 ± 16.37 ^{b}	519.59 ± 20.08 ^{A}	858.86 ± 56.40 ^{a}	711.13 ± 34.96 ^{A}	754.37 ± 56.73 ^{c}	594.92 ± 36.06 ^{A}	541.17 ± 54.43 ^{b}	310.60 ± 20.67 ^{AB}	489.93 ± 36.86 ^{cd}	398.49 ± 17.21 ^{AB}	338.99 ± 25.37 ^{bc}	237.84 ± 22.38 ^{A}
T2	2022	392.79 ± 27.31 ^{c}	422.60 ± 62.23 ^{C}	821.88 ± 10.34 ^{a}	658.44 ± 25.23 ^{AB}	686.90 ± 15.05 ^{a}	551.66 ± 25.27 ^{A}	404.83 ± 14.40 ^{a}	389.21 ± 12.90 ^{AB}	528.87 ± 7.48 ^{a}	461.57 ± 21.36 ^{ABC}	283.65 ± 6.03 ^{a}	255.53 ± 13.57 ^{A}
	2023	502.61 ± 61.38 ^{a}	567.43 ± 38.82 ^{A}	761.91 ± 10.39 ^{a}	696.90 ± 86.44 ^{A}	676.35 ± 15.58 ^{abc}	576.42 ± 83.04 ^{A}	473.44 ± 11.58 ^{b}	302.10 ± 39.44 ^{B}	508.06 ± 5.57 ^{cd}	335.00 ± 38.52 ^{A}	326.09 ± 13.27 ^{bc}	233.59 ± 37.51 ^{A}
T3	2022	369.32 ± 19.71 ^{c}	365.17 ± 29.54 ^{ABC}	927.79 ± 48.87 ^{ab}	809.15 ± 29.54 ^{C}	795.28 ± 45.99 ^{ab}	681.19 ± 24.39 ^{B}	487.04 ± 38.85 ^{ab}	481.78 ± 13.68 ^{C}	641.47 ± 32.06 ^{b}	505.40 ± 31.30 ^{C}	339.14 ± 23.47 ^{a}	326.26 ± 18.78 ^{B}
	2023	552.30 ± 24.23 ^{ab}	572.05 ± 14.70 ^{A}	866.86 ± 64.57 ^{ab}	780.92 ± 47.35 ^{A}	775.76 ± 46.69 ^{c}	695.70 ± 25.98 ^{A}	546.43 ± 44.70 ^{b}	419.47 ± 17.13 ^{C}	570.33 ± 41.16 ^{d}	501.40 ± 13.07 ^{C}	359.61 ± 12.42 ^{c}	270.92 ± 13.85 ^{A}
T4	2022	379.81 ± 17.87 ^{c}	333.11 ± 13.05 ^{ABC}	919.28 ± 79.90 ^{ab}	722.97 ± 93.80 ^{AB}	767.90 ± 71.56 ^{ab}	599.12 ± 90.69 ^{AB}	487.28 ± 40.82 ^{ab}	405.39 ± 47.94 ^{AB}	589.99 ± 45.26 ^{ab}	414.97 ± 56.57 ^{AB}	308.34 ± 25.95 ^{a}	285.04 ± 28.63 ^{AB}
	2023	547.03 ± 37.10 ^{ab}	565.86 ± 32.05 ^{A}	827.04 ± 38.39 ^{a}	681.96 ± 55.06 ^{A}	690.96 ± 72.44 ^{bc}	592.76 ± 61.64 ^{A}	441.86 ± 104.35 ^{b}	374.09 ± 48.56 ^{ABC}	434.37 ± 76.64 ^{bc}	438.91 ± 44.54 ^{BC}	298.65 ± 30.27 ^{b}	244.94 ± 34.23 ^{A}
T5	2022	395.68 ± 28.92 ^{c}	320.85 ± 19.08 ^{AB}	934.54 ± 39.06 ^{ab}	760.98 ± 35.24 ^{BC}	772.69 ± 33.34 ^{ab}	620.78 ± 23.33 ^{AB}	465.58 ± 27.42 ^{ab}	454.72 ± 29.06 ^{BC}	614.93 ± 15.67 ^{ab}	463.65 ± 24.71 ^{BC}	303.01 ± 21.81 ^{a}	287.15 ± 3.73 ^{AB}
	2023	514.08 ± 15.97 ^{a}	569.99 ± 47.21 ^{A}	772.80 ± 22.81 ^{a}	686.79 ± 62.30 ^{A}	592.51 ± 17.66 ^{ab}	608.93 ± 59.99 ^{A}	215.55 ± 43.07 ^{a}	389.57 ± 24.38 ^{BC}	355.38 ± 23.09 ^{ab}	427.22 ± 30.68 ^{BC}	227.15 ± 19.52 ^{a}	237.80 ± 22.69 ^{A}
T6	2022	291.00 ± 13.85 ^{a}	270.85 ± 34.93 ^{A}	853.34 ± 61.81 ^{a}	651.44 ± 26.92 ^{AB}	717.18 ± 48.28 ^{ab}	546.88 ± 21.94 ^{B}	437.34 ± 28.70 ^{ab}	404.83 ± 15.93 ^{AB}	599.71 ± 49.04 ^{ab}	376.11 ± 20.84 ^{A}	297.87 ± 26.15 ^{a}	271.48 ± 10.52 ^{A}
	2023	510.36 ± 28.57 ^{a}	506.65 ± 47.57 ^{A}	791.19 ± 26.50 ^{a}	716.84 ± 43.92 ^{A}	569.56 ± 7.34 ^{a}	627.69 ± 33.09 ^{A}	169.84 ± 2.93 ^{a}	357.80 ± 29.26 ^{ABC}	291.21 ± 8.98 ^{a}	444.47 ± 27.11 ^{BC}	209.88 ± 5.24 ^{a}	247.40 ± 20.97 ^{A}
p(T)	2022	<0.001		<0.01		<0.05		<0.001		<0.001		<0.001	
	2023	<0.05		<0.01		<0.001		<0.001		<0.001		<0.001	
p(Cv)	2022	>0.05		<0.001		<0.001		<0.001		<0.001		<0.001	
	2023	>0.05		<0.001		<0.001		<0.001		<0.05		<0.001	
p(T × Cv)	2022	<0.05		<0.001		<0.01		<0.01		<0.01		<0.05	
	2023	<0.01		<0.01		<0.001		<0.001		<0.001		<0.001	

Values are means ± SD (n = 3). Lowercase letters indicate significant differences ($p < 0.05$) between 'Duke' treatments and uppercase letters indicate significant differences ($p < 0.05$) between 'Draper' treatments by Tukey's test.

Table 2. Average levels of individual polyphenols (μg g^{-1}) (Part 2), including the sum of anthocyanins and the sum of polyphenols (mg g^{-1}) of fruits from 'Duke' and 'Draper' blueberry cultivars (Cv) depending on the treatments (T) used in the years 2022 and 2023.

Treatments	Years	Malvidin-3-O-arabinoside 'Duke'	Malvidin-3-O-arabinoside 'Draper'	Rutin 'Duke'	Rutin 'Draper'	Hyperoside 'Duke'	Hyperoside 'Draper'	Sum of Anthocyanins 'Duke'	Sum of Anthocyanins 'Draper'	Sum of Individual Polyphenols 'Duke'	Sum of Individual Polyphenols 'Draper'
T0	2022	458.56 ± 50.19 ^{b}	358.13 ± 4.24 ^{A}	334.50 ± 45.63 ^{b}	181.33 ± 4.66 ^{AB}	46.69 ± 8.02 ^{b}	26.21 ± 0.84 ^{A}	3.89 ± 0.43 ^{b}	2.52 ± 0.05 ^{A}	4.58 ± 0.48 ^{b}	3.03 ± 0.05 ^{A}
	2023	368.81 ± 4.06 ^{ab}	366.52 ± 7.77 ^{AB}	303.58 ± 13.80 ^{cd}	211.04 ± 11.16 ^{A}	51.69 ± 2.58 ^{c}	35.04 ± 2.51 ^{A}	3.38 ± 0.08 ^{c}	2.64 ± 0.08 ^{AB}	4.24 ± 0.08 ^{c}	3.44 ± 0.11 ^{AB}
T1	2022	455.32 ± 4.04 ^{b}	465.24 ± 17.07 ^{C}	277.36 ± 0.19 ^{ab}	191.69 ± 7.89 ^{ABC}	40.77 ± 0.26 ^{ab}	27.62 ± 1.70 ^{A}	3.48 ± 0.03 ^{ab}	3.06 ± 0.16 ^{BC}	4.15 ± 0.03 ^{ab}	3.66 ± 0.21 ^{B}
	2023	449.66 ± 42.17 ^{bc}	386.63 ± 20.45 ^{AB}	314.30 ± 24.92 ^{d}	223.24 ± 6.14 ^{A}	49.85 ± 5.49 ^{bc}	36.38 ± 1.26 ^{A}	3.44 ± 0.27 ^{c}	2.64 ± 0.15 ^{AB}	4.43 ± 0.31 ^{c}	3.42 ± 0.17 ^{AB}
T2	2022	381.63 ± 5.83 ^{a}	407.18 ± 21.93 ^{ABC}	262.13 ± 4.25 ^{a}	179.29 ± 11.06 ^{A}	35.69 ± 1.77 ^{a}	26.62 ± 1.70 ^{A}	3.10 ± 0.04 ^{a}	2.72 ± 0.12 ^{AB}	3.79 ± 0.05 ^{a}	3.35 ± 0.17 ^{AB}
	2023	445.03 ± 7.29 ^{bc}	313.27 ± 45.91 ^{A}	252.47 ± 2.37 ^{ab}	199.74 ± 23.20 ^{A}	40.61 ± 0.65 ^{ab}	32.99 ± 4.49 ^{A}	3.19 ± 0.06 ^{c}	2.45 ± 0.33 ^{A}	3.99 ± 0.11 ^{bc}	3.25 ± 0.39 ^{A}
T3	2022	478.61 ± 30.11 ^{b}	448.31 ± 22.55 ^{BC}	296.45 ± 13.57 ^{ab}	248.35 ± 10.20 ^{D}	40.88 ± 1.98 ^{ab}	36.09 ± 1.08 ^{B}	3.67 ± 0.22 ^{ab}	3.25 ± 0.14 ^{C}	4.38 ± 0.25 ^{ab}	3.90 ± 0.11 ^{B}
	2023	489.92 ± 45.05 ^{c}	490.00 ± 13.34 ^{C}	266.78 ± 20.12 ^{abc}	226.21 ± 11.80 ^{A}	43.77 ± 5.14 ^{abc}	38.56 ± 2.25 ^{A}	3.61 ± 0.25 ^{c}	3.16 ± 0.08 ^{B}	4.47 ± 0.30 ^{c}	4.00 ± 0.08 ^{B}
T4	2022	425.98 ± 29.23 ^{ab}	386.79 ± 46.08 ^{AB}	313.38 ± 24.39 ^{ab}	218.98 ± 18.69 ^{CD}	42.24 ± 4.59 ^{ab}	30.37 ± 3.08 ^{A}	3.50 ± 0.29 ^{ab}	2.81 ± 0.35 ^{ABC}	4.23 ± 0.33 ^{ab}	3.40 ± 0.38 ^{AB}
	2023	356.46 ± 49.79 ^{a}	396.14 ± 41.32 ^{AB}	283.69 ± 8.41 ^{bcd}	203.81 ± 21.38 ^{A}	46.19 ± 2.00 ^{bc}	33.21 ± 5.63 ^{A}	3.05 ± 0.36 ^{bc}	2.73 ± 0.28 ^{AB}	3.93 ± 0.37 ^{bc}	3.53 ± 0.34 ^{AB}
T5	2022	448.53 ± 17.92 ^{ab}	407.06 ± 18.41 ^{ABC}	302.33 ± 14.05 ^{ab}	211.18 ± 11.47 ^{BC}	40.71 ± 1.53 ^{ab}	29.40 ± 0.54 ^{A}	3.54 ± 0.15 ^{ab}	2.99 ± 0.13 ^{ABC}	4.28 ± 0.16 ^{ab}	3.56 ± 0.15 ^{AB}
	2023	416.40 ± 7.02 ^{abc}	379.47 ± 27.45 ^{AB}	255.46 ± 11.93 ^{ab}	201.86 ± 15.97 ^{A}	40.33 ± 1.21 ^{ab}	32.22 ± 4.24 ^{A}	2.58 ± 0.12 ^{ab}	2.73 ± 0.22 ^{AB}	3.39 ± 0.12 ^{ab}	3.53 ± 0.29 ^{AB}
T6	2022	430.81 ± 26.91 ^{ab}	359.18 ± 22.07 ^{A}	265.45 ± 20.06 ^{a}	181.23 ± 6.87 ^{AB}	35.01 ± 2.71 ^{a}	25.80 ± 1.56 ^{A}	3.34 ± 0.24 ^{ab}	2.61 ± 0.11 ^{AB}	3.93 ± 0.27 ^{ab}	3.09 ± 0.14 ^{AB}
	2023	403.93 ± 7.32 ^{ab}	409.25 ± 31.93 ^{BC}	226.16 ± 21.84 ^{a}	199.82 ± 11.49 ^{A}	35.13 ± 3.97 ^{a}	32.67 ± 1.16 ^{A}	2.44 ± 0.04 ^{a}	2.80 ± 0.18 ^{AB}	3.21 ± 0.09 ^{a}	3.54 ± 0.22 ^{AB}

https://doi.org/10.3390/plants15010092

Table 2. *Cont.*

Treatments	Years	Malvidin-3-*O*-arabinoside		Rutin		Hyperoside		Sum of Anthocyanins		Sum of Individual Polyphenols	
		'Duke'	'Draper'	'Duke'	'Draper'	'Duke'	'Draper'	'Duke'	'Draper'	'Duke'	'Draper'
p(T)	2022	<0.001		<0.001		<0.001		<0.01		<0.01	
	2023	<0.001		<0.001		<0.001		<0.001		<0.001	
p(Cv)	2022	<0.001		<0.001		<0.001		<0.001		<0.001	
	2023	<0.01		<0.001		<0.001		<0.001		<0.001	
p(T × Cv)	2022	<0.01		<0.01		<0.01		<0.01		<0.01	
	2023	<0.001		<0.01		<0.05		<0.001		<0.001	

Values are means ± SD ($n = 3$). Lowercase letters indicate significant differences ($p < 0.05$) between 'Duke' treatments and uppercase letters indicate significant differences ($p < 0.05$) between 'Draper' treatments by Tukey's test.

https://doi.org/10.3390/plants15010092

Results showed that both the sum of anthocyanins and the sum of individual polyphenols were affected by treatment ($p < 0.01$), cultivar ($p < 0.001$), and their interaction ($0.001 < p < 0.01$) during both years, as well as by the three-way interaction between treatment, cultivar, and year ($p < 0.001$) (Table A1 in Appendix A). Regarding pre-harvest foliar treatments, application of the low dose of EM (T2) decreased ($p < 0.01$) the sum of anthocyanins and individual polyphenols in 'Duke' blueberries by 20% and 17%, respectively, compared to the control (T0). In 2023, EM + GB treatments also decreased these compounds by 24% and 20%, respectively, for T5, and 28% and 24% for T6, when compared to T0. Conversely, EM (T1) and GB (T3) increased ($p < 0.01$) the sum of anthocyanins and the sum of individual polyphenols for cv. 'Draper' in 2022 by 21% and 29%, respectively.

The contrasting responses among cultivars regarding anthocyanin accumulation might be a result of environmental conditions and genotypic differences, since both abiotic and biotic stresses have been shown to influence anthocyanin biosynthesis and modulate the expression of key anthocyanins such as malvidin and delphinidin glycosides [31]. Consequently, the observed decrease in anthocyanin and total individual polyphenol contents following biostimulant application may indicate that, under low environmental stress, these treatments reduce the requirement for increased antioxidant production [22]. In addition, previous research has found a negative correlation between blueberry weight and anthocyanin concentration [5]. Thus, the increase in blueberry size observed in our recent study following the spraying of EM- and GB-based biostimulants application [16] might explain these results.

On the other hand, the increment of anthocyanin and individual polyphenol contents in 'Draper' fruits following the application of EM and GB in 2022 could also be attributed to a cultivar-specific response to these treatments. In line with our findings, previous work has shown that treatments with marine algae improve the content of delphinidin-3-O-galactoside and total polyphenols in 'Bluecrop' blueberries [7]. Similar effects were reported by Frioni et al. [32] on anthocyanins and phenolic contents of grapevine skin when a seaweed-based biostimulant was used. Such effects have been linked to the upregulation of genes responsible for anthocyanin transport by seaweed-based biostimulants, thereby supporting increases in anthocyanin content [26]. However, other investigations on seaweed-based biostimulant applications produced different results. An example of this is the study by Kapur et al. [27], which found that treatment with seaweed extract did not significantly alter the monomeric anthocyanin content of strawberries, suggesting that anthocyanin accumulation could be regulated by several factors. Regarding the external application of GB, transcriptome analysis suggests that this compound stimulates flavonoid biosynthesis by upregulating genes associated with flavonoid metabolic pathways [33]. In line with this, it has been shown that post-harvest application of GB led to an increase in anthocyanin content in 'Brightwell' blueberries [17], while similar effects were also observed in other crops, particularly in sweet cherries, in which the concentrations of cyanidin-3-*O*-rutinoside and cyanidin-3-*O*-glucoside increased after pre-harvest application of GB [34].

2.3. Vitamin C

Over the two experimental years, the average vitamin C content (Figure 2) for both cultivars ranged from 0.13 mg g^{-1} in cv. 'Draper' in 2023 to 1.05 mg g^{-1} in cv. 'Duke' in 2022. Similarly, a study conducted in Portugal by [6] with four highbush blueberry cultivars reported vitamin C levels between 0.06 and 1.7 mg g^{-1}. Consistent with our results, Gündüz et al. [5] found higher vitamin C levels in 'Duke' blueberries compared to 'Draper' over three years.

https://doi.org/10.3390/plants15010092

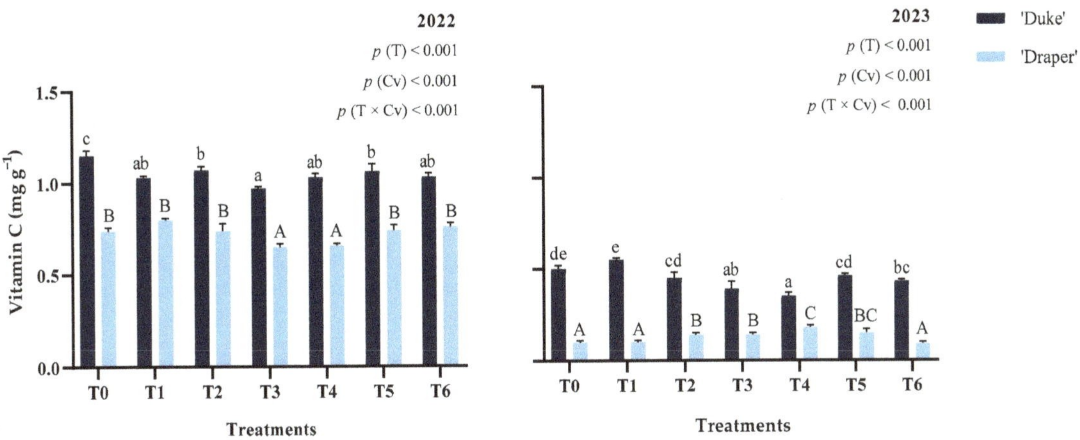

Figure 2. Vitamin C content of fruits from 'Duke' and 'Draper' blueberry cultivars (Cv) depending on the treatments (T) used in the years 2022 and 2023. Lowercase letters indicate significant differences ($p < 0.05$) between the 'Duke' treatments and uppercase letters indicate significant differences ($p < 0.05$) between the 'Draper' treatments by Tukey's test; means ($n = 3$) ± SD.

Although vitamin C content is primarily determined by genotypic factors, climatic conditions, such as light intensity and water availability, can also influence its concentration [35]. These same authors claim that crops that are exposed to less frequent irrigation may contain more concentrated vitamin C. In our study, the higher temperatures and reduced precipitation during the stage of active fruit development in 2022, particularly in May, when solar radiation showed comparable levels in both years, may have contributed to an increase in vitamin C accumulation. This effect has been evidenced in other species, for example, in sweet cherries [36].

Vitamin C content was affected by treatment, cultivar, and their interaction ($p < 0.001$). This compound was also influenced by year and by the interaction treatment × cultivar × year ($p < 0.001$) (Table A1 in Appendix A). Results showed that 'Duke' fruits treated with GB (T3 and T4) presented lower vitamin C content in 2022 (16% and 10%, respectively) and in 2023 (22% and 30%, respectively), compared with T0. These treatments also reduced the vitamin C content by 12% and 11%, respectively, for 'Draper' blueberries. Additionally, lower ($p < 0.001$) vitamin C content was obtained in T6 treated 'Duke' blueberries, with reductions of 10% in 2022 and 14% in 2023. Similarly, in 2022, 'Duke' blueberries treated with EM (T1 and T2) and EM + GB (T5) showed a 10%, 7%, and 8%, respectively, lower ($p < 0.001$) concentration of this compound. Conversely, in 2023, the use of 2 L ha^{-1} EM (T2), both GB doses (T3 and T4) and 4 L ha^{-1} EM + 4 kg ha^{-1} GB (T5) increased ($p < 0.001$) the vitamin C concentration of 'Draper' blueberries by 40%, 40%, 80%, and 50%, respectively.

These contrasting effects suggest that, because vitamin C is produced as a plant response to stress, the application of GB-based biostimulants may be involved in plant internal defense mechanisms, resulting in the stabilization of proteins and cellular membranes, as well as in the interaction with stress-responsive genes. Therefore, GB may lower oxidative stress, thereby helping to neutralize reactive oxygen species (ROS), and the need for antioxidants like vitamin C [37,38]. Also, Gündüz et al. [5] found a negative correlation between fruit weight and its respective vitamin C concentration after studying several northern highbush cultivars over different years and locations, which suggests that the increased fruit weight and size following the application of biostimulants, as reported in our previous study [16], may have contributed to the decrease in vitamin C concentration.

 https://doi.org/10.3390/plants15010092

Similar effects were reported by Correia et al. [36], who found a decrease in vitamin C in two cherry cultivars following foliar application of GB, compared with the untreated control. Regarding seaweed extract application, these authors observed different responses between cultivars, one of which showed increased vitamin C concentration, while in the other it decreased. Koort et al. [11] also observed a lower vitamin C content in 'Northblue' blueberries fertilized with organic fertilizer made from chicken manure and seaweed, as compared with mineral fertilization. Similarly, Ashour et al. [39] showed that vitamin C content was reduced in bell peppers after foliar application with seaweed extracts.

On the other hand, an increase in vitamin C content with the application of seaweed extract or GB through a foliar application method has been documented for several fruits [18,34,39–41]. In this context, the higher levels of vitamin C obtained after the application of these biostimulants might be linked to the improvement of plant health and metabolism, promoting the synthesis of various secondary metabolites, including bioactive compounds such as vitamin C. These results may indicate a potential genetic difference in response to vitamin C accumulation between the two blueberry cultivars, with genotype being a potential modifier of the response to biostimulants.

2.4. Antioxidant Capacity

The AC measured by the $ABTS^{\bullet+}$, CUPRAC, $DPPH^{\bullet}$, and FRAP methods can be observed in Table 3. In 2022, the $ABTS^{\bullet+}$, CUPRAC, and FRAP methods registered significant differences between treatments ($p < 0.001$), whereas in 2023, all AC methods were affected by treatments ($0.001 < p < 0.01$). Additionally, all AC parameters were statistically different among cultivars during both years ($p < 0.001$), confirming that genotype is an important factor that determines blueberry antioxidant activity [42]. Regarding the interaction treatment × cultivar, there was a statistically significant effect ($0.001 < p < 0.01$) in both AC methods for 2022. In the following year, the interaction was influenced only by the $DPPH^{\bullet}$ and FRAP methods ($p < 0.001$). All AC assays were also affected ($p < 0.001$) by the interaction treatment × cultivar × year (Table A1 in Appendix A).

Table 3. Antioxidant capacity (AC) of fruits from 'Duke' and 'Draper' blueberry cultivars (Cv) depending on the treatments (T) used in the years 2022 and 2023.

Treatments	Years	$ABTS^{\bullet+}$ (μmol L^{-1} TE g^{-1})		CUPRAC (μmol L^{-1} TE g^{-1})		$DPPH^{\bullet}$ (μmol L^{-1} TE g^{-1})		FRAP (μmol L^{-1} $FeSO_4$ g^{-1})	
		'Duke'	'Draper'	'Duke'	'Draper'	'Duke'	'Draper'	'Duke'	'Draper'
T0	2022	149.89 ± 13.12^{ab}	86.03 ± 3.40^{A}	119.48 ± 5.02^{b}	88.49 ± 0.35^{A}	82.60 ± 9.29^{a}	77.21 ± 0.19^{B}	178.69 ± 0.11^{a}	151.25 ± 1.36^{AB}
	2023	176.50 ± 5.38^{abc}	169.53 ± 1.43^{AB}	72.60 ± 4.13^{b}	60.57 ± 10.75^{A}	118.47 ± 2.61^{a}	95.57 ± 2.86^{A}	96.01 ± 2.83^{c}	81.19 ± 3.57^{A}
T1	2022	154.24 ± 0.38^{b}	100.90 ± 1.83^{BC}	118.27 ± 2.10^{b}	101.27 ± 2.77^{BC}	91.86 ± 3.42^{a}	66.12 ± 8.85^{AB}	187.49 ± 0.54^{abc}	145.64 ± 3.94^{A}
	2023	196.32 ± 0.78^{d}	176.16 ± 3.77^{B}	72.57 ± 0.16^{b}	64.89 ± 3.77^{A}	146.65 ± 5.83^{b}	104.18 ± 5.04^{A}	91.47 ± 1.69^{abc}	81.53 ± 1.58^{A}
T2	2022	128.10 ± 4.94^{a}	91.36 ± 0.75^{AB}	100.88 ± 2.20^{a}	100.70 ± 4.05^{BC}	97.09 ± 3.14^{a}	71.44 ± 6.63^{AB}	180.64 ± 1.61^{ab}	154.43 ± 4.58^{AB}
	2023	176.18 ± 8.87^{ab}	172.40 ± 4.65^{AB}	72.92 ± 2.49^{b}	69.99 ± 0.86^{A}	133.90 ± 6.13^{ab}	102.42 ± 11.57^{A}	95.05 ± 1.80^{bc}	83.95 ± 2.68^{A}
T3	2022	142.35 ± 8.15^{ab}	110.87 ± 1.62^{CD}	128.39 ± 4.35^{c}	107.24 ± 6.45^{C}	97.11 ± 0.70^{a}	73.90 ± 5.54^{AB}	183.69 ± 3.17^{abc}	159.03 ± 0.24^{B}
	2023	189.58 ± 0.16^{bcd}	176.85 ± 7.42^{B}	81.01 ± 0.43^{c}	66.82 ± 7.00^{A}	119.55 ± 18.15^{a}	109.91 ± 2.88^{A}	92.34 ± 1.13^{abc}	85.49 ± 1.83^{A}
T4	2022	134.85 ± 11.02^{ab}	117.69 ± 7.86^{D}	123.92 ± 1.16^{bc}	99.33 ± 4.45^{BC}	95.71 ± 4.58^{a}	62.90 ± 2.14^{AB}	181.95 ± 5.68^{ab}	147.00 ± 6.43^{A}
	2023	190.03 ± 5.51^{cd}	170.83 ± 8.80^{AB}	62.81 ± 0.16^{a}	57.50 ± 4.66^{A}	108.73 ± 9.16^{a}	96.21 ± 6.24^{A}	87.10 ± 1.74^{a}	84.71 ± 4.67^{A}
T5	2022	143.80 ± 3.61^{ab}	107.51 ± 3.38^{CD}	130.29 ± 1.76^{c}	101.08 ± 2.13^{BC}	98.13 ± 3.48^{a}	60.20 ± 5.34^{A}	194.76 ± 4.29^{c}	147.07 ± 4.04^{A}
	2023	167.15 ± 4.98^{a}	157.08 ± 4.77^{A}	80.42 ± 0.71^{c}	55.61 ± 13.36^{A}	109.48 ± 4.44^{a}	104.43 ± 6.67^{A}	90.29 ± 2.12^{ab}	93.37 ± 1.95^{B}
T6	2022	137.02 ± 9.75^{ab}	91.16 ± 3.00^{AB}	129.54 ± 3.50^{c}	92.59 ± 0.71^{AB}	81.89 ± 16.96^{a}	75.06 ± 2.42^{B}	190.64 ± 7.16^{bc}	158.60 ± 1.07^{B}
	2023	182.30 ± 2.19^{bc}	169.35 ± 4.92^{AB}	75.44 ± 5.16^{bc}	72.41 ± 2.88^{A}	108.83 ± 8.07^{a}	106.21 ± 2.03^{A}	88.74 ± 1.42^{a}	84.33 ± 0.58^{A}
p(T)	2022	<0.001		<0.001		>0.05		<0.001	
	2023	<0.001		<0.01		<0.001		<0.01	
p(Cv)	2022	<0.001		<0.001		<0.001		<0.001	
	2023	<0.001		<0.001		<0.001		<0.001	
p(T × Cv)	2022	<0.001		<0.001		<0.01		<0.001	
	2023	>0.05		<0.05		<0.001		<0.001	

Values are means ± SD (n = 3). Lowercase letters indicate significant differences ($p < 0.05$) between 'Duke' treatments and uppercase letters indicate significant differences ($p < 0.05$) between 'Draper' treatments by Tukey's test.

The average AC values determined by the CUPRAC and FRAP assays were significantly higher ($p < 0.001$) in 2022, whereas the $ABTS^{\bullet+}$ and $DPPH^{\bullet}$ assays had higher ($p < 0.001$) values in 2023. In 2022, the application of EM + GB (T5 and T6) showed a 9%

https://doi.org/10.3390/plants15010092

and 8% increase in the CUPRAC method, respectively, compared with T0. In addition, FRAP values of 'Duke' blueberries recorded 9% and 7% increases, respectively, after the application of T5 and T6. For cv. 'Draper', the treatment with 4 L ha^{-1} EM (T1), both doses of GB (T3 and T4), and 4 L ha^{-1} EM + 4 kg ha^{-1} GB (T5) showed higher ($p < 0.001$) AC as determined by the $ABTS^{\bullet+}$ assay by 17%, 29%, 37%, and 25%, respectively, and by the CUPRAC by 14%, 21%, 12%, and 14%, respectively.

Taken together, these results might be attributed to the concentration of bioactive compounds in treated plants, which correlate closely with the AC measured by the methods assayed, as revealed by the results of Pearson's correlation analysis (Figure A1 in Appendix A) and principal component analysis (PCA) (Figure A2 in Appendix A). The PCA, which incorporated all antioxidant and phenolic parameters for each year, explained 79.9% and 67.0% of total variance in 2022 and 2023, respectively. Thus, in both years, PC1 was closely related to TPC, flavonoids, rutin, vitamin C, and to all antioxidant capacity assays (CUPRAC, FRAP, $ABTS^{\bullet+}$, and $DPPH^{\bullet}$), illustrating the general antioxidant potential of the fruit. Also, PC2 was associated with the individual phenolic compounds, particularly anthocyanin derivatives, reflecting differences in specific phenolic composition across the samples.

Regarding cv. 'Draper', the results of T1 and T3 application on AC can be explained by the positive correlation ($p < 0.001$) of total anthocyanins and total individual polyphenols with CUPRAC ($r = 0.782$ and $r = 0.784$, respectively) and $ABTS^{\bullet+}$ ($r = 0.770$ and $r = 0.776$, respectively) (Figure A1 in Appendix A). T3 sprays increased TPC, which, in turn, showed a positive correlation ($p < 0.001$) with $ABTS^{\bullet+}$ ($r = 0.765$) and CUPRAC ($r = 0.724$). Likewise, we observed that T4 increased rutin content, which correlated positively ($p < 0.001$) with the $ABTS^{\bullet+}$ ($r = 0.842$) and CUPRAC ($r = 0.792$) methods. Additionally, T5-treated fruits had higher delphinidin-3-*O*-galactoside and delphinidin-3-*O*-arabinoside contents, which had a positive correlation ($p < 0.001$) with CUPRAC ($r = 0.766$ and $r = 0.586$, respectively) and $ABTS^{\bullet+}$ ($r = 0.809$ and $r = 0.534$, respectively).

In contrast, T2 decreased ($p < 0.001$) the CUPRAC values in cv. 'Duke' by 16% in 2022, while this treatment increased ($p < 0.001$) the AC values in 'Draper' fruits by 14%. $DPPH^{\bullet}$ results indicate that the high dose of EM + GB ($p < 0.01$) decreased by 22% the AC of 'Draper' blueberries as compared with T0. For cv. 'Duke', the decrease in AC measured by the CUPRAC method following the application of T2 might be related to the reduction in the sum of anthocyanins, the sum of individual polyphenols, and vitamin C, which showed strong positive correlations with this method ($r = 0.782$, $r = 0.784$, and $r = 0.704$, respectively; $p < 0.001$). In both years, results showed that the application of 4 kg ha^{-1} GB (T3) increased ($p < 0.001$) the AC of 'Duke' blueberries according to the CUPRAC method by 7% in 2022 and 12% in 2023. In 2023, regarding cv. 'Duke', the application of the high dose of EM (T1) increased ($p < 0.001$) the AC by 11% according to the $ABTS^{\bullet+}$ method and by 24% according to the $DPPH^{\bullet}$. Similarly, 'Duke' blueberries treated with T5 showed 11% higher ($p < 0.001$) CUPRAC values, with a similar effect ($p < 0.01$) regarding FRAP results (15% increase) in the cultivar 'Draper'. Exogenous T3 application increased AC levels as measured by the CUPRAC method in both years. This increase might be explained by the higher concentrations of malvidin-3-*O*-arabinoside in 2023, which also showed a positive correlation ($r = 0.323$; $p < 0.05$) with this assay. In the case of 'Draper' blueberries treated with T5, the increase in AC measured by the FRAP can be attributed to the higher vitamin C concentration, positively correlated with this method ($r = 0.717$; $p < 0.001$).

Seaweed-based biostimulants can improve fruit development and quality by enhancing its antioxidant properties [43]. According to the literature, the efficiency of EM in increasing AC might be related to its composition in bioactive compounds, affecting plant metabolism and promoting the activity of several antioxidant enzymes that lower oxida-

https://doi.org/10.3390/plants15010092

tive stress [43–45]. On the other hand, GB also controls the production of antioxidant enzymes, such as ascorbate peroxidase, catalase, peroxidase, and superoxide dismutase, in crops [24,26]. As a consequence, this can lead to higher AC levels, as observed in the present study. Secondly, GB could also improve antioxidant systems indirectly by the improvement of sugar and organic acid content [46]. Overall, these biostimulants may enhance the phenylpropanoid pathway and improve the activity of antioxidant enzymes, enabling plants to better tolerate several environmental stresses [47].

Regarding the pre-harvest application of a marine algae-based biostimulant, Lenart et al. [7] found higher levels of AC in 'Bluecrop' blueberries assessed by the DPPH$^•$ method. According to this assay, post-harvest application of GB in 'Brightwell' blueberries also resulted in higher AC values compared with the control group [17]. The authors attribute these findings to increased enzymatic and non-enzymatic antioxidants. Similar results were observed following the pre-harvest foliar application of GB, as determined by the FRAP method in grapes [26] and by the DPPH$^•$ method in tomatoes [40] and bell peppers [39]. Similar effects were found for the application of a foliar seaweed-based biostimulant in cherries, where the treatment increased AC (DPPH$^•$ and FRAP) compared with the control [19]. Furthermore, it has been shown that seaweed- and yeast-based biostimulants enhanced the antioxidant properties of tomatoes, measured by the DPPH$^•$ and ABTS$^{•+}$ assays [48].

However, a reduction in the AC was also observed in our study, depending on the antioxidant test used, following biostimulant application to blueberries. In 2023, 'Duke' blueberries treated with 2 kg ha^{-1} GB (T4) had 13% and 9%, respectively, lower ($p < 0.001$) CUPRAC and FRAP values compared with T0. Lower vitamin C content was observed in treated blueberries, which positively correlated with CUPRAC ($r = 0.525$; $p < 0.001$) and FRAP ($r = 0.717$; $p < 0.001$) assays. Fruits sprayed with T4 also showed lower delphinidin-3-*O*-galactoside, which correlated with CUPRAC ($r = 0.413$; $p < 0.01$) as well as FRAP ($r = 0.496$; $p < 0.001$). Both doses of EM + GB (T5 and T6) decreased ($p < 0.001$) FRAP values of 'Duke' blueberries in 2023 by 6% and 8%, respectively. Based on our results, it was possible to observe that T5 and T6 both reduced the levels of compounds correlated with the FRAP method, such as the sum of anthocyanins ($r = 0.450$; $p < 0.01$) and the sum of polyphenols ($r = 0.442$; $p < 0.01$), while T6 also reduced concentrations of vitamin C ($r = 0.717$; $p < 0.001$) and flavonoids ($r = 0.457$; $p < 0.01$).

Based on our results, foliar applications of EM and GB are therefore effective tools for improving the nutraceutical quality of blueberries. However, their effects are influenced by the interaction of genotype and environmental conditions. Thus, future research should focus on elucidating the mechanisms underlying these responses by quantifying the activity of antioxidant enzymes and enzymes of the phenylpropanoid pathway, as well as by expressing key genes responsible for flavonoid biosynthesis. Such investigations will clarify whether the observed increases in bioactive compounds and AC result from the transcriptional activation of these metabolic pathways, and hence the establishment of a direct link between biostimulant application, molecular responses, and improved fruit quality.

3. Materials and Methods

3.1. Plant Material and Sampling

Experiments were carried out in a commercial orchard located in Vilarandelo, Valpaços municipality (41°40′8.38″ N, 7°19′22.81″ W, 593 m asl), north of Portugal, for two years (2022 and 2023). 'Duke' and 'Draper' (*Vaccinium corymbosum* L.) northern highbush blueberry cultivars, planted in 2012, were selected due to their commercial importance and widespread cultivation in Portugal. Plants were spaced 3 × 1 m apart in a north–south

 https://doi.org/10.3390/plants15010092

orientation and irrigated using a drip irrigation system with two lines per ridge. The bushes were treated with the same fertilization program based on soil analysis during the trial. The soil contained 4.71% organic matter, moderate acidity (pH 5.6), medium texture, high phosphorus content (105 mg P_2O_5 kg^{-1}), and very high potassium content (426 mg K_2O kg^{-1}).

Ecklonia maxima (EM) macroalgae (Kelpak®, Daymsa, Zaragoza, Spain) and glycine betaine (GB) (Greenstim®, Massó Agro Department, Barcelona, Spain) were used. Kelpak® is a natural concentrate of 100% EM algae with a pH of 4.4, 20 mS/cm conductivity, and 0.55% K_2O content (*w*/*w*), while Greenstim® is made up of 97% GB obtained from sugar beet with a C/N ratio of 4.9, 56% organic C, 11.5% organic N, and 12% total N (*w*/*w*). A manual backpack sprayer equipped with a single adjustable nozzle was used for the application of biostimulants on groups of nine plants per treatment for each cultivar at three phenological stages: full bloom (BBCH 65), early green fruit (BBCH 71), and fruit coloring (BBCH 81) [49]. At each growth stage, the following treatments were applied: 4 L ha^{-1} (T1) and 2 L ha^{-1} (T2) of EM-based biostimulant; 4 kg ha^{-1} (T3) and 2 kg ha^{-1} (T4) of GB-based biostimulant; a combination of 4 L ha^{-1} EM + 4 kg ha^{-1} GB (T5) and 2 L ha^{-1} EM + 2 kg ha^{-1} GB (T6); and water as a control treatment (T0). The dosages for GB and EM were applied in accordance with the manufacturer's recommendations. Applications were made in the morning, the canopy was fully covered, and no precipitation event was forecasted for the next 24 h.

'Duke' blueberries were picked at commercial maturity, according to cultivar traits, on 14 June (2022) and 12 June (2023), whereas 'Draper' blueberries were picked on 27 June (2022) and 20 June (2023). For every treatment, hundreds of fruits with uniform size were randomly collected and taken to the laboratory for analysis. The fruits were frozen in liquid nitrogen and stored at −80 °C until quantification of bioactive compounds and AC assays. Prior to analysis, blueberries were freeze-dried, ground into a powder using a commercial blender, and properly identified before being used in the analytical procedure. All determinations for chemical and biochemical analyses were performed in triplicate.

3.2. Climatic Conditions

The plantation area has a temperate climate according to the Köppen classification, with wet winters and dry, hot summers, classified as "Csa." Temperature and precipitation data (Figure 3) were obtained from reanalysis datasets based on nearby meteorological stations, specifically the E-OBS dataset, whilst mean solar radiation was provided by the EMA network property of DRAPN whose data are controlled through mySense Platform. The average air temperature throughout the growing season, from March to June, was approximately 1.0 °C lower in 2022 compared to 2023. Nevertheless, a temperature about 2.0 °C higher was observed in May during the first season. Compared to the long-term air temperature of Vila Real (1981–2010), March and April 2022 were below average, while May and June exceeded the long-term mean. In contrast, all months in 2023 recorded temperatures above the climatological normals. The values of precipitation during March (54.1 mm) and April (34.2 mm) were higher in 2022, whereas in 2023 higher values were presented during May (42.8 mm) and June (78.3 mm). Overall, both years showed lower precipitation levels than the long-term, except for June 2023. In addition, the mean solar radiation was higher in 2023 than in 2022.

https://doi.org/10.3390/plants15010092

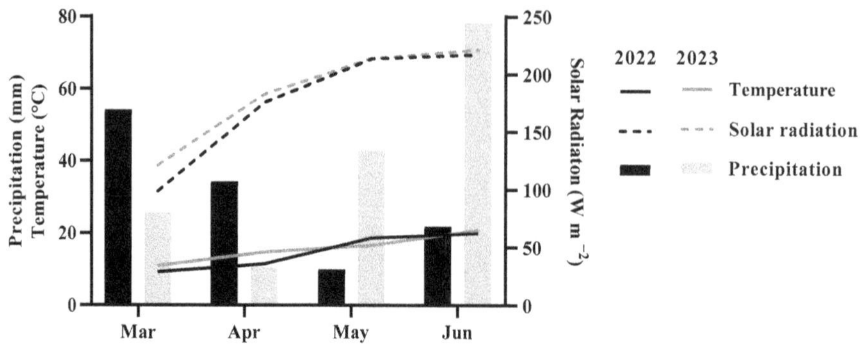

Figure 3. Mean temperature (°C), monthly precipitation (mm), and mean solar radiation (W m^{-2}) for Vilarandelo in the period from March to June in 2022 and 2023.

3.3. Determination of Phenolic Compounds

Prior to the analysis of TPC, flavonoids, *ortho*-diphenols, and antioxidant capacity (AC), a blueberry extract was prepared. For this, 40 mg dry weight (DW) of each sample was mixed with 1 mL of 70% methanol (v/v) in 2 mL centrifuge tubes. The mixture was homogenized using a vortex and then heated at 70 °C for 30 min. Afterwards, the samples were centrifuged at 11,000 rpm and 1 °C for 15 min (Centrifuge 5804R, Eppendorf, Hamburg, Germany). The supernatant was then collected and filtered with 0.2 µm Spartan filters into amber vials.

3.3.1. Total Phenolic Compounds (TPC)

TPC were analyzed via the colorimetric Folin–Ciocalteu method of Singleton and Rossi [50] and Dewanto et al. [51], in a 96-well microplate. A total of 20 µL extract was mixed with 100 µL diluted Folin–Ciocalteu reagent (1:10 in bidistilled water) and 80 µL of 7.5% Na_2CO_3. The microplates were subsequently incubated in the dark at 45 °C for 15 min. Absorbance was measured at 765 nm using a microplate reader (Multiskan GO Microplate Spectrophotometer, Thermo Scientific, Vantaa, Finland). Results were expressed as milligrams of gallic acid equivalents per gram (mg GAE g^{-1}), on a DW basis.

3.3.2. *Ortho*-Diphenols

The determination of *ortho*-diphenols was made according to the methodologies described by Gutfinger [52] and Garcia et al. [53]. In each well of a 96-well microplate, 20 µL of sample was added, followed by the addition of 100 µL ultrapure water, 80 µL of phosphate buffer pH 6.5 (0.1 mol L^{-1}), and 160 µL of 5% sodium molybdate solution $Na_2MoO_4 \cdot 2H_2O$. Then, the microplate was incubated in the dark at room temperature for 15 min. A calibration curve was prepared using caffeic acid ($C_9H_8O_4$) as standard and absorbance values were recorded at 350 nm with a microplate reader (Multiskan GO Microplate Spectrophotometer, Thermo Scientific, Vantaa, Finland). Results were expressed in milligrams of caffeic acid equivalents per gram (mg CAE g^{-1}), on a DW basis.

3.3.3. Flavonoids

The determination of flavonoid content was performed using the colorimetric method adopted by Dewanto et al. [51], using 96-well microplates. In each well of the microplate, 25 µL of extract was mixed with 100 µL of ultrapure water and 10 µL of a 5% sodium nitrite ($NaNO_2$) solution. The solution was then homogenized and incubated at room temperature in the dark for 5 min. Then, 15 µL of aluminum chloride ($AlCl_3$) 10% was added and the microplate was placed again for 6 min at room temperature in the dark. Sodium hydroxide

https://doi.org/10.3390/plants15010092

(NaOH) 1 mol L^{-1} (50 µL) and ultrapure water (50 µL) were added to each well. A calibration curve was prepared using catechin as standard and absorbances were recorded at 510 nm using a microplate reader (Multiskan GO Microplate Spectrophotometer, Thermo Scientific, Vantaa, Finland). The results were reported as milligrams of catechin equivalents per gram (mg CE g^{-1}), on a dry weight (DW) basis.

3.4. Individual Polyphenols

Individual polyphenols were determined via HPLC-DAD-UV/VIS according to Aires et al. [54]. First, 40 mg of dried powder of blueberry sample was extracted using 950 µL of 70% methanol (*v*/*v*) and 50 µL of naringin (1 mg mL^{-1}) as internal standard. The mixture was subjected to a warm bath at 70 °C for 30 min with intermittent stirring every 5 min and extracts were centrifuged at 11,000 rpm, 4 °C during 15 min (Centrifuge 5804R, Eppendorf, Hamburg, Germany). The supernatant was filtered using 0.20 µm cellulose ester filters (WhatmanTM, Spartan 13/0.2 RC, Maidstone, UK) and transferred to amber HPLC vials to prevent light degradation. After this process, vials were stored at −20 °C until analysis. The HPLC-DAD-UV/VIS analysis was carried out using a reverse-phase column (C18 Spherisorb ODS2, 250 mm × 4.6 mm), and eluents of 0.1% trifluoroacetic acid (TFA) in water (solvent A) and 0.1% TFA in acetonitrile (solvent B), in a run of 60 min, starting with 100% solvent A, while the flow rate was set to 1 mL min^{-1}, and an injection volume of 10 µL. Chromatograms were recorded at 280, 320, and 370 nm for phenolic compounds and at 520 nm for anthocyanins, and identification was based on retention time, UV spectra of each compound, and comparison with commercial external standards. Quantification was performed using external calibration curves, internal standards, and the response factor for each detected polyphenol. The chromatographic profile of polyphenols is presented in Supplementary Figure S1. Results for individual polyphenols are expressed in µg g^{-1}, whereas the sums of individual polyphenols and anthocyanins are expressed in mg g^{-1}, on a DW basis.

3.5. Determination of Vitamin C Content

The vitamin C content was quantified using HPLC-DADUV/VIS according to Hernandez et al. [55], adapted by Aires et al. [56]. For the extraction, 200 mg of lyophilized blueberry sample was weighed and mixed with 5.0 mL extraction solvent comprising 3.0% metaphosphoric acid and 8% acetic acid as well as 1.0 mmol L^{-1} tert butylhydroquinone (TBHQ) (Sigma-Aldrich, Tauferkichen, Germany). The samples were homogenized using a disperser (T 25 digital ULTRA-TURRAX®, Staufen, Germany) and then centrifuged at 4000 rpm for 5 min at 4 °C (Centrico 250, UniEquip, Munich, Germany). The supernatant was then filtered through PTFE 0.2 µm filters (Whatman™ Spartan 13/0.2 RC, Maidstone, UK) into amber HPLC vials and injected immediately. The analysis was conducted in an HPLC system equipped with a C18 reverse-phase column (Spherisorb ODS2, 250 × 4.6 mm, 5 µm) applying a mobile phase composed of 0.2% *ortho*-phosphoric acid in an isocratic gradient. The flow rate was set to 1.2 mL min^{-1}. The injection volume was 20 µL, and the recording of chromatograms was performed at 245 nm. The identity of vitamin C was ascertained by the retention time of chromatographic peaks by comparing the same with an external standard of vitamin C (Sigma-Aldrich, Tauferkichen, Germany). Finally, the contents were presented in mg g^{-1}, on a DW basis.

3.6. Antioxidant Capacity Assays

3.6.1. $ABTS^{\bullet+}$ Radical-Scavenging Activity

$ABTS^{\bullet+}$ (2,2′-azino-bis (3-ethylbenzothiazoline-6-sulphonic acid)) discoloration assay was performed using the method of Stratil et al. [57] with modifications. A working solution was made by mixing 1 mg mL^{-1} $ABTS^{\bullet+}$ with 0.68 mg mL^{-1} potassium persulfate ($K_2S_2O_8$)

 https://doi.org/10.3390/plants15010092

in distilled water. It was incubated at room temperature in the dark for 12 to 16 h. After this period, the solution was diluted with absolute ethanol to reach an absorbance between 0.8 and 1.0 at 734 nm. For quantification, 15 μL of the extract previously prepared was added in each well of the 96-well microplate, considering that the blank was made with 70% methanol (*v*/*v*). Then, 285 μL of $ABTS^{\bullet+}$ working solution was added, and the microplate was incubated in the dark for 10 min. A calibration curve was prepared using Trolox (Sigma-Aldrich, Tauferkichen, Germany) as standard and absorbances were recorded at 734 nm using a microplate reader (Multiskan GO Microplate Spectrophotometer, Thermo Scientific, Vantaa, Finland). Results were presented as micromoles Trolox equivalent (TE) per gram (μmol TE g^{-1}, on a DW basis).

3.6.2. Cupric-Reducing Antioxidant Capacity (CUPRAC)

The CUPRAC assay was made according to the method described by Apak et al. [58]. For the determination, solutions of 10 mmol L^{-1} $CuCl_2$ in water, 7.4 mmol L^{-1} neocuproine (Sigma-Aldrich, Tauferkichen, Germany) in 96% ethanol, and 1 mmol L^{-1} ammonium acetate buffer solution at pH 7.0 in water were prepared. Then, 50 μL of $CuCl_2$ solution, 50 μL of neocuproine solution, 50 μL of buffer solution, 25 μL of sample, and 25 μL of bidistilled H_2O were added sequentially to each well of the 96-well microplate. The blank was composed by adding all reagents except CuCl2. The microplate was left in the dark at room temperature for 30 min and the absorbance was then measured at 450 nm using a microplate reader (Multiskan GO Microplate Spectrophotometer, Thermo Scientific, Vantaa, Finland). A calibration curve was performed with Trolox (Sigma-Aldrich, Tauferkichen, Germany) as standard at different concentrations. The results were expressed as μmol L^{-1} Trolox equivalents (TE) per gram of sample (μmol L^{-1} TE g^{-1}), on a dry weight (DW) basis.

3.6.3. $DPPH^{\bullet}$ Radical-Scavenging Capacity

$DPPH^{\bullet}$ radical scavenging capacity was assessed following the method of Siddhraju and Becker [59] adapted to 96-well microplates. An amount of 285 μL of $DPPH^{\bullet}$ solution prepared by dissolving 4 mg of 2,2-diphenyl-1-picrylhydrazyl radical in 100 mL of 95% ethanol was added to each microplate well, to which 15 μL of extract was added. A blank sample was prepared by adding all reagents except the extract, which was replaced with the solvent used for extraction. The microplate was left in the dark at room temperature for 30 min. A calibration curve was performed using Trolox (Sigma-Aldrich, Tauferkichen, Germany) as standard at different concentrations, and absorbances were recorded at 517 nm in a microplate reader (Multiskan GO Microplate Spectrophotometer, Thermo Scientific, Vantaa, Finland). Values were expressed as μmol L^{-1} Trolox equivalents per gram (μmol L^{-1} TE g^{-1}), on a DW basis.

3.6.4. Ferric-Reducing Antioxidant Power (FRAP)

The FRAP assay was performed according to Stratil et al. [57]. Initially, FRAP reagent was prepared by combining, in a ratio of 10:1:1 (*v*/*v*/*v*), 300 mmol L^{-1} acetate buffer at pH 3.6, 10 mmol L^{-1} TPTZ (2,4,6-tri(2-pyridyl)-s-triazine) in 40 mmol L^{-1} HCl, and 20 mmol L^{-1} $FeCl_3 \cdot 6H_2O$. A calibration curve was also prepared using a solution of $FeSO_4$ as standard at different concentrations. To each well of the microplate, 25 μL of sample or standard was added, followed by 275 μL of the FRAP reagent. The extraction solvent (70% methanol) was the blank. The microplates were therefore incubated in the dark at room temperature for 5 min. Then, the absorbance values were recorded at 593 nm in a microplate reader (Multiskan GO Microplate Spectrophotometer, Thermo Scientific, Vantaa, Finland). Results were expressed as μmol of $FeSO_4$ equivalents per gram (μmol $FeSO_4$ g^{-1}), on a DW basis.

https://doi.org/10.3390/plants15010092

3.7. Statistical Analysis

Data were subjected to analysis of variance (ANOVA), followed by Tukey's HSD post hoc test with a significance level of 5% for mean comparison. All the above tests were carried out using SPSS software version 27.0 (SPSS-IBM, Orchard Road, Armonk, New York, NY, USA). One-way, two-way, and three-way ANOVA were applied to compare the treatment and cultivar effect in a particular year, as well as the year effect. Pearson's correlation test and a principal component analysis (PCA) were also performed to evaluate the relationships between bioactive compounds and AC.

4. Conclusions

This two-year study reveals that the foliar application of biostimulants, particularly those based on EM and GB, strongly affects the phenolic compounds and vitamin C content and, correspondingly, the antioxidant capacity (AC) of blueberries. Significant differences were recorded between the responses of 'Duke' and 'Draper' cultivars to treatments and between the years. In both years, the application of GB (2 kg ha^{-1}) generally resulted in a significant increase in the total phenolic content (TPC) of 'Duke' blueberries by 22–33% compared with the control. Furthermore, regarding cv. 'Draper', the higher dose of this biostimulant (4 kg ha^{-1} GB) contributed to a 16–29% increase in total anthocyanins and individual polyphenols. Accordingly, 'Draper' blueberries subjected to this treatment generally had 21–29% higher AC measured by the $ABTS^{•+}$ and CUPRAC methods in 2022. However, both GB doses and the combined application of 2 L ha^{-1} EM + 2 kg ha^{-1} GB reduced vitamin C concentrations in 'Duke' blueberries by 10–30% during both years. Conversely, in 2023, both GB doses increased vitamin C content by 40–80% in fruits of the cv. 'Draper'. In addition, the application of 4 kg ha^{-1} GB and the combination of 4 L ha^{-1} EM + 4 kg ha^{-1} GB increased the AC in blueberries of the cv. 'Duke' by 7–12% as measured via the CUPRAC assay in both years. The treatment with 4 L ha^{-1} EM in 2022 increased by 14–21% total anthocyanins, individual polyphenols, and AC according to $ABTS^{•+}$ and CUPRAC methods in 'Draper' blueberries. It also increased TPC in 'Duke' fruits by 21%. In another way, the application of 2 L ha^{-1} EM reduced anthocyanins, individual phenolics, and AC by the CUPRAC assay by 16–20%. Our findings underscore that the effect of biostimulants is influenced by the applied dose, genotype, and climatic conditions of the year. Therefore, dosages can be modulated based on each cultivar's specific response and environmental conditions. This study showed the potential of EM- and GB-based biostimulants to enhance the nutritional quality and antioxidant capacity of blueberries.

Supplementary Materials: The following supporting information can be downloaded at https://www.mdpi.com/article/10.3390/plants15010092/s1; Figure S1: Representative HPLC chromatogram of polyphenols in blueberry extract recorded at 280, 320, 370, and 520 nm.

Author Contributions: Conceptualization, T.L., A.P.S., A.A.V., and B.G.; methodology, T.L., A.P.S., A.A.V., and B.G.; investigation, T.L., R.C., M.F., and A.A.; writing—original draft preparation, T.L.; writing—review and editing, T.L., A.P.S., A.A., A.A.V., and B.G.; supervision, A.P.S., A.A.V., and B.G. All authors have read and agreed to the published version of the manuscript.

 https://doi.org/10.3390/plants15010092

Funding: This work was supported by National Funds from FCT—Portuguese Foundation for Science and Technology, MCTES, and FSE, grant number 2021.07171.BD (https://doi.org/10.54499/2021.07171.BD (accessed on 28 August 2024), and under the projects UID/04033/2025 (CITAB), LA/P/0126/2020 (https://doi.org/10.54499/LA/P/0126/2020 (accessed on 28 August 2024), UIDB/04469/2020 (CEB), and LA/P/0029/2020 (LABBELS—Associate Laboratory in Biotechnology, Bioengineering, and Microelectromechanical Systems).

Data Availability Statement: The original contributions presented in this study are included in the article/supplementary material. Further inquiries can be directed to the corresponding author.

Acknowledgments: The author Tiago Lopes acknowledges FCT, MCTES, and FSE for the Fellowship (2021.07171.BD). The authors also thank João Santos, André Fonseca, and Raul Santos for providing the climate data.

Conflicts of Interest: The authors declare no conflicts of interest. The funders had no role in the design of the study; in the collection, analyses, or interpretation of data; in the writing of the manuscript; or in the decision to publish the results.

Appendix A

Table A1. Statistical effect of treatment (T), cultivar (Cv), year (Y), and their interactions on the analyzed blueberry parameters.

Parameters	p (T)	p (Cv)	p (Y)	p (T × Cv)	p (T × Y)	p (C × Y)	p (T × Cv × Y)
Total Phenolic Compounds	<0.001	<0.001	>0.05	<0.001	>0.05	<0.001	<0.001
Flavonoids	<0.001	<0.001	<0.001	>0.05	<0.05	<0.001	<0.05
Ortho-diphenols	<0.01	<0.01	<0.001	>0.05	>0.05	>0.05	>0.05
Sum of Anthocyanins	<0.001	<0.001	<0.001	<0.001	<0.05	<0.01	<0.001
Sum of Individual Polyphenols	<0.001	<0.001	>0.05	<0.001	>0.05	<0.01	<0.001
Vitamin C	<0.001	<0.001	<0.001	<0.001	<0.001	>0.05	<0.001
$ABTS^{\bullet+}$	<0.001	<0.001	<0.001	<0.001	<0.001	<0.001	<0.001
CUPRAC	<0.001	<0.001	<0.001	<0.001	<0.001	<0.001	<0.001
$DPPH^{\bullet}$	<0.001	<0.001	<0.001	<0.001	<0.01	>0.05	<0.001
FRAP	<0.001	<0.001	<0.001	<0.01	<0.01	<0.001	<0.001

https://doi.org/10.3390/plants15010092

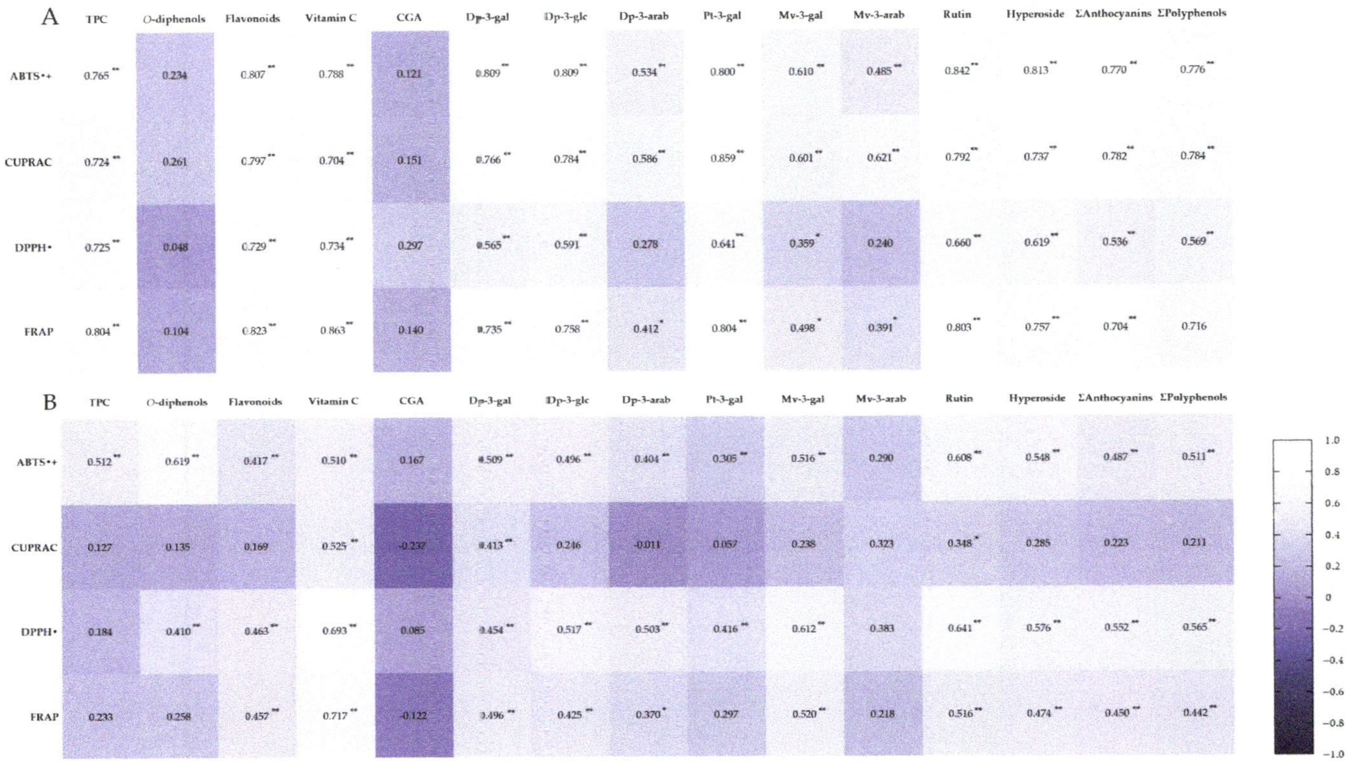

Figure A1. Pearson's correlation analysis between bioactive compounds and AC across the different methods of blueberries from both cultivars in the years 2022 (**A**) and 2023 (**B**). Abbreviations: TPC—total phenolic compounds; O-diphenols—*ortho*-diphenols; CGA—chlorogenic acid; Dp-3-gal—delphinidin-3-*O*-galactoside; Dp-3-glc—delphinidin-3-*O*-glucoside; Dp-3-arab—delphinidin-3-*O*-arabinoside; Pt-3-gal—petunidin-3-*O*-galactoside; Mv-3-gal—malvidin-3-*O*-galactoside; Mv-3-arab—malvidin-3-*O*-arabinoside; ΣAnthocyanins—sum of anthocyanins; ΣPolyphenols—sum of individual polyphenols. * means correlation is significant at $p < 0.05$; ** correlation is significant at $p < 0.01$.

https://doi.org/10.3390/plants15010092

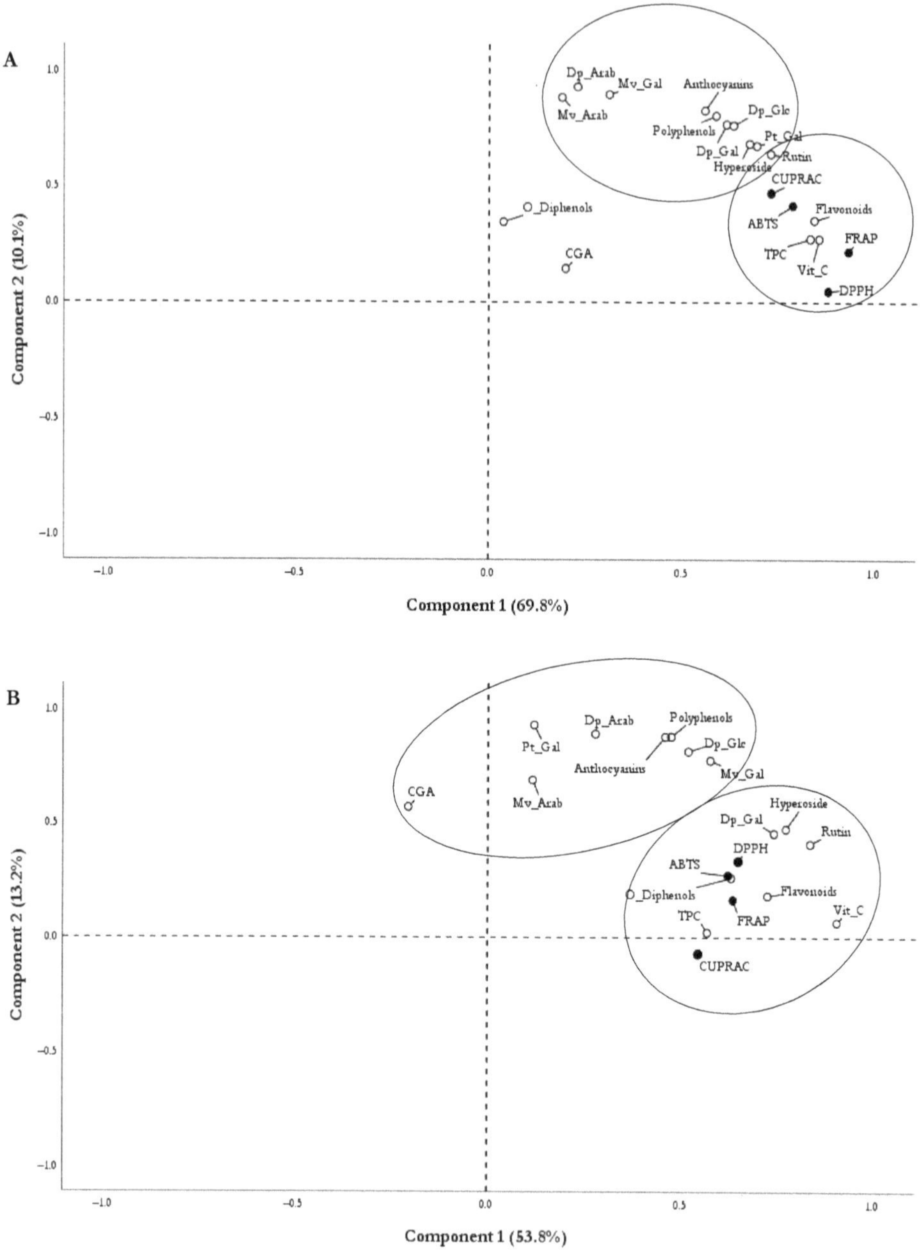

Figure A2. Principal component analysis (PCA) of bioactive compounds and AC of blueberries from both cultivars in the years 2022 (**A**) and 2023 (**B**). Ellipses indicate variable associations based on the loadings of PC1 and PC2. Abbreviations: TPC—total phenolic compounds; O_Diphenols—*ortho*-diphenols; Vit_C—vitamin C; CGA—chlorogenic acid; Dp_Gal—delphinidin-3-*O*-galactoside; Dp_Glc—delphinidin-3-*O*-glucoside; Dp_Arab—delphinidin-3-*O*-arabinoside; Pt_Gal—petunidin-3-*O*-galactoside; Mv_Gal—malvidin-3-*O*-galactoside; Mv_Arab—malvidin-3-*O*-arabinoside; Anthocyanins—sum of anthocyanins; Polyphenols—sum of individual polyphenols.

 https://doi.org/10.3390/plants15010092

References

1. Gilbert, J.L.; Olmstead, J.W.; Colquhoun, T.A.; Levin, L.A.; Clark, D.G.; Moskowitz, H.R. Consumer-assisted selection of blueberry fruit quality traits. *HortScience* **2014**, *49*, 864–873. [CrossRef]
2. Silva, S.; Costa, E.M.; Veiga, M.; Morais, R.M.; Calhau, C.; Pintado, M. Health-promoting properties of blueberries: A review. *Crit. Rev. Food Sci. Nutr.* **2020**, *60*, 181–200. [CrossRef]
3. Sater, H.; Ferrão, L.F.V.; Olmstead, J.; Munoz, P.R.; Bai, J.; Hopf, A.; Plotto, A. Exploring environmental and storage factors affecting sensory, physical, and chemical attributes of six southern highbush blueberry cultivars. *Sci. Hortic.* **2021**, *289*, 110468. [CrossRef]
4. Duan, Y.; Tarafdar, A.; Chaurasia, D.; Singh, A.; Bhargava, P.C.; Yang, J.; Li, Z.; Ni, X.; Tian, Y.; Li, H.; et al. Blueberry fruit valorization and valuable constituents: A review. *Int. J. Food Microbiol.* **2022**, *381*, 109890. [CrossRef] [PubMed]
5. Gündüz, K.; Serçe, S.; Hancock, J.F. Variation among highbush and rabbiteye cultivars of blueberry for fruit quality and phytochemical characteristics. *J. Food Compos. Anal.* **2015**, *38*, 69–79. [CrossRef]
6. Correia, S.; Gonçalves, B.; Aires, A.; Silva, A.; Ferreira, L.; Carvalho, R.; Fernandes, H.; Freitas, C.; Carnide, V.; Paula Silva, A. Effect of harvest year and altitude on nutritional and biometric characteristics of blueberry cultivars. *J. Chem.* **2016**, *2016*, 8648609. [CrossRef]
7. Lenart, A.; Wrona, D.; Krupa, T. Health-promoting properties of highbush blueberries depending on type of fertilization. *Agriculture* **2022**, *12*, 1741. [CrossRef]
8. Radünz, A.L.; Herter, F.G.; Radünz, M.; Silva, V.N.D.; Cabrera, L.D.C.; Radünz, L.L. Bud position influences the fruit quality attributes of three blueberry (*Vaccinium ashei* Reade) cultivars. *Rev. Ceres* **2022**, *69*, 131–138. [CrossRef]
9. Du Jardin, P.; Xu, L.; Geelen, D. Agricultural functions and action mechanisms of plant biostimulants (PBS): An introduction. In *The Chemical Biology of Plant Biostimulants*, 1st ed.; Geelen, D., Xu, L., Eds.; Wiley: Hoboken, NJ, USA, 2020; pp. 1–30. [CrossRef]
10. Lenart, A.; Wrona, D.; Krupa, T. Biostimulators with marine algae extracts and their role in increasing tolerance to drought stress in highbush blueberry cultivation. *PLoS ONE* **2024**, *19*, e0306831. [CrossRef] [PubMed]
11. Koort, A.; Starast, M.; Põldma, P.; Moor, U.; Mainla, L.; Maante-Kuljus, M.; Karp, K. Sustainable fertilizer strategies for *Vaccinium corymbosum* × *V. angustifolium* under abandoned peatland conditions. *Agriculture* **2020**, *10*, 121. [CrossRef]
12. Stirk, W.A.; Rengasamy, K.R.R.; Kulkarni, M.G.; Van Staden, J. Plant biostimulants from seaweed: An overview. In *The Chemical Biology of Plant Biostimulants*, 1st ed.; Geelen, D., Xu, L., Eds.; Wiley: Hoboken, NJ, USA, 2020; pp. 31–55. [CrossRef]
13. Chen, T.H.H.; Murata, N. Glycinebetaine protects plants against abiotic stress: Mechanisms and biotechnological applications. *Plant Cell Environ.* **2011**, *34*, 1–20. [CrossRef]
14. Monteiro, E.; Baltazar, M.; Pereira, S.; Correia, S.; Ferreira, H.; Alves, F.; Cortez, I.; Castro, I.; Gonçalves, B. *Ascophyllum nodosum* extract and glycine betaine preharvest application in grapevine: Enhancement of berry quality, phytochemical content and antioxidant properties. *Antioxidants* **2023**, *12*, 1835. [CrossRef]
15. Afonso, S.; Oliveira, I.; Ribeiro, C.; Vilela, A.; Meyer, A.S.; Gonçalves, B. Exploring the role of biostimulants in sweet cherry (*Prunus avium* L.) fruit quality traits. *Agriculture* **2024**, *14*, 1521. [CrossRef]
16. Lopes, T.; Silva, A.P.; Ribeiro, C.; Carvalho, R.; Aires, A.; Vicente, A.A.; Gonçalves, B. *Ecklonia maxima* and glycine–betaine-based biostimulants improve blueberry yield and quality. *Horticulturae* **2024**, *10*, 920. [CrossRef]
17. Zhang, Y.; Kou, X.; Zhang, G.; Luo, D.; Cao, S. Exogenous glycine betaine maintains postharvest blueberry quality by modulating antioxidant capacity and energy metabolism. *LWT* **2024**, *212*, 116976. [CrossRef]
18. Consentino, B.B.; Vultaggio, L.; Iacuzzi, N.; La Bella, S.; De Pasquale, C.; Rouphael, Y.; Ntatsi, G.; Virga, G.; Sabatino, L. Iodine biofortification and seaweed extract-based biostimulant supply interactively drive the yield, quality, and functional traits in strawberry fruits. *Plants* **2023**, *12*, 245. [CrossRef] [PubMed]
19. Zhi, H.; Dong, Y. Seaweed-based biostimulants improve quality traits, postharvest disorders, and antioxidant properties of sweet cherry fruit in response to gibberellic acid treatment. *Sci. Hortic.* **2024**, *336*, 113454. [CrossRef]
20. Kumar, K.; Debnath, P.; Singh, S.; Kumar, N. An Overview of Plant Phenolics and Their Involvement in Abiotic Stress Tolerance. *Stresses* **2023**, *3*, 570–585. [CrossRef]
21. Saini, N.; Anmol, A.; Kumar, S.; Wani, A.W.; Bakshi, M.; Dhiman, Z. Exploring Phenolic Compounds as Natural Stress Alleviators in Plants—A Comprehensive Review. *Physiol. Mol. Plant Pathol.* **2024**, *133*, 102383. [CrossRef]
22. Naikoo, M.I.; Dar, M.I.; Raghib, F.; Jaleel, H.; Ahmad, B.; Raina, A.; Khan, F.A.; Naushin, F. Role and regulation of plant phenolics in abiotic stress tolerance. In *Plant Signaling Molecules*; Elsevier: Amsterdam, The Netherlands, 2019; pp. 157–168. [CrossRef]
23. Wang, L.; Shan, T.; Xie, B.; Ling, C.; Shao, S.; Jin, P.; Zheng, Y. Glycine betaine reduces chilling injury in peach fruit by enhancing phenolic and sugar metabolisms. *Food Chem.* **2019**, *272*, 530–538. [CrossRef] [PubMed]
24. Chen, L.-L. Postharvest application of glycine betaine ameliorates chilling injury in cold-stored banana fruit by enhancing antioxidant system. *Sci. Hortic.* **2021**, *287*, 110264. [CrossRef]
25. Ali, O.; Ramsubhag, A.; Jayaraman, J. Biostimulatory activities of *Ascophyllum nodosum* extract in tomato and sweet pepper crops in a tropical environment. *PLoS ONE* **2019**, *14*, e0216710. [CrossRef]

https://doi.org/10.3390/plants15010092

26. Monteiro, E.; De Lorenzis, G.; Ricciardi, V.; Baltazar, M.; Pereira, S.; Correia, S.; Ferreira, H.; Alves, F.; Cortez, I.; Gonçalves, B.; et al. Exploring seaweed and glycine betaine biostimulants for enhanced phenolic content, antioxidant properties, and gene expression of *Vitis vinifera* cv. "Touriga Franca" berries. *Int. J. Mol. Sci.* **2024**, *25*, 5335. [CrossRef]
27. Kapur, B.; Karaca, C.; Sarıdaş, M.A.; Ağçam, E.; Çeliktopuz, E.; Kargı, S.P. Enhancing secondary compounds in strawberry fruit through optimized irrigation and seaweed application. *Sci. Hortic.* **2024**, *324*, 112609. [CrossRef]
28. Samuels, L.J.; Papageorgiou, A.E.; Setati, M.E.; Blancquaert, E.H. Effects of *Ecklonia maxima* seaweed extract on the fruit, wine quality, and microbiota in *Vitis vinifera* L. cv. Cabernet Sauvignon. *S. Afr. J. Bot.* **2024**, *172*, 647–662. [CrossRef]
29. Kuntz, S.; Kunz, C.; Rudloff, S. Inhibition of pancreatic cancer cell migration by plasma anthocyanins isolated from healthy volunteers receiving an anthocyanin-rich berry juice. *Eur. J. Nutr.* **2017**, *56*, 203–214. [CrossRef] [PubMed]
30. Kalt, W.; Cassidy, A.; Howard, L.R.; Krikorian, R.; Stull, A.J.; Tremblay, F.; Zamora-Ros, R. Recent research on the health benefits of blueberries and their anthocyanins. *Adv. Nutr.* **2020**, *11*, 224–236. [CrossRef] [PubMed]
31. Yang, W.; Guo, Y.; Liu, M.; Chen, X.; Xiao, X.; Wang, S.; Gong, P.; Ma, Y.; Chen, F. Structure and function of blueberry anthocyanins: A review of recent advances. *J. Funct. Foods* **2022**, *88*, 104864. [CrossRef]
32. Frioni, T.; Sabbatini, P.; Tombesi, S.; Norrie, J.; Poni, S.; Gatti, M.; Palliotti, A. Effects of a biostimulant derived from the brown seaweed *Ascophyllum nodosum* on ripening dynamics and fruit quality of grapevines. *Sci. Hortic.* **2018**, *232*, 97–106. [CrossRef]
33. Zheng, Z.; Wang, T.; Liu, M.; Xu, X.; Wang, J.; Sun, G.; He, S.; Liao, L.; Xiong, B.; Wang, X.; et al. Effects of exogenous application of glycine betaine treatment on 'Huangguoggan' fruit during postharvest storage. *Int. J. Mol. Sci.* **2023**, *24*, 14316. [CrossRef]
34. Gonçalves, B.; Morais, M.C.; Sequeira, A.; Ribeiro, C.; Guedes, F.; Silva, A.P.; Aires, A. Quality preservation of sweet cherry cv. "Staccato" by using glycine-betaine or *Ascophyllum nodosum*. *Food Chem.* **2020**, *322*, 126713. [CrossRef]
35. Lee, S.K.; Kader, A.A. Preharvest and postharvest factors influencing vitamin C content of horticultural crops. *Postharvest Biol. Technol.* **2000**, *20*, 207–220. [CrossRef]
36. Correia, S.; Aires, A.; Queirós, F.; Carvalho, R.; Schouten, R.; Silva, A.P.; Gonçalves, B. Climate conditions and spray treatments induce shifts in health-promoting compounds in cherry (*Prunus avium* L.) fruits. *Sci. Hortic.* **2020**, *263*, 109147. [CrossRef]
37. Khan, M.S.; Yu, X.; Kikuchi, A.; Asahina, M.; Watanabe, K.N. Genetic engineering of glycine betaine biosynthesis to enhance abiotic stress tolerance in plants. *Plant Biotechnol.* **2009**, *26*, 125–134. [CrossRef]
38. Sami, A.; Han, S.; Haider, M.Z.; Khizar, R.; Ali, Q.; Shafiq, M.; Tabassum, J.; Khalid, M.N.; Javed, M.A.; Sajid, M.; et al. Genetic aspects of vitamin C (ascorbic acid) biosynthesis and signaling pathways in fruits and vegetable crops. *Funct. Integr. Genom.* **2024**, *24*, 73. [CrossRef]
39. Ashour, M.; Hassan, S.M.; Elshobary, M.E.; Ammar, G.A.G.; Gaber, A.; Alsanie, W.F.; Mansour, A.T.; El-Shenody, R. Impact of commercial seaweed liquid extract (TAM®) biostimulant and its bioactive molecules on growth and antioxidant activities of hot pepper (*Capsicum annuum*). *Plants* **2021**, *10*, 1045. [CrossRef]
40. Abad, M.S.F.; Abedi, B.; Ne'emati, S.H.; Arouiee, H. Studying the effects of foliar spraying of seaweed extract as a biostimulant on the increase in the yield and quality of tomato (*Lycopersicon esculentum* L.). *World J. Environ. Biosci.* **2014**, *8*, 14–17.
41. Adak, N. Effects of glycine betaine concentrations on the agronomic characteristics of strawberry grown under deficit irrigation conditions. *Appl. Ecol. Environ. Res.* **2019**, *17*, 3753–3767. [CrossRef]
42. Cardeñosa, V.; Girones-Vilaplana, A.; Muriel, J.L.; Moreno, D.A.; Moreno-Rojas, J.M. Influence of genotype, cultivation system and irrigation regime on antioxidant capacity and selected phenolics of blueberries (*Vaccinium corymbosum* L.). *Food Chem.* **2016**, *202*, 276–283. [CrossRef] [PubMed]
43. Pohl, A.; Kalisz, A.; Sękara, A. Seaweed extracts' multifactorial action: Influence on physiological and biochemical status of Solanaceae plants. *Acta Agrobot.* **2019**, *72*, 1758. [CrossRef]
44. Nanda, S.; Kumar, G.; Hussain, S. Utilization of seaweed-based biostimulants in improving plant and soil health: Current updates and future prospective. *Int. J. Environ. Sci. Technol.* **2022**, *19*, 12839–12852. [CrossRef]
45. Sujeeth, N.; Petrov, V.; Guinan, K.J.; Rasul, F.; O'Sullivan, J.T.; Gechev, T.S. Current insights into the molecular mode of action of seaweed-based biostimulants and the sustainability of seaweeds as raw material resources. *Int. J. Mol. Sci.* **2022**, *23*, 7654. [CrossRef]
46. Habibi, F.; Valero, D.; Serrano, M.; Guillén, F. Exogenous application of glycine betaine maintains bioactive compounds, antioxidant activity, and physicochemical attributes of blood orange fruit during prolonged cold storage. *Front. Nutr.* **2022**, *9*, 873915. [CrossRef]
47. Kumar, R.; Trivedi, K.; Anand, K.G.V.; Ghosh, A. Science behind biostimulant action of seaweed extract on growth and crop yield: Insights into transcriptional changes in roots of maize treated with *Kappaphycus alvarezii* seaweed extract under soil moisture stressed conditions. *J. Appl. Phycol.* **2020**, *32*, 599–613. [CrossRef]
48. Mannino, G.; Campobenedetto, C.; Vigliante, I.; Contartese, V.; Gentile, C.; Bertea, C.M. The application of a plant biostimulant based on seaweed and yeast extract improved tomato fruit development and quality. *Biomolecules* **2020**, *10*, 1662. [CrossRef] [PubMed]
49. Wichura, M.A.; Koschnick, F.; Jung, J.; Bauer, S.; Wichura, A. Phenological growth stages of highbush blueberries (*Vaccinium* spp.): Codification and description according to the BBCH scale. *Botany* **2024**, *102*, 428–437. [CrossRef]

https://doi.org/10.3390/plants15010092

50. Singleton, V.L.; Rossi, J.A. Colorimetry of total phenolics with phosphomolybdic–phosphotungstic acid reagent. *Am. J. Enol. Vitic.* **1965**, *16*, 144–158. [CrossRef]
51. Dewanto, V.; Wu, X.; Adom, K.K.; Liu, R.H. Thermal Processing Enhances the Nutritional Value of Tomatoes by Increasing Total Antioxidant Activity. *J. Agric. Food Chem.* **2002**, *50*, 3010–3014. [CrossRef]
52. Gutfinger, T. Polyphenols in olive oils. *J. Am. Oil Chem. Soc.* **1981**, *58*, 966–968. [CrossRef]
53. Garcia, B.; Coelho, J.; Costa, M.; Pinto, J.; Paiva-Martins, F. A simple method for the determination of bioactive anti-oxidants in virgin olive oils. *J. Sci. Food Agric.* **2013**, *93*, 1727–1732. [CrossRef]
54. Aires, A.; Carvalho, R.; Rosa, E.A.S.; Saavedra, M.J. Phytochemical characterization and antioxidant properties of baby-leaf watercress produced under organic production system. *CyTA J. Food* **2013**, *11*, 343–351. [CrossRef]
55. Hernandez, Y.; Lobo, M.G.; Gonzalez, M. Determination of vitamin C in tropical fruits: A comparative evaluation of methods. *Food Chem.* **2006**, *96*, 654–664. [CrossRef]
56. Aires, A.; Carvalho, R.; Matos, M.; Carnide, V.; Silva, A.P.; Gonçalves, B. Variation of chemical constituents, anti-oxidant activity, and endogenous plant hormones throughout different ripening stages of highbush blueberry (*Vaccinium corymbosum* L.) cultivars produced in centre of Portugal. *J. Food Biochem.* **2017**, *41*, e12414. [CrossRef]
57. Stratil, P.; Klejdus, B.; Kubán, V. Determination of total content of phenolic compounds and their antioxidant activity in vegetables—Evaluation of spectrophotometric methods. *J. Agric. Food Chem.* **2006**, *54*, 607–616. [CrossRef]
58. Apak, R.; Güçlü, K.; Özyürek, M.; Karademir, S.E. Novel total antioxidant capacity index for dietary polyphenols and vitamins C and E, using their cupric ion reducing capability in the presence of neocuproine: CUPRAC method. *J. Agric. Food Chem.* **2004**, *52*, 7970–7981. [CrossRef]
59. Siddhuraju, P.; Becker, K. Antioxidant properties of various solvent extracts of total phenolic constituents from three different agroclimatic origins of drumstick tree (*Moringa oleifera* Lam) leaves. *J. Agric. Food Chem.* **2003**, *51*, 2144–2155. [CrossRef]

https://doi.org/10.3390/plants15010092

plants

MDPI

Article

Effect of Pyroligneous Acid on Needle Retention and Certain Stress-Related Phytochemicals in Balsam Fir (*Abies balsamea*)

Niruppama Senthilkumar, Ravalika Kasu, Raphael Ofoe , Lord Abbey and Mason T. MacDonald *

Faculty of Agriculture, Dalhousie University, Bible Hill, NS B2N 5E3, Canada; nr977159@dal.ca (N.S.); rv260387@dal.ca (R.K.); raphael.ofoe@dal.ca (R.O.); labbey@dal.ca (L.A.)
* Correspondence: mason.macdonald@dal.ca; Tel.: +1-902-893-6626

Abstract

Balsam fir is an important specialty horticultural crop in eastern North America and commonly harvested for use as Christmas trees. Postharvest quality is a major challenge for producers, who are particularly concerned about postharvest needle retention. It was hypothesized that pyroligneous acid (PA) would help increase postharvest needle retention in balsam fir when supplied via xylem or foliage. This project first identified foliar spraying as the best application method, then designed a multivariate experiment with two factors. The first factor was foliar treatment (control, water, 1% PA, 2% PA, and 4% PA). The second factor was time, where branches were evaluated for needle abscission at 0, 2, 4, 6, and 8 weeks after harvest. The experiment was replicated 5 times and needle abscission, water uptake, chlorophyll, carotenoids, flavonoids, total phenolics, membrane injury, proline, and H_2O_2 production were all measured in response. Postharvest abscission reached 100% over the 8-week experiment and water uptake decreased by over 80%. Chlorophyll, proline, membrane injury, and H_2O_2 production all increased over time. Although PA did not improve needle retention compared to the control under the tested conditions, 4% PA spray increased proline concentration by 40% while decreasing membrane injury by 26%. Ultimately, PA did not consistently improve needle retention but did induce proline accumulation and membrane protection.

Keywords: abscission; carotenoids; chlorophyll; conifer; flavonoids; membrane injury; phenolics; proline; reactive oxygen species

1. Introduction

Christmas trees are a culturally and economically important specialty crop. Conifers were used in Christmas celebrations beginning in the 16th century and have since evolved to require millions of trees each year [1]. The Christmas tree industry in Canada alone was worth $163.3 million in 2021 [2]. Larger Christmas tree-producing areas, such as the United States and Europe, have over $1 billion in annual sales [3]. However, postharvest needle retention is a major challenge for the industry, with estimated economic losses exceeding $600 million [4].

Postharvest needle retention is a complex physiological process that presents a major challenge for Christmas tree producers. A multitude of factors have been shown to influence needle retention [5–10]. However, the driving factor for postharvest needle abscission appears to be dehydration and/or water uptake deficit through embolism despite the provision of water [11–13]. Freshly cut branches uptake water at 0.20 mL g^{-1} d^{-1} However, water uptake, stomatal conductance, and water potential all decrease within

https://doi.org/10.3390/plants15020261

the first 2 weeks postharvest, which directly coincides with the commencement of leaf abscission [8,13].

The original premise behind postharvest dehydration was that if water could be conserved by blocking stomata, then dehydration would decrease, and needle retention would increase. However, anti-transpirants have generally been only effective at reducing transpiration, with little effect on water status and needle retention [14,15]. Another strategy was delivering water to trees via an intravenous type of device, which was also ineffective [16]. Several species face postharvest challenges in water uptake due to bacterial accumulation in the water and xylem of the cut end [13,17]. Xylem blockages (embolism) can be mitigated to some extent by antibacterial leachates from the plant into the water, practices such as changing water daily, or addition of low-concentration silver nitrate to standing water [13,18]. However, such strategies are inconvenient and not guaranteed to work.

Pyroligneous acid (PA) is a by-product of pyrolysis where plant biomass undergoes thermal degradation in the absence or near absence of oxygen. The condensed organic vapors form an aqueous liquid fraction, PA, that has several desirable properties [19]. PA has been an effective organic biostimulant, fertilizer, insecticide, fungicide, bactericide, and antioxidant [20–23]. As such, PA has effectively increased growth, maintained chlorophyll membranes, reduced the impact of stress, and inhibited fungal and bacterial growth in agricultural species. Furthermore, the use of PA in agriculture is desirable because it can be produced sustainably and has virtually no environmental footprint [24,25]. The combination of PA's beneficial properties has high potential to improve postharvest characteristics of balsam fir, specifically by inhibiting growth of contaminants as a bactericide/fungicide and membrane protection as an antioxidant. It is hypothesized that PA could inhibit bacterial blockage in balsam fir xylem, inhibit fungal contamination of stomata, and protect chloroplast membranes. The specific objectives of this research are to determine (1) the effectiveness of xylem-fed PA on balsam fir needle retention, (2) the effectiveness of foliar applied PA on balsam fir needle retention, and (3) any effect on phytochemical composition. Although this study investigates physiological responses only and does not test antifungal or antibacterial mechanisms directly, all the three hypothesized mechanisms could lead to improved postharvest needle retention.

2. Results

2.1. Experiment 1: Determination of Application Method and Concentration Range

2.1.1. Xylem-Fed Delivery of PA

There was a significant ($p < 0.001$) difference in all water quality response variables due to PA concentration (Table 1). The significant differences in water quality parameters, specifically pH, were expected since PA is an acidic aqueous solution. There was considerable variation in pH, salinity, total dissolved solids (TDS), and electrical conductivity (EC). It is noteworthy that there is a clear trend that pH greatly decreases with even a small percentage of PA, while salinity, TDS, and EC all increase.

Visual differences in abscission due to xylem uptake are shown in Figure 1. There was a significant ($p < 0.001$) difference in needle abscission due to PA concentration (Table 1), but not the trend we had expected. The 0.25% PA solution had no impact on abscission compared to the control, but every other concentration significantly ($p < 0.05$) increased abscission. It is possible that PA altered water chemistry too much at concentration above 0.25%. The pH was very low compared to distilled water, while salinity, TDS, and EC had all significantly increased.

 https://doi.org/10.3390/plants15020261

Table 1. Water quality parameters and abscission due to PA added to stand water. Data is expressed as mean ± standard error. Different letters within columns indicate significant differences between treatments at 5% significance using Fisher's least significant difference test.

Concentration (%)	pH	Salinity (ppm)	TDS (ppm)	EC (μS)	Abscission (%)
0	6.15 ± 0.09 A	441.0 ± 0.6 A	634.7 ± 1.2 A	908.3 ± 1.3 A	1.0 ± 0.5 A
0.25	3.54 ± 0.06 B	490.0 ± 1.2 B	703.4 ± 1.2 B	1008.3 ± 1.76 B	1.3 ± 0.5 A
0.50	3.29 ± 0.02 C	505.7 ± 0.3 C	724.7 ± 0.9 C	1035.3 ± 0.3 C	44.7 ± 5.7 B
1.00	3.23 ± 0.03 C	550.7 ± 0.9 D	784.7 ± 0.7 D	1122.0 ± 1.0 D	67.9 ± 9.5 C
2.00	3.06 ± 0.01 D	577.0 ± 0.6 E	821.0 ± 1.0 E	1174.7 ± 2.3 E	75.2 ± 9.9 C
p-value	<0.001	<0.001	<0.001	<0.001	<0.001

(a) (b) (c) (d) (e)

Figure 1. Visual inspection of branches after finger run test when displayed in (**a**) 0%, (**b**) 0.25%, (**c**) 0.5%, (**d**) 1%, or (**e**) 2% pyroligneous acid after 6 weeks of xylem feeding.

2.1.2. Foliar Application of PA

There was a significant difference ($p = 0.035$) in needle abscission due to foliar spray treatments (Figure 2). The no-spray control lost significantly more needles than any other treatment, while 1% and 4% PA sprays lost significantly less needles. The 1% PA treatment lost 78% fewer needles than the no-spray control and 69% fewer needles than the water-only spray. The 4% PA treatment lost 66% fewer needles than the no-spray control and 55% fewer needles than the water-only spray. However, it is also important to state that the raw numbers are all very small percentages of the total needles. Ultimately, all branches retained needles reasonably well throughout the 5 weeks of this experiment, but the 1% and 4% PA foliar treatments were statistically the best. The 1%, 2%, and 4% treatments were selected for future experiments.

2.2. Experiment 2: Effect of Foliar PA on Postharvest Needle Retention

There were no significant differences in needle abscission or water uptake due to foliar PA treatments ($p = 0.961$ and $p = 0.230$, respectively) when branches were monitored until complete abscission. However, needle abscission and water uptake each increased significantly ($p < 0.001$) over time (Figure 3). When data from PA treatments was pooled to clearly depict time effects, needle abscission began slowly with only 5% of needles abscised within the first 2 weeks. Needle abscission then accelerated, increasing 37% over

https://doi.org/10.3390/plants15020261

the next 2 weeks. Abscission continued to 100% over the remaining 4 weeks almost linearly. Meanwhile, water uptake decreased rapidly in the first 2 weeks by 56% (Figure 3). Water uptake did not decrease significantly again until week 6.

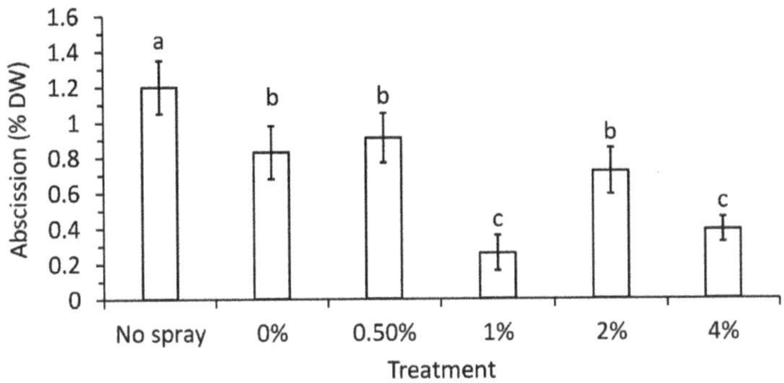

Figure 2. Needle abscission percentage in balsam fir seedlings after foliar spray with various concentrations of pyroligneous acid after 8 weeks. Bars represent the means, and the error bars indicate standard error. Each mean was calculated from 5 replicates from independent experiments and those bars with different letter groupings are significantly different ($p < 0.05$).

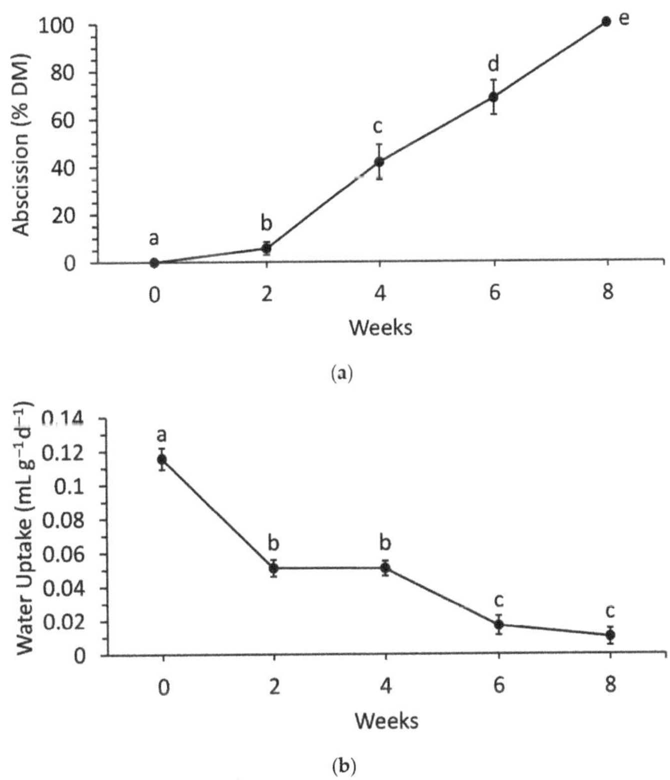

Figure 3. (**a**) Needle abscission and (**b**) water uptake of balsam fir branches over time. In each, data points denote the means and error bars denote standard error from a pooled 25 branches. Since treatment effects were not significant, displayed means are pooled from all treatments and show only time effects. Means with different letters are significantly different as determined by Fisher's least significant difference at 5% significance.

https://doi.org/10.3390/plants15020261

2.3. Experiment 3: Effect of Foliar PA on Stress Response

All pigments significantly ($p < 0.001$) changed over time throughout the experiment (Figure 4). Chl a and chl b increased by 41% and 55%, respectively, within the first 2 weeks before decreasing to near initial concentrations by week 4 (Figure 4a,b). Chl a and Chl b increased throughout the rest of the experiment and ultimately 66% and 82% higher than initial concentrations, respectively. Carotenoid concentration had a different trend, where it increased by 23% by week 4 from initial concentrations. The carotenoid concentration decreased for the rest of the experiment to a level not significantly different than the initial concentration (Figure 4c).

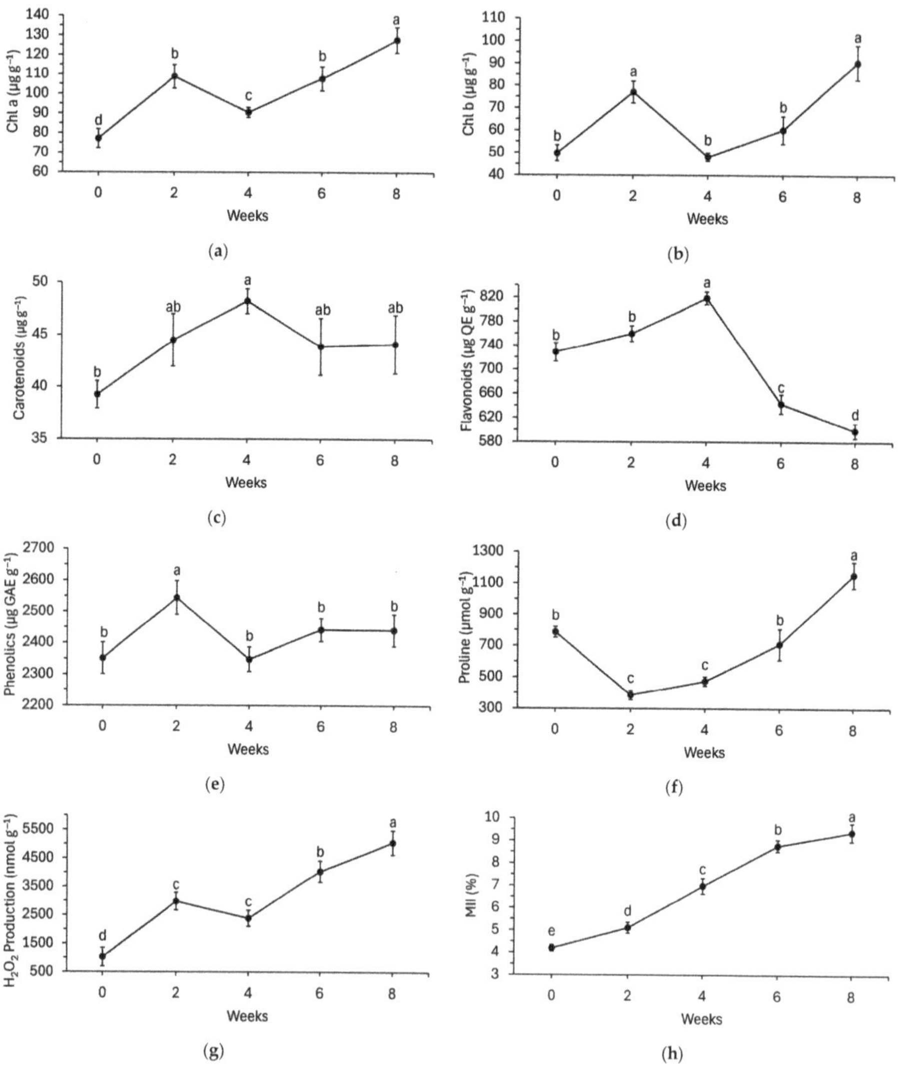

Figure 4. (**a**) Chlorophyll a, (**b**) chlorophyll b, (**c**) carotenoids, (**d**) flavonoids, (**e**) phenolics, (**f**) proline, (**g**) H_2O_2 production, and (**h**) membrane injury of balsam fir branches over time. In each, data points denote the means and error bars denote standard error from a pooled 25 branches. Time effects were pooled across control, water, 1% PA, 2% PA, and 4% PA treatments. Means with different letters are significantly different as determined by Fisher's least significant difference at 5% significance.

https://doi.org/10.3390/plants15020261

Flavonoids and total phenolics significantly changed over time ($p < 0.001$ and $p = 0.031$, respectively). Flavonoids increased by 12% by week 4 compared to initial concentrations, but then rapidly decreased (Figure 4d). By week 8, flavonoids were 18% lower than initial concentrations. Total phenolics only increased during the second week but otherwise did not differ significantly from initial concentrations (Figure 4e).

Stress indicators included proline, H_2O_2 production, and membrane injury index and all eventually increased by the end of the experiment. Proline was originally present at 785.0 µmol g^{-1} but decreased by 51% in the first 2 weeks (Figure 4f). Proline concentration recovered by week 6 before increasing by 47% by week 8. H_2O_2 production and membrane injury followed a more linear progression over time. H_2O_2 production increased 191% by week 2, 133% by week 4, 295% by week 6, and 395% by week 8 compared to initial levels (Figure 4g). The membrane injury index increased 22% by week 2, 66% by week 4, 109% by week 6, and 124% by week 8 compared to initial levels (Figure 4h).

There was a significant foliar spray treatment effect on chl a and chl b ($p = 0.001$ for each), where the water-sprayed treatment had significantly lower chlorophyll (Table 2). Chl a and chl b were 18% and 33% lower than the control, respectively. Chlorophyll did not differ from the control in any PA spray treatment. Similarly, total phenolics were 14% lower in the water-sprayed branches compared to the control, but spraying with PA had no significant effect (Table 2). Carotenoids and flavonoids were not significantly different in any treatment ($p > 0.05$).

Table 2. Chlorophyll pigment and stress-related phytochemicals of balsam fir branches due to PA added to the stand water. Data is expressed as mean ± standard error. Different letters within columns indicate significant differences between treatments at 5% significance.

Treatment	Chl a (µg g^{-1})	Chl b (µg g^{-1})	Carotenoids (µg g^{-1})	Flavonoids (µg GAE g^{-1})	Phenolics (µg QE g^{-1})
Control	104.7 ± 7.0 A	73.3 ± 8.8 A	41.2 ± 2.7	737.7 ± 22.0	2493.2 ± 71.4 A
Water	85.4 ± 2.7 B	45.9 ± 1.8 B	46.3 ± 1.4	718.3 ± 21.1	2132.8 ± 87.4 B
1% PA	103.5 ± 6.3 A	63.9 ± 5.2 A	46.1 ± 2.4	704.1 ± 18.7	2546.3 ± 116.1 A
2% PA	111.1 ± 6.9 A	75.1 ± 8.6 A	43.0 ± 3.1	692.0 ± 18.1	2550.9 ± 82.7 A
4% PA	107.3 ± 6.5 A	68.1 ± 7.0 A	43.1 ± 1.4	699.7 ± 18.9	2407.2 ± 85.8 A
p-value	0.001	0.001	0.55	0.357	<0.001

There was a significant treatment effect on proline ($p = 0.014$). Water-sprayed branches had 19% less proline than the control, while branches sprayed with 4% PA had 40% more proline than the control (Table 3). There was a significant treatment effect on H_2O_2 production ($p = 0.032$), though the effect was only observed in one treatment. Water-sprayed branches had 36% less H_2O_2 production (Table 3). There was also a significant treatment effect on the membrane injury index ($p < 0.001$). Membrane injury was 6% lower in the 1% PA treatment, 29% lower in the 2% PA treatment, and 26% lower in the 4% PA treatment compared to the control (Table 3).

Table 3. Stress marker contents of balsam fir branches due to PA added to the stand water. Data is expressed as mean ± standard error. Different letters within columns indicate significant differences between treatments at 5% significance.

Treatment	Proline (µmol g^{-1})	H_2O_2 Production (nmol g^{-1})	MII (%)
Control	705.5 ± 102.3 B	3465.5 ± 448.8 A	7.8 ± 0.6 A
Water	570.0 ± 49.3 C	2224.8 ± 322.5 B	8.2 ± 0.6 A

https://doi.org/10.3390/plants15020261

Table 3. *Cont.*

Treatment	Proline	H_2O_2 Production	MII
	(μmol g^{-1})	(nmol g^{-1})	(%)
1% PA	610.7 ± 61.2 BC	3302.8 ± 504.0 A	7.3 ± 0.5 B
2% PA	611.4 ± 44.2 BC	3617.0 ± 663.9 A	5.5 ± 0.4 C
4% PA	991.2 ± 122.9 A	2903.2 ± 329.6 AB	5.8 ± 0.3 C
p-value	0.035	0.038	<0.001

3. Discussion

3.1. PA and Needle Retention

The progression of needle abscission and decline in water uptake observed in this study mirrors established patterns for balsam fir and other conifer species. As shown in Figure 3a, needle abscission was minimal in the first two weeks postharvest, accelerated sharply between weeks two and six, and reached completion by the eighth week. Water uptake (Figure 3b) dropped significantly within the first two weeks, then stabilized at low levels as abscission increased. These findings are consistent with previous studies [8], that highlights dehydration and xylem embolism as key drivers of postharvest needle loss. Like earlier studies, this current study confirms that water status is a critical determinant of needle retention in balsam fir, with water uptake deficits directly preceding and likely triggering abscission. These findings suggest that xylem dysfunction and embolism formation can occur on the branch and whole-tree level under stressed conditions and that future work should consider the effects of these conditions on whole trees under commercial postharvest conditions.

3.2. PA Effects on Pigments and Antioxidant Metabolites

All major pigments and stress-related phytochemicals changed significantly over time in balsam fir branches (Figure 4). Chlorophyll a and b increased in the first two weeks, then declined near baseline by week four and ultimately rose again above initial levels by week eight. Carotenoids peaked in week four before returning to baseline. These results align with other species, where PA showed a positive effect in increasing chlorophyll pigment in tomato leaves [26] and Indian mustard [26]. This increase in chlorophyll pigments may suggest that PA could have delayed chlorophyll degradation by suppressing senescence-related pathways [27] or by providing antioxidant compounds that stabilize pigment metabolism [28]. Interestingly, the PA used in this study was shown to contain high levels of N, K and Ca as well as other bioactive compounds including phenolics and flavonoids [29]. Also, in this study, flavonoids and phenolics showed transient increases, with flavonoids peaking at week four and dropping below initial levels by week eight. While these increased phenolic compounds are fundamental for maintaining metabolic stability by scavenging ROS, their levels have also been associated with improved chlorophyll contents in leaf tissues by suppressing oxidative stress induced photorespiration [30]. Hence, the increased chlorophyll content with PA application can be ascribed to its rich bioactive contents that modulated pigment breakdown, possibly via improved chlorophyll biosynthesis and reduction in oxidative stress damage.

Proline, H_2O_2 production, and membrane injury all increased over the course of the experiment. These trends differ from those in other species. In apples and tomatoes, PA treatments led to sustained increases in antioxidant capacity and suppression of ROS, delaying abscission and reducing the postharvest losses [31,32]. In this experiment, however, ROS and membrane injury increased steadily despite the increase in antioxidants like proline and flavonoids, suggesting that PA did not mitigate oxidative stress. This

https://doi.org/10.3390/plants15020261

may reflect species-specific differences in cuticle structure, stress signaling or the relative importance of hydraulic versus biochemical factors in abscission.

Branches treated with 4% PA had more proline than controls, while water-sprayed branches had less (Table 3). Proline is a well-established osmoprotectant, accumulating in plant tissues under drought and other abiotic stresses [33]. Increased proline here indicates that PA may have triggered a stress response, potentially aiding in osmotic adjustment and temporary stabilization of cellular structures. However, proline accumulation alone was not sufficient to prevent needle loss, likely because the primary cause of abscission was hydraulic failure rather than cellular dehydration [34]. This apparent coexistence of both increased proline and hydraulic failure represents a layered stress physiology where proline accumulation reflects a short-term cellular response that may delay visible senescence but cannot prevent the structural failure of the vascular system. PA treatments reduced membrane injury compared to controls (Table 3) suggesting that PA may offer some protection to cellular membranes, possibly by enhancing antioxidant defenses or stabilizing lipid bilayers [35,36]. In other crops, such as apples, similar effects are associated with delayed senescence and improved postharvest quality [31]. A study in canola seed showed that PA has a positive effect in increasing the SOD activity and reducing oxidative stress [36]. However, in balsam fir, these improvements did not translate into better needle retention, again highlighting the dominant role of hydraulic factors over biochemical stress responses in this species.

3.3. *Hydraulic Versus Biochemical Control of Needle Abscission*

Even though PA increased proline and reduced membrane injury, needle retention was not consistently improved by PA because of several reasons. First, needle abscission in balsam fir is driven primarily by water uptake deficits and xylem embolism [34]; improvements in cellular stress tolerance could not overcome systemic hydraulic failure. Second, while not directly measured in this study, ethylene is a known trigger of abscission in conifers [4,34,37,38]. PA may not have affected ethylene signaling pathways or may have encouraged ethylene production. While PA can modulate biochemical stress markers, these are not the primary determinants of abscission in balsam fir, unlike in some angiosperms, where oxidative stress is more central to senescence and tissue loss [39]. Finally, conifer needles have a thick, waxy cuticle [40], which likely limited PA penetration and efficacy. Unlike apples and tomatoes, PA may not have reached the target tissues in balsam fir.

This study shows that whereas PA can modulate biochemicals linked to stress and decrease membrane damage in balsam fir, these modifications do not result in better needle retention in postharvest settings. The results highlight the significance of hydraulic parameters in conifer abscission and recommend that future research concentrate on methods that directly address embolism and water intake. Also, since light conditions were not controlled, the influence of photoperiod and light quality on postharvest physiology could not be assessed. Future studies should examine these factors under defined photoperiods. Future research may provide more encouraging outcomes if alternative surfactants are used to improve PA penetration, replication, and combinatorial treatments that target both physiological and biochemical processes.

4. Materials and Methods

4.1. *Experiment 1: Determination of Application Method and Concentration Range*

4.1.1. Sample Collection

Branches were collected from the Christmas tree clonal orchard in Plumdale Research Facility at Dalhousie University. Branches were cut to include 2 full years of growth at a height of 1 m. Branches were first visually inspected and any with obvious signs of stress

https://doi.org/10.3390/plants15020261

(e.g., nutrient stress discoloration, pest damage, etc.) were excluded from the study. A total of 55 branches were required for the first experiment, so 5 branches were collected from 11 trees of the same genotype within the clonal orchard.

Samples were taken to the lab, provided a fresh cut approximately 1 cm from the previously cut end, and then placed in glass vases to simulate household conditions. All cuts were performed with the cut section submerged in deionized water to reduce the risk of introducing air bubbles to trigger embolism. Branches were then randomly assigned to treatments via a random number generator.

4.1.2. Xylem-Fed Delivery of PA

This experiment followed a completely randomized design where the factor of interest was PA concentration diluted in deionized water. Treatment levels consisted of 0% (a water-only control), 0.25%, 0.5%, 1%, and 2% PA solutions. Each branch was provided with 1 L of its respective treatment solutions. The base of each branch was wrapped in aluminum foil at the mouth of the vase to reduce evaporation and provide stability to the branch. Since PA is a naturally derived product containing multiple organic compounds, specific water quality parameters pH, salinity, total dissolved solids (TDS), and electrical conductivity (EC) of solutions were measured at the beginning of the experiment. This experiment was replicated 5 times.

4.1.3. Foliar Application of PA

This experiment followed a completely randomized design where the factor of interest was PA concentration. Treatment levels consisted of a control (no spraying), 0% (water only spray), 0.5% PA, 1% PA, 2% PA, and 4% PA. Each branch was provided with 1 L of deionized water, and the base of the branch was wrapped in aluminum foil at the mouth of the vase to reduce evaporation and provide stability to the branch. Branches were sprayed with their respective treatment at the beginning of the experiment and then once per week throughout the remainder of the experiment. The experiment was conducted under ambient laboratory conditions at room temperature in typical indoor lighting with 40% relative humidity. This condition was selected to represent a worst-case scenario of water stress, thereby allowing us to evaluate treatment efficacy under challenging circumstances. It is worth noting that no controlled photoperiod, artificial light/dark cycle, or supplemental light sources were applied. As above, spray solutions were evaluated for pH, salinity, TDS, and EC. This experiment was replicated 5 times.

4.1.4. Measuring Needle Abscission

Since these experiments were preliminary in nature, the major response variable was only needle abscission. Needle abscission was determined by performing a finger run test [5,11]. The first finger run test for both experiments was conducted on 4 January 2023. The water additive experiment was terminated at this point. A second finger run test was conducted 23 January 2023, on the remaining foliar spray experiment, after which the foliar spray experiment was also terminated. In each case, abscised needles and branches were collected, dried in an oven at 80 °C for 24 h, and then weighed. Needle abscission was then expressed a percentage of total dry mass of a branch.

Data from each experiment was submitted to an analysis of variance to determine significant differences. The null hypothesis was that there was no difference between treatments, and the alternative was that there was at least one difference. Specific differences were then determined through Fisher's least significant difference at 5% significance. Statistical assumption of normality, homogeneity, and independence were all valid.

 https://doi.org/10.3390/plants15020261

4.2. Experiment 2: Effect of Foliar PA on Postharvest Needle Retention

4.2.1. Experimental Design

This experiment had two factors. The first factor was foliar treatment, which included a control (no spraying), water, 1% PA, 2% PA, and 4% PA. The second factor was time, where branches were evaluated for needle abscission at 0, 2, 4, 6, and 8 weeks after harvest. The experiment was replicated 5 times. Branch collection and display protocol were the same as Experiment 1. Needle abscission and water uptake were the only response variables. Experiment 2 required 25 branches, where 5 branches were collected from 5 trees and then randomly assigned to treatments.

4.2.2. Needle Abscission

Needle abscission was measured gravimetrically after a finger run test, like Experiment 1. Needles were dried in an oven at 80 °C for 24 h. Branches were displayed until complete abscission occurred, and all branches had completed abscission within the 8-week experiment duration. All needles were weighed at the end of the experiment to determine the total dry mass of needles on each branch. Needle abscission was then expressed as a cumulative percentage.

4.2.3. Water Uptake

Water uptake was determined gravimetrically as suggested by Lada et al. [13]. To determine water uptake the branches were gently removed from the amber flasks and the remaining contents (water and flask together) were weighed every alternate day. Since bottles were sealed, any loss in mass indicated water uptake by the branches. The water uptake was reported in mL g^{-1} d^{-1}.

4.2.4. Statistical Analysis

The data collected from the experiment was analyzed statistically using Minitab 22 software (Minitab, LLC, State College, PA, USA). Analysis of variance (ANOVA) from the general linear model was used to determine if there is a difference in needle retention or water uptake due to foliar PA in balsam fir. The residuals from needle retention data were nearly normal ($p = 0.034$) from an Anderson-Darling test; normality was achieved with a mild square root transformation. Mean separation was determined using Fisher's method with a 95% confidence interval at $p \leq 0.05$ after a significant ANOVA result was obtained.

4.3. Experiment 3: Effect of Foliar PA on Stress Response

4.3.1. Experimental Design

Like Experiment 2, this experiment had two factors. The first factor was foliar treatment, which included a control (no spraying), water, 1% PA, 2% PA, and 4% PA. The second factor was time, where branches were evaluated for needle abscission at 0, 2, 4, 6, and 8 weeks after harvest. Display conditions were identical to those described earlier in Experiment 1. The branches used in Experiment 3 were distinct from those used in previous experiments. and the experiment was replicated 6 times. The destructive nature of sampling for biochemical analysis required all needles on each branch to be removed every 2 weeks, ensuring that biweekly sampling was independent As such, 150 branches were needed for Experiment 3, which were obtained by collecting 15 branches from 10 trees.

Response variables were chlorophyll a, chlorophyll b, total carotenoids, total phenols, total flavonoids, proline concentration, reactive oxygen species, and membrane injury index.

https://doi.org/10.3390/plants15020261

4.3.2. Chlorophyll and Carotenoids

Chlorophyll a, b and carotenoids were measured in reference to Lichtenthaler [41]. A 0.2 g of ground samples were transferred into a sterile 50 mL falcon tube and 10 mL of 80% acetone was added. The mixture was vortexed for 1 min and centrifuged at 12,000× *g* for 15 min. 1 mL of supernatant was transferred into a cuvette. The absorbance was measured at 646.8 and 663.2 nm with a UV-Vis spectrophotometer against 80% acetone as blank. The concentration of chlorophyll a, b, and carotenoids was calculated as μg g^{-1} FW using the Formulas (1)–(3).

$$\text{Chlorophyll a}\left(\frac{\mu g}{mL}\right) = (12.25 \times A663.2) - (2.79 \times A646.8) \quad (1)$$

$$\text{Chlorophyll b}\left(\frac{\mu g}{mL}\right) = (21.50 \times A646.8) - (5.1 \times A663.2) \quad (2)$$

$$\text{Carotenoids}\left(\frac{\mu g}{mL}\right) = \frac{(1000 \times A470) - (1.8 \times \text{chl a}) - (85.02 \times \text{chl b})}{198} \quad (3)$$

4.3.3. Phenolics and Flavonoids

The total phenolic content was determined by the Folin–Ciocalteu assay described by Ainsworth & Gillespie [42] with a slight modification. A 0.2 g of sample was homogenized in a 2 mL of ice-cold methanol and incubated in the dark at room temperature for 48 h. The mixture was centrifuged at 13,000× *g* for 5 min and 100 μL of supernatant was transferred into a new microfuge tube 200 μL of 10% Folin-Ciocalteau reagent was added and vortexed for 5 min. 800 μL of 700 nM sodium carbonate (Na_2CO_3) was added, vortexed for a min, and incubated at room temperature (25 °C) for 2 h. The absorbance of the resultant mixture was measured at 765 nm. Total phenolic content was estimated using gallic acid equivalents standard curve and was expressed as mg gallic acid equivalents per g of the sample.

Total flavonoids was determined as described by Chang et al. [43] with some modifications. A 0.2 g of ground samples were homogenized with 2.5 mL of 95% methanol. The mixture was centrifuged at 13,000× *g* for 10 min and 500 μL of the supernatant was transferred into a new tube. To each tube, 1.5 mL of 95% methanol, 0.1 mL of 10% $AlCl_3$, 0.1 mL of 1 M potassium acetate, and 2.8 mL of distilled water was added. The mixture was vortexed and incubated for 30 min at room temperature and absorbance was measured at 415 nm against a blank. Flavonoids content was estimated using a quercetin standard curve and expressed in μg of quercetin per g of sample.

4.3.4. Proline

A 0.5 g of ground samples were mixed with 1 mL of 70% ethanol. Then the mixture was centrifuged at 12,000× *g* for 15 min at 4 °C and 500 μL of supernatant was added to 1 mL of the reaction mixture (ninhydrin 1% (*w*/*v*) dissolved in 60% acetic acid (*v*/*v*) and 20% ethanol (*v*/*v*). The tubes were sealed and vortexed for 30 s. The mixture was incubated in a water bath at 95 °C for 30 min. After cooling at room temperature, the absorbance was measured at 520 nm using a spectrophotometer against a blank containing ethanol and reaction mixture. Proline content was estimated using the L-Proline standard curve. The absorbance was plotted against L-Proline concentration to obtain the standard curve. Total proline content can be calculated using Equation (4).

$$\text{Proline}\left(\frac{\mu mol}{gFW}\right) = \frac{Abs_{ext} - blank}{Slope} \times Vol_{ext}/Vol_{aliquot} \times \left(\frac{1}{FW}\right) \quad (4)$$

https://doi.org/10.3390/plants15020261

In Equation (4), where Abs_{ext} is the absorbance of plant extract, blank is the absorbance of clear extraction solution, the slope is determined by linear regression of a calibration curve, Vol_{ext} is the total extract volume, $Vol_{aliqout}$ is the extract volume used for the assay, FW is the weight of the plant material.

4.3.5. Reactive Oxygen Species

ROS was estimated to use the H_2O_2 concentration by following the procedure described by Patterson et al. [44]. A total of 0.1 g of ground samples were mixed with 1 mL of 0.1% (*w*/*v*) TC. The mixture was vortexed and centrifuged at 16,000 rpm for 10 min. After that 0.2 mL of supernatant was added to 200 µL of 100 mM potassium phosphate buffer (pH 7.0) and 800 µL of 1 M KI. The mixture was incubated in the dark for 1 h. The absorbance was measured at 390 nm against the blank (0.1 TCA in the absence of needle extract). The H_2O_2 concentration was calculated according to the standard curve (H_2O_2 dissolved in 0.1% TCA) and expressed in µM/g of fresh weight.

4.3.6. Membrane Injury Index

MII was performed in reference to the study by MacDonald and Lada [8]. The percentage of electrolyte that leaks into the solution was used by the membrane injury index (MII) to measure membrane integrity. Approximately 30 mL of distilled water was added to test tubes, which were then let to warm up to room temperature (25 °C). Using a CDM 2e Conductivity Meter (Vernon Hills, IL, USA), the electrical conductivity of the distilled water (EC_w) alone was measured. Then, 0.4 g of needles were taken from each branch and immersed entirely in a centrifuge tube. After being sealed, the tubes were kept at room temperature for 24 h. To estimate how much of the electrolytes are leached into the solution, the initial conductivity (EC_0) was measured. After 4 h at 90 °C in a forced-air oven to kill tissues, sealed tubes were then cooled to room temperature. To estimate the maximum leakage, final conductivity measurements (EC_f) were made after the system had reached equilibrium at 25 °C. The following formula was used to determine MII.

$$MII = EC_0 - EC_w / EC_f - EC_w \times 100 \quad (5)$$

4.3.7. Statistical Analysis

Data were submitted to general linear model using Minitab 22 (Minitab, LLC, State College, PA, USA). Foliar application and time were entered as explanatory variables and a 2-way interaction between spraying and time was also included in the model. Means were compared using Fisher's least significant difference when significant differences were found from the ANOVA at $p \leq 0.05$. Statistical assumptions of normality, homogeneity, and independence were confirmed.

Author Contributions: Conceptualization, all authors.; methodology, all authors; software, N.S., R.K.; validation, all authors.; formal analysis, N.S., R.K., R.O.; investigation, N.S., R.K., R.O.; resources, M.T.M., L.A.; data curation, M.T.M., N.S., R.K.; writing—original draft preparation, N.S., R.K.; writing—review and editing, M.T.M., L.A., R.O.; visualization, M.T.M.; supervision, M.T.M., L.A., R.O.; project administration, M.T.M., L.A.; funding acquisition, M.T.M., L.A. All authors have read and agreed to the published version of the manuscript.

Funding: This research was funded by the Real Christmas Tree Board, grant number 22-10-DU.

Data Availability Statement: The datasets presented in this article are not readily available because the data are part of an ongoing study. Requests to access the datasets should be directed to Mason MacDonald (mason.macdonald@dal.ca).

https://doi.org/10.3390/plants15020261

Acknowledgments: The authors thank the Department of Plant, Food, and Environmental Sciences for permission to obtain samples from the balsam fir clonal orchard.

Conflicts of Interest: The authors declare no conflicts of interest.

Abbreviations

The following abbreviations are used in this manuscript:

EC	Electrical conductivity
MII	Membrane injury index
PA	Pyroligneous acid
ROS	Reactive oxygen species
TDS	Total dissolved solids

References

1. Albers, H.H.; Davis, A.K. *The Wonderful World of Christmas Trees*, 1st ed.; Woolverton Printing: Cedar alls, IA, USA, 1997.
2. Garside, M. Farm Cash Receipts from Christmas Trees Canada. Available online: https://www.statista.com/statistics/449337/farm-cash-receipts-of-christmas-trees-canada/ (accessed on 19 November 2025).
3. Chastagner, G.A.; Benson, D.M. The Christmas Tree: Traditions, Production, and Diseases. *Plant Health Prog.* **2000**, *1*, 15. [CrossRef]
4. Thiagarajan, A.; MacDonald, M.T.; Lada, R. Environmental and Hormonal Physiology of Postharvest Needle Abscission in Christmas Trees. *Crit. Rev. Plant Sci.* **2016**, *35*, 1–17. [CrossRef]
5. MacDonald, M.T.; Lada, R.R.; MacDonald, G.E.; Caldwell, C.D.; Udenigwe, C.C. Changes in Polar Lipid Composition in Balsam Fir during Seasonal Cold Acclimation and Relationship to Needle Abscission. *Int. J. Mol. Sci.* **2023**, *24*, 15702. [CrossRef] [PubMed]
6. Bartas, M. Abiotic Stresses in Plants: From Molecules to Environment. *Int. J. Mol. Sci.* **2024**, *25*, 8072. [CrossRef] [PubMed]
7. Amponsah, I.G.; Comeau, P.G.; Brockley, R.P.; Lieffers, V.J. Effects of Repeated Fertilization on Needle Longevity, Foliar Nutrition, Effective Leaf Area Index, and Growth Characteristics of Lodgepole Pine in Interior British Columbia, Canada. *Can. J. For. Res.* **2005**, *35*, 440–451. [CrossRef]
8. MacDonald, M.T.; Lada, R.R. Biophysical and Hormonal Changes Linked to Postharvest Needle Abscission in Balsam Fir. *J. Plant Growth Regul.* **2014**, *33*, 602–611. [CrossRef]
9. Addicott, F.T. Environmental Factors in the Physiology of Abcission. *Plant Physiol.* **1968**, *43*, 1471–1479.
10. Manter, D.K.; Bond, B.J.; Kavanagh, K.L.; Stone, J.K.; Filip, G.M. Modelling the Impacts of the Foliar Pathogen, Phaeocryptopus Gaeumannii, on Douglas-Fir Physiology: Net Canopy Carbon Assimilation, Needle Abscission and Growth. *Ecol. Model.* **2003**, *164*, 211–226. [CrossRef]
11. Chastagner, G.A.; Riley, K.L. Postharvest Quality of Noble and Nordmann Fir Christmas Trees. *HortSci* **2003**, *38*, 419–421. [CrossRef]
12. Tremblay, J.D.; Smith, R.F.; D'Orangeville, L. Integrating the Rate of Moisture Loss into Needle Retention Testing to Improve the Selection of Balsam Fir (*Abies balsamea*) for Use as Christmas Trees. *Forests* **2023**, *14*, 1626. [CrossRef]
13. Lada, R.R.; MacDonald, M.T.; West, R.R. Physiology of Postharvest Needle Abscission in Balsam Fir: Water Quality Modulates Postharvest Needle Abscission. *Acta Hortic.* **2014**, *1119*, 111–120. [CrossRef]
14. Hinesley, L.E.; Snelling, L.K.; Goodman, S. "Crop-Life" Does Not Slow Postharvest Drying of Fraser Fir and Eastern Red Cedar Christmas Trees. *HortSci* **1993**, *28*, 1054. [CrossRef]
15. Duck, M.W.; Cregg, B.M.; Cardoso, F.F.; Fernandez, F.T.; Behe, B.K.; Heins, R.D. Can Antitranspirants Extend the Shelf-Life of Tabletop Christmas Trees. *Acta Hortic.* **2003**, *618*, 153–161. [CrossRef]
16. Chastagner, G.A.; Hinesley, E.; Riley, K. Effectiveness of I–V Watering Devices in Maintaining Postharvest Freshness and Quality of Cut Christmas Trees. *Postharvest Biol. Technol.* **2007**, *43*, 178–181. [CrossRef]
17. Hoogerwerf, A.; Van Doorn, W.G. Numbers of Bacteria in Aqueous Solutions Used for Postharvest Handling of Cut Flowers. *Postharvest Biol. Technol.* **1992**, *1*, 295–304. [CrossRef]
18. Hinesley, L.E.; Snelling, L.K. Postharvest Drying of Leyland Cypress, Eastern Red Cedar, and Fraser Fir Christmas Trees. *HortSci* **1995**, *30*, 1427–1428. [CrossRef]
19. Mathew, S.; Zakaria, Z.A. Pyroligneous Acid—The Smoky Acidic Liquid from Plant Biomass. *Appl. Microbiol. Biotechnol.* **2015**, *99*, 611–622. [CrossRef]
20. Grewal, A.; Abbey, L.; Gunupuru, L.R. Production, Prospects and Potential Application of Pyroligneous Acid in Agriculture. *J. Anal. Appl. Pyrolysis* **2018**, *135*, 152–159. [CrossRef]

 https://doi.org/10.3390/plants15020261

21. Lee, S.H.; H'ng, P.S.; Lee, A.N.; Sajap, A.S.; Tey, B.T.; Salmiah, U. Production of Pyroligneous Acid from Lignocellulosic Biomass and Their Effectiveness Against Biological Attacks. *J. Appl. Sci.* **2010**, *10*, 2440–2446. [CrossRef]
22. Loo, A.; Jain, K.; Darah, I. Antioxidant Activity of Compounds Isolated from the Pyroligneous Acid, Rhizophora Apiculata. *Food Chem.* **2008**, *107*, 1151–1160. [CrossRef]
23. Abinandan, S.; Kuppan, P.; Venkateswarlu, K.; Mukunthan, K.; Megharaj, M. Pyroligneous Acid as a Multifunctional Biostimulant Enhances Microalgal Growth and Soil Beneficial Metabolites for Sustainable Agriculture. *World J. Microbiol. Biotechnol.* **2025**, *41*, 306. [CrossRef]
24. Jindo, K.; Goron, T.L.; Kurebito, S.; Matsumoto, K.; Masunaga, T.; Mori, K.; Miyakawa, K.; Nagao, S.; Tokunari, T. Sustainable Plant Growth Promotion and Chemical Composition of Pyroligneous Acid When Applied with Biochar as a Soil Amendment. *Molecules* **2022**, *27*, 3397. [CrossRef]
25. Mmojieje, J.; Hornung, A. The Potential Application of Pyroligneous Acid in the UK Agricultural Industry. *J. Crop Improv.* **2015**, *29*, 228–246. [CrossRef]
26. Benzon, H.R.L.; Lee, S.C. Potential of Wood Vinegar in Enhancing Fruit Yield and Antioxidant Capacity in Tomato. *Korean J. Plant Resour.* **2016**, *29*, 704–711. [CrossRef]
27. Dominguez, F.; Cejudo, F.J. Chlorophyll Dismantling in Leaf Senescence. *J. Exp. Bot.* **2021**, *72*, 5905–5918. [CrossRef]
28. Tan, S.; Sha, Y.; Sun, L.; Li, Z. Abiotic Stress-Induced Leaf Senescence: Regulatory Mechanisms and Application. *Int. J. Mol. Sci.* **2023**, *24*, 11996. [CrossRef]
29. Ofoe, R.; Gunupuru, L.R.; Abbey, L. Metabolites, Elemental Profile and Chemical Activities of *Pinus Strobus* High Temperature-Derived Pyroligneous Acid. *Chem. Biol. Technol. Agric.* **2022**, *9*, 85. [CrossRef]
30. Meyer, S.; Cerovic, Z.; Goulas, Y.; Montpied, P.; Demontes-Mainard, S.; Bidel, L.; Moya, I.; Dreyer, E. Relationships between Optically Assessed Polyphenols and Chlorophyll Contents, and Leaf Mass per Area Ratio in Woody Plants: A Signature of the Carbon–Nitrogen Balance within Leaves? *Plant Cell Environ.* **2006**, *29*, 1338–1348. [CrossRef] [PubMed]
31. Liu, X.; Li, J.; Cui, X.; Ji, D.; Xu, Y.; Chen, T.; Tian, S. Exogenous Bamboo Pyroligneous Acid Improves Antioxidant Capacity and Primes Defense Responses of Harvested Apple Fruit. *LWT* **2020**, *134*, 110191. [CrossRef]
32. Rivera, F.R.; Valida, A.D.; Alcedo, A.L., Jr. Effects of Wood Vinegar on Tomato Fruit Quality and Shelf Life at Ambient and Low Temperatures. *Acta Hortic.* **2015**, *1088*, 145–148. [CrossRef]
33. Kaur, G.; Asthir, B. Proline: A Key Player in Plant Abiotic Stress Tolerance. *Biol. plant.* **2015**, *59*, 609–619. [CrossRef]
34. Lada, R.R.; MacDonald, M.T. Understanding the Physiology of Postharvest Needle Abscission in Balsam Fir. *Front. Plant Sci.* **2015**, *6*, 1069. [CrossRef]
35. Cândido, N.R.; Pasa, V.M.D.; de Oliveira Vilela, A.; Campos Â, D.; de Fátima, Â.; Modolo, L.V. Understanding the Multifunctionality of Pyroligneous Acid from Waste Biomass and the Potential Applications in Agriculture. *Sci. Total Environ.* **2023**, *881*, 163519. [CrossRef]
36. Yang, J.-F.; Yang, C.-H.; Liang, M.-T.; Gao, Z.-J.; Wu, Y.-W.; Chuang, L.-Y. Chemical Composition, Antioxidant, and Antibacterial Activity of Wood Vinegar from Litchi Chinensis. *Molecules* **2016**, *21*, 1150. [CrossRef]
37. Fuhrer, J. Ethylene Production and Premature Senescence of Needles from Fir Trees (*Abies alba*). *Eur. J. For. Pathol.* **1985**, *15*, 227–236. [CrossRef]
38. Yamanaka, K. Leaf Abscission and Ethylene Production in Chamaecyparis Obtusa S. et Z. *J. Jpn. For. Soc.* **1984**, *66*, 297–305. [CrossRef]
39. Sedigheh, H.G.; Mortazavian, M.; Norouzian, D.; Atyabi, M.; Akbarzadeh, A.; Hasanpoor, K.; Ghorbani, M. Oxidative Stress and Leaf Senescence. *BMC Res. Notes* **2011**, *4*, 477. [CrossRef]
40. Zhang, Y.; Wang, A.; Li, J.; Wu, J. Water Content Estimation of Conifer Needles Using Leaf-Level Hyperspectral Data. *Front. Plant Sci.* **2024**, *15*, 1428212. [CrossRef] [PubMed]
41. Lichtenthaler, H.K. Chlorophylls and Carotenoids: Pigments of Photosynthetic Membranes. In *Methods in Enzymology*; Elsevier: Amsterdam, The Netherlands, 1987; Volume 148, pp. 350–382.
42. Ainsworth, E.A.; Gillespie, K.M. Estimation of Total Phenolic Content and Other Oxidative Substrates in Plant Tissues Using Folin-Ciocalteu Reagent. *Nat. Protoc.* **2007**, *2*, 875–877.
43. Chang, C.C.; Yang, M.H.; Wen, H.M.; Chern, J.C. Estimation of Total Flavanoid Content in Propolis by Two Complimentary Colorimetric Methods. *J. Food Drug Anal.* **2002**, *10*, 178–182.
44. Patterson, B.D.; MacRae, E.A.; Ferguson, I.B. Estimation of Hydrogen Peroxide in Plant Extracts Using Titanium(IV). *Anal. Biochem.* **1984**, *139*, 487–492. [CrossRef] [PubMed]

 https://doi.org/10.3390/plants15020261

MDPI AG
Grosspeteranlage 5
4052 Basel
Switzerland
Tel.: +41 61 683 77 34

Plants Editorial Office
E-mail: plants@mdpi.com
www.mdpi.com/journal/plants

www.ingramcontent.com/pod-product-compliance
Lightning Source LLC
Chambersburg PA
CBHW040817250926
51850CB00007BA/1197